FERTILIZERS AND CROP PRODUCTION

DR. LUCIUS L. VAN SLYKE

FERTILIZER AND CROP PRODUCTION

DR. LUCIUS L. VAN SLYKE, Ph.D.
Formerly Chemist of the
New York Agricultural Experiment Station

AGROBIOS (INDIA)

Published by:
AGROBIOS (INDIA)
Agro House, Behind Nasrani Cinema
Chopasani Road, Jodhpur 342 002
Phone: 91-0291-2642319, 2643994, Fax: 2643993
E. mail: agrobios@sify.com

AGROBIOS (INDIA)

First Indian Impression 2003
Reprinted: 2012

ISBN
Pric

Published by: Dr. Updesh Purohit for Agrobios (India), Jodhpur
Laser typeset at: Yashee Computers, Jodhpur
Cover Design by: Reena
Printed at: Bharat Printers (Press), Jodhpur

PREFACE

This book is the successor to the author's "Fertilizers and Crops," published in 1912. The intervening years have been filled with the ever-increasing activity of agricultural research workers in the study of problems relating to plant nutrition and crop production, covering all of the various phases,—chemical, physical, and biological. Many definite facts have thus been added to our knowledge.

Like its predecessor, this book has for its purpose not merely to give information but also to present it in such a systematic way as to make clear fundamental relations and to emphasize the reasons that underlie every practice suggested. The conditions are studied under which plant-foods, whether in the form of soil compounds, of farm-produced materials, or of commercial fertilizers, can be conserved and, at the same time, utilized with the greatest efficiency and economy in the production of crops.

Not only is it desirable but essential that a book that treats of the practical use of fertilizers on the farm be something more than a mere collection of formulas or recipes, that give the number of pounds of nitrogen, phosphorus, potassium or other compounds to apply to individual crops. Important also is it to make as prominent as possible other vital factors that influence or control the effectiveness with which a crop utilizes the nutrient materials supplied to it.

It has been the privilege of the writer for a period of forty years to be in touch with work relating to plant nutrition. Through personal contact he has been enabled to learn the practices, difficulties, and needs of farmers in relation to many of the problems of crop growing with the use of fertilizers. In no small degree, this book is, therefore, the outgrowth of an appreciation of the farmer's point of view. It seeks to assemble and arrange many of the facts and principles the farmer desires and needs to know.

As Chemist of the New York Agricultural Experiment Station, the author in 1890 had the responsibility of organizing the New York State fertilizer inspection work, with which he continued in close official relations until his retirement in 1929. He was thus for many years brought into intimate contact with the various aspects connected with the distribution and use of commercial fertilizers. He has been able to know at first hand the changes and developments that have taken place in the fertilizer trade. There is thus made possible a full discussion of the new synthetic fertilizer compounds now coming into increasing use and also of the highly concentrated mixtures of commercial fertilizers now in the market.

In the preparation of this book, it has been the author's aim to keep in mind not only the requirements of students in agricultural colleges and high schools but as well the needs of student farmers who, in increasing numbers, are directly and vitally concerned in the profitable production of crops. The book can be used to advantage in the educational work promoted by granges and other farmers' organizations.

For the marked favor which has been accorded its predecessor, as shown by its extensive use during nearly twenty years, appreciation can not be expressed in any way more appropriate than by writing the present book and bringing the fertilizer subject up to date.

LUCIUS L. VAN SLYKE

Geneva, N. Y.

It is greatly to be regretted that Dr. Van Slyke did not live to see this book through the press. He died only a few weeks after completing the manuscript. Fortunately his son, Dr. Donald D. Van Slyke, Chemist, Member of Rockefeller Institute for Medical Research, has edited and proofread the text which is thus given the further stamp of authority.

PUBLISHER'S EDITOR

Ammo-phos-ko. After this book was in plate form the United States Court of Customs and Patent Appeals cancelled the registration of this trade mark name on the grounds that it is an infringement of the previously established trade mark term "nitrophoska." At the date of publication it is not known whether the former product has been renamed.—(See page 180.)

CONTENTS

Part I

FACTORS OF SOIL FERTILITY

Part II

FUNCTIONS AND PHYSICAL PROPERTIES OF SOILS

Part III

SOURCES AND COMPOSITION OF MATERIALS USED AS FERTILIZERS

Part IV

FACTORS THAT GUIDE THE SELECTION OF FERTILIZER MATERIALS

Part V

PRACTICAL USE OF FERTILIZERS IN GROWING INDIVIDUAL CROPS

ILLUSTRATIONS

TABLES

PART I

FACTORS OF SOIL FERTILITY

I

INTRODUCTION

CONSERVATION OF PLANT-FOOD—RELATIONS OF PLANT-FOOD TO SOIL FERTILITY

There has never been in this country so widespread and so deep an interest in the subject of conservation of plant-food as at present; and never before have so many men, trained for investigation, been engaged in studying the subject in all its manifold phases and varied relations.

The condition of many American farmers is aptly illustrated in the following experience described by Dr. Holmes in "The Professor at the Breakfast Table": "Let me tell you what happened to me once. I put a little money into a bank, and bought a check-book, so that I might draw it as I wanted, in sums to suit. Things went on nicely for a time; scratching with a pen was as easy as rubbing Aladdin's Lamp; and my blank check-book seemed a dictionary of possibilities, in which I could find all the synonyms of happiness, and realize any one of them on the spot. A check came back to me at last with these two words on it—*No funds*."

In its main features, the application of the foregoing experience is so highly suggestive as to require no special explanation. American farmers have practised the wasteful if not exhaustive drawing upon the readily available plant-food supplies deposited for their use by the accumulations of long ages. The main problem before them is, therefore, the utilization of plant-food under such conditions and methods as will stop needless waste and at the same time increase the average yield of crops without a proportionate increase of cost. It may seem paradoxical to suggest that it is possible to produce larger crops without using up more rapidly the soil's supply of plant-food. Speaking in general terms, it becomes a matter of preventing unnecessary losses of plant-food materials, and, at the same time, of replacing unavoidable losses in the most economical way, while maintaining such other conditions requisite for plant growth as will permit crops to utilize to greatest advantage the food supplies within reach.

While much attention has been publicly directed to the conservation of our phosphate deposits, and most wisely so, the real field for conservation, the one that furnishes the most extensive and important opportunity for the effective practice of conservation of plant-food, is the American farm. Each farmer has at his own door opportunities for the study and practice of conservation, and each must for himself apply in his work the principles which will aid in the solution of his immediate problems. Fortunately, trained investigators have done much to point the way in whch true conservaton lies, though some details yet remain to be worked out.

Taken in the aggregate, the controllable losses of plant-food on American farms are fairly appalling. To give, as an illustration, the loss from only a single source, it is a moderate statement, well within the limits of actual truth, that one-third of the plant-food value of the manure produced by the various kinds of farm animals in the United States is lost by carelessness, a loss equivalent to $700,000,000 a year. Most of this enormous waste, equal in value to an average annual wheat crop, *is preventable.*

Wasteful methods of crop raising and neglect of ordinary precautions in saving and utilizing farm resources have led to increasing dependence upon the blind use of commercial fertilizers. This practice in too many cases has developed into a habit not unlike the patent-medicine habit, so far as relates to knowledge used and intelligent reasons underlying use of materials. Millions of dollars have been thrown away because the conditions which make the use of commercial fertilizers most effective have not been appreciated or understood by farmers.

Generally speaking, the fundamental purpose of this book is to study the conditions under which plant-foods, whether in the form of soil compounds, of farm-produced materials, or of commercial fertilizers, can be conserved and at the same time utilized with the greatest efficiency and economy in the production of crops.

PLANT-FOOD

Plant-feeding is not a simple problem. Its solution depends upon much more than the liberal use of farm manure, commercial fertilizers or other materials. Intelligent and successful use of plant-foods in the growing of crops is based upon knowledge of definite facts and of the relations of these facts to one another. As we shall learn later, the beneficial action of fertilizers upon crops depends upon, and is limited by, many different conditions, each of which must be taken into careful consideration if the cost of crop production is to be satisfactorily regulated. The main purpose of this introductory chapter is to make prominent at the beginning some of the important fundamental facts which underlie the practical problem of feeding crops.

We shall have occasion to use frequently such terms as *plant-*

food, available, unavailable, soil fertility, etc., so it is desirable that we define them in the beginning; although the details of our definitions in their various applications will be discussed further on, and their meaning made clearer by more complete development.

As commonly used, the term, *plant-food,* includes compounds that contain nitrogen, phosphorus or potassium. This use has arisen mainly because these constituents are the only ones usually considered in connection with the application of fertilizers in crop growing. This application of the term answers some practical purposes, but is not sufficiently broad to be accurate. For the objects of this treatise, *a plant-food is a substance which supplies any constituent necessary for the nourishment, and in a form suited to promote the development of plants or capable of being changed by natural processes into such a form.*

This swamp area, drained in 1925, has given a real crop every year since. The first crop paid for the drainage expenses.

We shall later learn in detail what are the various constituents essential to plant nutrition and also what are the particular forms or compounds containing these essential constituents that plants can utilize for their growth. For the present it is sufficient to state that the air and the earth's surface are the original sources of all plant-foods.

It will be shown later (p. 123) that a substance must enter a plant before it can be used as food. Gases and liquids are able to enter plants readily, but solids must be made soluble in water before they can do so. The solid compounds that contain plant-food constituents vary greatly in respect to the quickness and completeness with which they dissolve in water. Some compounds must first be changed into other forms of combination. It is, however, not sufficient merely that compounds containing plant-food constituents should be soluble in water in order to serve as plant-food. Many such soluble substances, even after they get inside the plant, are not used as a source of nourishment, though they contain abundance of some plant-food constituent. The number of substances in forms suited to promote plant development at once, or capable of being changed into such forms, is comparatively small, especially those practically or commercially important. To distinguish substances that contain plant-food constituents in forms suited to promote plant development, or capable of being more or less readily changed into such forms, from those not in, nor readily changeable into such forms we use the terms *available* and *unavailable*. The common usage of these terms has been somewhat confusing, and necessarily so, because in many cases, the line of division is obscure and must be more or less arbitrary. The property of availability is dependent on so many complicated conditions, it is well nigh impossible to give a definition that will cover all possible cases. Moreover, full comprehension of any adequate definition can be gained only by study of subsequent chapters.

Available plant-food.—*A plant-food, or a plant-food constituent, is available when it is in such form of combination that plants can immediately utilize it, or when it is in such form that, though not suited to immediate use, it becomes so more or less readily under favorable conditions.*

Unavailable plant-food.—*A plant-food, or a plant-food constituent, is unavailable when it is in such form of combination that plants cannot utilize it under any natural conditions, or when it becomes available so slowly under favorable conditions that it does not furnish appreciable amounts of material utilizable by growing crops.*

The meaning of these definitions will be made clear by specific illustrations.

(1) *Plant-food immediately available.*—Nitrogen in the combination known as *nitrate* (p. 30) furnishes a typical case of *immediately available* plant-food. Of the large number of compounds that contain the essential plant-food constituent, nitrogen, the one form of chemical combination which plants use most quickly and extensively as food is that of nitrate nitrogen. It is a significant fact also that all nitrate combinations present in soils are easily soluble in water. Similarly, the *phosphorus* and *calcium* in acid calcium phosphate (p. 35) and the *potassium* in potassium chloride

(muriate of potash), or in potassium sulfate or carbonate (p. 38), are available, because these compounds readily dissolve in water and can be taken into plants at once for use as food without having to be changed into other forms of chemical combination.

(2) *Plant-food not immediately available.*—Organic materials that contain nitrogen, dried blood, fish-scrap, cottonseed-meal, castor pomace, etc., are insoluble in water and can not be used at once as plant-food; but under the action of micro-organisms (p. 139), and under favorable conditions their nitrogen is changed into nitrate nitrogen, the rapidity of the change depending upon temperature, moisture, kind and number of micro-organisms, fineness of plant-food material, etc. Under usual conditions, when applied to soils in spring, the nitrogen in such substances becomes completely available in time to be used by crops during the growing season. While not immediately available, this nitrogen is classed as available, and is actually so from the practical standpoint of supplying the needs of growing crops.

As a further illustration under this head, farm manure, when strictly fresh, contains no immediately available plant-food. As it passes through various processes of fermentation changes under the action of micro-organisms, its plant-food constituents gradually pass into immediately available forms, beginning with the liquid portion and in time extending to the solid part. When applied to soils in sufficient amount, farm manure usually continues to show beneficial effects on crop growth for three or four years, corresponding with the change of slowly available plant-food constituents into forms immediately available.

(3) *Compounds wholly unavailable.*—Many compounds that contain constituents essential to plant growth are readily soluble in water but, if taken into the plant, are not used as food; in some cases compounds of this kind, even in minute amounts, act as poisons and seriously interfere with the normal activities of plants. When such substances, even though they contain constituents essential to plant nutrition and readily soluble in water, are not capable of change under ordinary conditions into compounds that plants can use as food, they are wholly unavailable—in fact, they are not plant-foods at all. Instances are potassium cyanide, calcium or magnesium chloride.

(4) *Plant-food constituents largely unavailable.*—When the term ***unavailable*** is applied to materials that contain plant-food constituents, it is usually in the sense of *very slowly available*. In illustration of this class of substances are leather and hair, which show great power of resistance to the action of micro-organisms. Their nitrogen requires many years for complete conversion into available form. So they are of little value in furnishing plant-

food for growing crops. For practical purposes, they are regarded as unavailable or inert. Similarly, iron and aluminum phosphates (p. 37), which constitute a large portion of the phosphate compounds in many soils, are so slowly changed under ordinary conditions into forms that plants can use they are classed as unavailable.

Relative influence of alfalfa and timothy on succeeding crop of fodder corn. *Left, one year of alfalfa; right, one year of timothy.*

Strictly speaking, a plant-food constituent in unavailable form is not *actual* plant-food at all, but simply raw material which under proper conditions may become actual plant-food in time. Unavailable plant-food is, therefore, regarded as *potential* plant-food; that is, an existing source of future supply of *actual* plant-food. Unavailable plant-foods are constantly becoming available, so control of the processes of converting potential into actual plant-food constitutes one of the fundamental problems of successful crop production. The conditions under which unavailable is changed into available plant-food constitute an important line of study in future chapters.

Factors that affect availability of plant-food.—Since many factors must be taken into consideration, the subject of availability of plant-food is in many cases even more complicated than might appear from the preceding discussion (p. 283). We will anticipate to the extent of stating without enlargement that, prominent among the factors that affect the availability of plant-foods, are the following: (1) Chemical character of the plant-food materials, (2) fineness of division, (3) various soil conditions, especially those that affect micro-organisms, and (4) the character of the crop grown.

FACTORS OF SOIL FERTILITY

In common usage, a soil is fertile when it produces large crops. The expression, soil fertility, is used to include the factors that make soils productive; definitions vary in respect to the factors included. The most common meaning given to soil fertility is simply abundance of available plant-food in the soil. This is too narrow, since a soil may contain available plant-food sufficient to meet the requirements of large crops and yet be unproductive. The broadest possible use of the term, soil fertility, includes all conditions that affect crop production, not only those more immediately centered in the soil itself, but also those outside, such as rainfall, temperature, adaptation of crops to soil and climate, selection of seed, and all other factors that may be regarded as external.

While the factors of soil fertility are divided into two general classes, internal and external, they are so closely bound together in crop production they can be separated only in theory. When we speak of a soil as fertile, we always assume that the external factors, rain, sunshine, seed, etc., are present to furnish favorable conditions for crop growth, and to supplement those of the soil.

As a general statement, *under the term, soil fertility, are included all the soil factors that contribute to plant growth when supplemented by favorable external conditions. Ability to produce crops under conditions favorable to plant growth is a measure of soil fertility.*

The conditions needed in a soil to promote plant growth are: (1) Abundance of available plant-food. (2) A physical structure (*a*) which combines mellowness and firmness, permits plant-roots to extend their growth freely; (*b*) which enables the soil to receive water easily, distribute it promptly, hold it with sufficient tenacity, and give it up as needed by plants ; (*c*) which permits some circulation of air; and (*d*) which enables the soil to absorb heat and maintain a degree of warmth suited to plant growth. (3) The presence of and conditions favorable to beneficial micro-organisms. (4) Absence of injurious amounts of substances poisonous to plants.

The factors of soil fertility are conveniently grouped under three general divisions, as follows:

(1) *Chemical composition,* or the relations of soils to (*a*) the kinds and amounts of plant-food present and (*b*) substances injurious to plant growth.

(2) *Physical condition,* the state of fine or coarse division and other physical properties of soils, and also the relations of soils to air, moisture, heat and light.

(3) *Biochemical condition,* or the relations of soils to the action of micro-organisms.

RELATION OF PLANT-FOOD TO OTHER FACTORS OF SOIL FERTILITY

The fact that soil fertility is so generally regarded as synonymous with abundance of available plant-food has some justification, since it is a fact worthy of notice that plant-food appears to be the one factor to which others are, in a sense, more or less largely subordinated. Someone has defined an animal as a stomach with appendages necessary for feeding it. In a somewhat similar way, a plant can be regarded as an organism furnished with an environment whose factors combine to prepare and supply food for it. The insoluble plant-food constituents of the rock-crust of the earth have been pulverized by various natural agencies (p. 69), and also been made soluble as the result of the chemical action of water, carbon dioxide (carbonic acid) and oxygen (p. 70), together with the help of micro-organisms (p. 153). Various agencies combine to make the soil mellow so that the plant-roots can find their way to the supplies of plant-food (p. 77). Water holds plant-food in solution ready for absorption by plant-roots, carrying it into and throughout the plant (p. 101). And so, chemical, physical and biological factors, all combine to prepare the plant-food supply. Those conditions of soil recognized as best suited to crop production are the ones which best furnish available plant-food. It follows, therefore, that the preparation and utilization of plant-food are dependent upon many factors, all of which must be considered in any practical discussion of plant-feeding. We must, then, keep in mind that certain conditions are essential before a crop can use plant-food to advantage, whether in the form of an applied fertilizer or as derived directly from the soil. Various conditions limit and determine the effect of plant-food. Some of the factors are more or less largely beyond man's control; others are within his power to regulate. Among the most important specific factors, capable of control, are the following: (1) Soil organic matter, (2) soil moisture, (3) soil acidity, (4) soil structure, (5) soil microorganisms.

It is obvious that any treatment of the use of fertilizers which confines itself to a statement of how many pounds of nitrogen, phosphorus, or potassium compounds to apply to crops, must be narrow and superficial and in the nature of an ordinary recipe. Therefore, in the preparation of this book, it has seemed necessary to consider first the factors that determine effectiveness before we can consider to advantage the practical use of fertilizers in crop-growing.

II

THE CHEMICAL COMPOSITION OF PLANTS

IN our study of fertilizers, we shall take as our starting-point the constituents necessary for plant nutrition, their sources, compounds, properties and functions.

All forms of matter known to us are composed of over ninety different chemical elements. An element is a substance which cannot be separated by any known means into two or more different substances. For example, pure iron is an element because by no process can we get anything out of it but iron. Pure iron contains nothing but iron.

Chemical elements in plants.—Chemical analysis has shown that the many thousands of kinds of plants growing on the earth are made from a comparatively small number of the elements. While more than twenty elements have been found in plants, they are not all necessary to plant nutrition.

It is now generally agreed that the following ten elements are essential to the normal growth of plants: calcium, carbon, hydrogen, iron, magnesium, nitrogen, oxygen, phosphorus, potassium, sulfur.

Recent research indicates that four other elements, boron, copper, manganese and zinc, may be required. In more than minute amounts, compounds containing these elements are injurious to plants.

Some other elements (aluminum, arsenic, barium, chromium, cobalt, iodine, nickel, silicon, strontium, cæsium, vanadium and titanium) may be found to be needed by some plants or to meet some special physiological demand.

While sodium, chlorine and silicon compounds are always present in plants, it has not yet been satisfactorily shown that they perform any necessary function in plant nutrition. They may be regarded as simply being carried into a plant because they are present in solution in the vicinity of the roots of growing plants and are absorbed in company with the plant nutrients proper.

Air-derived and soil-derived elements.—The elements present in plants are divided into two distinct classes: *Air-derived* (carbon, hydrogen, nitrogen, oxygen) and *soil-derived* or *mineral* or *inorganic* (calcium, iron, magnesium, phosphorus, potassium, etc.). These two classes differ in three important respects:

(1) *Source.*—The elements of the first class come to plants exclusively from the air, either directly, or indirectly through the medium of the soil. Elements of the second class come exclusively from the soil.

(2) *Effect of burning.*—When a plant is completely burned, its air-derived elements disappear for the most part as gases; the soil-derived (mineral or inorganic) elements, usually much the smaller part, are left as an unburned residue or ash, upon which further heating has no effect. This distinction is not perfectly sharp, since some oxygen is always found in the ash, while some of the chlorine, phosphorus and sulfur may be vaporized by the burning.

(3) *Proportions in plants.*—Air-derived elements make up over 95 per cent. of plants, while the soil-derived elements occur in small amounts.

Elements and compounds.—The chemical elements do not usually exist separate from one another as pure elements. While we are familiar with impure carbon in the form of coal and charcoal, and with nitrogen and oxygen as mixed together in the air, we never find under ordinary conditions elements like hydrogen, phosphorus, potassium, etc., existing separate from other elements. Various elements combine to form compounds somewhat as the letters of the alphabet combine to form words. The chemical elements used in plant growth exist in air and soil in the form of compounds, some of which are used as food by plants. After being taken into the living plant, they are worked over into a great variety of new compounds; and these within the plant, are more or less regularly grouped or mixed together in a great variety of ways in the processes of plant-building. The elements may, therefore, be regarded as the raw materials from which plant-foods come.

Classes of chemical compounds in plants.—Some of the more familiar chemical compounds found in plants are:

(1) *Water,* the most abundant compound found in fresh, growing plants, forms 60 to 95 per cent. The relations of water to soils and crops will be discussed later (p. 101).

(2) *Carbohydrates.*—Under this general name is included one of the most important and widely distributed groups of compounds which form a large proportion of the dry matter of plants. Each carbohydrate contains carbon, together with hydrogen and oxygen in the proportion present in water, as suggested by the word. This extensive group includes the sugars and starches.

Celluloses, gums and pectins are more complex carbohydrates. They furnish the material of which the structural framework or skeleton of plants is made. The walls of plant-cells and the intervening spaces consist largely of these materials. While found in largest proportions in the stems of woody plants, these are also present as structural material in every part of plants. The several kinds of celluloses include *normal* cellulose, as represented by the fibers of cotton, flax, hemp, etc., and *ligno-cellulose* or lignin, commonly known as woody fiber. Familiar examples of *gums* are gum arabic, gum tragacanth, and cherry gum. *Pectins* are familiar in fruits as the materials which cause concentrated fruit juices to jelly.

(3) *Oils* are contained in all plants, many of which are of commercial importance, such as linseed, castor, olive, cottonseed, corn, etc. These contain the elements, carbon, hydrogen and oxygen, but in relative proportions different from those occurring in carbohydrates.

(4) *Proteins.*—This class of compounds may be regarded as the most important group of the organic constituents of plants. They make up the active portion of protoplasm, which is the living part of plant-cells, making possible the chemical changes that result in plant growth. Vegetable proteins are highly essential in animal nutrition. In addition to the three elements contained in carbohydrates, (carbon, hydrogen and oxygen), proteins contain nitrogen and frequently sulfur. Some of them contain phosphorus also. As a familiar representative of this class, we have "gluten," a mixture of proteins, which forms the sticky portion of wheat flour.

(5) *Acids.*—Many of the acids and their acid salts produced in plants are important articles of commerce. Their presence in many fruits accounts in part for the special flavors of such fruits. Among the more common vegetable acids are: (*a*) Tartaric acid, found in grapes; (*b*) citric acid, present in lemons, currants, cherries, strawberries and similar fruits; (*c*) malic acid, the characteristic acid of apples, tomatoes and several small fruits. These acids contain the elements, hydrogen, carbon and oxygen, combined in various proportions.

(6) *Aromatic substances.*—Most plants possess in their flowers, fruits, leaves and roots compounds having characteristic flavors. These are due frequently to volatile oils (consisting of carbon and hydrogen), such as oils of turpentine, lemon, cinnamon, almond, etc., or to other special plant compounds. The characteristic flavors of pineapple, apple, strawberry, banana, wintergreen, mint, etc., are due to aromatic substances. They usually consist of the three elements, carbon, hydrogen and oxygen.

(7) *Medicinal substances.*—There is a large variety of widely distributed substances used especially for medicinal purposes, among which, for example, are morphine (from the poppy), quinine

(from cinchona bark), nicotine (from tobacco), strychnine (from seeds of *Strychnos nux-vomica*), belladonna (from deadly nightshade), etc.

(8) *Mineral compounds.*—It is noticeable that the preceding classes of plant compounds are made up chiefly of air-derived elements and are mostly among the most important articles of commerce. The soil-derived or inorganic elements are present in the plant largely in the form of combinations known as salts, such as carbonates, chlorides, phosphates and sulfates of potassium, calcium, magnesium, iron, sodium, etc. (p. 80). These mineral compounds exist in the tissues of plants in three general forms: (1) In solution in the plant juice or sap, (2) as crystals in plant cells, and (3) as incrustations in the cell walls.

Relative agricultural importance of various elements.—While the absence of any one essential constituent seriously impairs or wholly prevents plant growth, there is a sense in which some of the soil-derived elements are of much greater agricultural importance than others. Certain ones are more extensively used by crops and sooner require special attention to increase the available supply, while others are used in such small amounts, relative to the available supply, that they rarely need attention. The elements of special importance are: Nitrogen, phosphorus, potassium, calcium, and, in some cases, sulfur.

REFERENCES

The Rôle of the Rarer Elements in Soils, Foods and Drugs. O. Schreiner. Jour. Ass'n. Off'l. Agr'l. Chemists. Vol. 12, p. 16, 1928.

Occurrence of Copper, Manganese, Zinc, Nickel and Cobalt in Soils, Plants and Minerals, and their Possible Functions as Vital Factors. J. S. McHargue. Jour. Agr. Res. Vol. 30, p. 193. (16 references.)

Chemistry of Plant Life. R. W. Thatcher, 1921.

III

ELEMENTARY FACTS OF CHEMISTRY[1]

As material for study of a few elementary chemical facts of importance, and as a source of reference, we will present briefly some of the simpler fundamentals of chemical combination, which have a direct bearing on the subject of plant-foods. It is desirable to get some idea, even though crude and limited, of what is meant by such terms as acid, alkali, base, salt, chemical neutralization, reaction, etc.

Symbols and combining proportions of elements.—As a matter of convenience in saving the labor of writing in full the names of elements and chemical compounds every time used, a kind of chemical shorthand system has been adopted, in which the name of each element is represented by a characteristic symbol, consisting of the first letter or letters of the element's name. Where the English name differs from the Latin name, the letters of the symbol are taken from the Latin form.

Elements combine to form compounds. It is a fact of great importance that in combining with one another, they unite in certain definite weights or proportions. Each chemical compound always contains the same elements in exactly the same proportions. By analyzing large numbers of the compounds of various elements, a number has been found which represents the proportion by weight in which each element enters into combination with others. In Table 1 we give the names of those elements in which we are interested, their symbols and their approximate combining weights.

The chemical symbol of an element is seen to stand for two things, (1) for the name of the element and (2) for the combining weight. The compound known chemically as sodium chloride,

[1] For those who do not desire any knowledge of chemistry, this chapter may be omitted in reading. For others who desire to become acquainted with some of the simpler fundamental facts of chemistry, such as will give them a clearer understanding of the various relations of plant-foods, this chapter is recommended for serious study, especially in connection with subsequent chapters. It will be well to refer to it from time to time until its contents are familiar. Teachers who use the book in class work with students who have not studied chemistry can exercise their judgment as to the way in which this chapter shall be used. For a complete understanding of parts of Chapter III, this chapter is necessary.

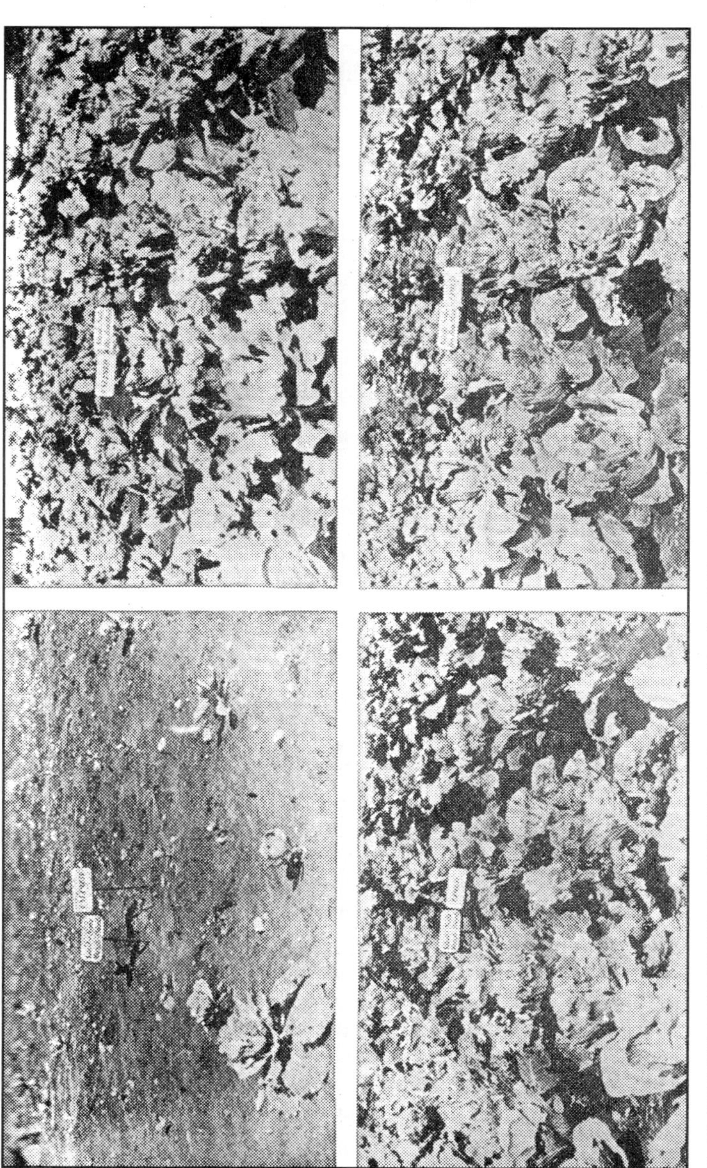

Effect of acidulated phosphate in neutralizing "soil acidity." *Upper left*, nitrate of soda and muriate of potash without lime, 160 pounds good cabbage, 840 poor (crop failure). The omission of phosphorus under acid soil conditions was fatal. *Upper right*, nitrate of soda and phosphoric acid without lime, 13,560 pounds good cabbage, 9,600 poor. *Lower left*, nitrate of soda, muriate of potash and lime, 21,560 pounds of good cabbage, 5,880 poor. *Lower right*, nitrate of soda, phosphoric acid and lime, 20,000 pounds good cabbage, 7,600 poor.

familiar to us as common salt, has its composition shown by the symbol or formula, NaCl, in which 23 parts by weight of sodium, its combining weight, are always combined with 35.5 parts by weight of chlorine, no more and no less, the combining weight of Cl being 35.5 In calcium oxide, commonly known as lime, the symbol or formula, CaO, means that 40 parts by weight of calcium are always united with exactly 16 parts by weight of oxygen. It should be stated further that the combining weight always represents the *smallest amount* of an element that enters into combination. In other words, we do not have half of a combining weight but always the full value given in the table. Taking 1 as the combining weight of hydrogen (H), the lightest element, the other combining weights are those stated in Table 1.

TABLE I—NAMES, SYMBOLS AND COMBINING WEIGHTS OF CHEMICAL ELEMENTS

Name of element	*Symbol*	*Combining weight*
Hydrogen	H	1
Carbon	C	12
Nitrogen	N	14
Oxygen	O	16
Sodium (Latin, *Natrium*)	Na	23
Magnesium	Mg	24
Aluminum	Al	27
Silicon	Si	28.5
Phosphorus	P	31
Sulfur	S	32
Chlorine	Cl	35.5
Potassium (Latin, *Kalium*)	K	39
Calcium	Ca	40
Manganese	Mn	55
Iron (Latin, *Ferrum*)	Fe	56

In many compounds, one or more elements may be combined in a proportion representing two or more times the combining weight given in the preceding table. For example, in the case of water we have 2 parts by weight of hydrogen (H) and 16 parts by weight of oxygen (O); this is conveniently expressed by the formula, H_2O (equal to HHO). Whenever an element is used in proportions representing more than one combining weight, this fact is expressed by writing the corresponding number below and at the right hand of the symbol of the element. This is further illustrated in the case of nitric acid, HNO_3 (equal to HNOOO), in which we have 1 part (1 combining weight) of hydrogen, 14 parts (1 combining weight) of nitrogen and 48 parts (3 combining weights, 3×16) of oxygen. Again, in the compound sodium phosphate, Na_3PO_4 (equal to NaNaNaPOOOO), we have 3 combining weights of sodium, 1 of phosphorus, and 4 of oxygen; or, expressed in parts by weight, 69 (23×3) parts of sodium, 31 parts of phosphorus and 64 (16×4) parts of oxygen.

In this connection, it is desirable to point out a mistake often made in respect to the meaning of chemical formulas by those who have not studied chemistry. For illustration: Na_3PO_4 by some would erroneously be taken to mean 3 parts of sodium, 1 part of phosphorus and 4 parts of oxygen, which is far from the truth, since the symbol of an element stands, not for equal parts, but for *combining weights*. No two elements have the same combining weight. If the combining weights of different elements were the same, it would be proper to speak of the symbols in the formula Na_3PO_4 as representing 3, 1 and 4 parts respectively of the elements in question; but since these symbols represent different weights, the compound contains by weight, as a matter of fact, 69 parts (3 combining weights) of sodium, 31 parts (1 combining weight) of phosphorus and 64 parts (4 combining weights) of oxygen.

How to calculate percentages of elements in a compound.—When we know (1) the chemical formula of any compound and (2) the combining weight of each element in the compound, we can easily calculate the percentage of each constituent element. For illustration, potassium chloride (muriate of potash) composition is expressed by KCl, containing, according to Table 1, 39 parts by weight of K and 35.5 parts of Cl. In order to find the percentage of each element in KCl, we add the combining weights of the K and Cl, multiply each by 100 and then divide each result by the sum of the combining weights of K and Cl as follows:

(1) Combining weight of K = 39
(2) " " " Cl = 35.5

(3) Sum of combining weights of K + Cl = 74.5
(4) (K) 39 × 100 ÷ 74.5 = 52.3 the percentage of K in KCl
(5) (Cl) 35.5 × 100 ÷ 74.5 = 47.7 " " " Cl " "

Three other cases are: (1) Sodium nitrate (nitrate of soda), (2) potassium sulfate and (3) calcium phosphate.

Example 1.—Find the precentages of each element in sodium nitrate, $NaNO_3$.

(1) Combining weight of Na = 23
(2) " " " N = 14
(3) " " " O_3 = 48 (16 × 3)

(4) Sum of combining weights of Na + N + O_3 = 85
(5) (Na) 23 × 100 ÷ 85 = 27, the percentage of Na in $NaNO_3$
(6) (N) 14 × 100 ÷ 85 = 16.5 " " " N " "
(7) (O_3) 48 × 100 ÷ 85 = 56.5 " " " O " "

Example 2.—Find the percentages of each element in potassium sulfate, K_2SO_4.

(1) Combining weight of $K_2 = 78$ (39×2)
(2) " " " $S = 32$
(3) " " " $O_4 = 64$ (16×4)

(4) Sum of combining weights of $K_2 + S + O_4 = 174$
(5) (K_2) $78 \times 100 \div 174 = 44.8$, the percentage of K in K_2SO_4
(6) (S) $32 \times 100 \div 174 = 18.4$ " " " S " "
(7) (O_4) $64 \times 100 \div 174 = 36.8$ " " " O " "

Example 3.—Find the percentages of each element in calcium phosphate, $Ca_3P_2O_8$.

(1) Combining weight of $Ca_3 = 120$ (40×3)
(2) " " " $P_2 = 62$ (31×2)
(3) " " " $O_8 = 128$ (16×8)

(4) Sum of combining weights of $Ca_3 + P_2 + O_8 = 310$
(5) (Ca_3) $120 \times 100 \div 310 = 38.7$, percentage of Ca in $Ca_3P_2O_8$
(6) (P_2) $62 \times 100 \div 310 = 20.0$ " " P " "
(7) (O_8) $128 \times 100 \div 310 = 41.3$ " " O " "

A careful study of the preceding examples will enable one easily to make use of the following rule: To find the percentage of any constituent of a chemical compound, when the chemical symbol or formula is known, *first find in Table 1 the combining weight for each element represented in the compound, and take the combining weight of each element as many times as indicated in the formula. Add the combining weights thus obtained. Then divide this sum into 100 times the combining weight or weights of each element present.*

Chemical action of acids and bases.—Caustic soda or soda lye, a material familiar in many common uses, is known in chemistry as *sodium hydroxide,* having the composition represented by the symbol Na(OH); it is also known as a *base,* an *alkali,* and an *alkali base.* (p. 22). This substance, when added to an acid, produces noticeable changes. For example, if we add together solutions of nitric acid (HNO_3) and of sodium hydroxide in certain proportions, the result will be neither nitric acid nor sodium hydroxide, but a new compound, *sodium nitrate* ($NaNO_3$), formed by the action of the acid and the alkali upon each other.

The nitric acid used in the experiment tastes sour and biting while the caustic soda solution has a peculiar odor and feels soapy on the hands, and, if concentrated enough, destroys or "eats" the skin. After these two compounds in solution are brought together in the right proportions, there can be no longer observed any sour taste of acid, or soapy feeling, or alkali odor of the caustic soda, because acid and alkali have *neutralized,* or in popular language "killed," each other, which means that they have combined to form new compounds; in this case the new compounds are water (H_2O) and sodium nitrate ($NaNO_3$), this latter belonging to a class of chemical compounds known as *salts.* The presence of the nitrate is readily perceived by its characteristic salty taste. This chemical change or reaction can be represented by means of a chemical equation, as follows:

Na [OH	+	H]	NO_3	=	$NaNO_3$	+	H_2O
Sodium hydroxide			Nitric acid		Sodium nitrate		Water

Many important and practical facts of chemistry are illustrated by this simple change. We are at once met by such questions as the following: What is a base or an alkali? What is an acid? What kind of a product is formed in bringing together a base or an alkali and an acid?

Acid, alkaline and neutral reactions.—How can we tell, in a practical way, the solution of a base from one of an acid? This is done by means of some coloring substance, called an indicator, which is so acted on by alkalis and acids as to undergo changes of color, producing one color with alkalis and a different color with acids. For example, one substance in common use as an indicator is the dyestuff known as *litmus,* which is blue. If we put a few drops of a litmus solution in some water and then add a drop or two of some acid, the color changes to red. If now the red solution is treated with a few drops of a solution of a base like sodium hydroxide, the red color disappears and the solution again becomes blue.

For ordinary uses, such as testing acidity or alkalinity of soils, litmus is used in the form of strips of paper colored red or blue with litmus dye. To test whether a substance is basic (alkaline) or acid, a strip each of blue litmus and red litmus paper is put into the solution tested; if the blue strip changes to red, the solution is acid; if the red changes to blue, the solution contains a base or alkali; if neither color changes, the solution contains neither acid nor alkali. The behavior of any solution to litmus, or to any other indicator used for acids and alkalis, is described by calling it (1) *acid,* (2) *alkaline* or *basic,* or (3) *neutral.* A reaction is called *strong* when the color change is quick and pronounced. Solutions reacting thus are called strongly acid or strongly alkaline. A reaction is called *weak* when the color change is slow or not very pronounced. Solutions acting thus are said to be weakly acid or weakly alkaline. The strong or the weak character of the reaction of a solution depends primarily upon the amount of the alkali or acid in solution.

Making application of these statements to the illustration given above in the action of sodium hydroxide and nitric acid, sodium hydroxide turns red litmus blue, so is alkaline or basic in reaction; nitric acid turns blue litmus red so its reaction is acid; the mixture, of water and sodium nitrate, which are the products formed by the action of sodium hydroxide and nitric acid, has no effect upon either red or blue litmus and is therefore neutral in reaction.

Bases *are compounds which have the power of neutralizing acids and thereby forming neutral salts.* The compounds that act most commonly as bases are *hydroxides* and *oxides* of certain elements known as metals, called *base-forming* elements. The bases of most interest to us are the following:

TABLE 2—NAMES AND SYMBOLS OF COMMON BASES

Name of base	*Symbol*	*Common name*
Sodium hydroxide	$NaOH$	Caustic soda
Potassium hydroxide	KOH	Caustic potash
Ammonium hydroxide	$(NH_4)OH$	Ammonia water
Calcium hydroxide	$Ca(OH)_2$	Slaked lime
Calcium oxide	CaO	Lime, quicklime, etc.
Magnesium hydroxide	$Mg(OH)_2$	Hydrated magnesia
Magnesium oxide	MgO	Magnesia

Attention should be called to the *ammonium* compound. The elements nitrogen and hydrogen, taken in the proportions represented by the symbol NH_4, act together as if one element, and resemble metals in forming basic and other compounds.

(1) *Alkalis.*—Of the various bases mentioned in the preceding list, potassium, sodium and ammonium hydroxides are called *alkalis* or *alkali bases;* their solutions have a strongly alkaline reaction, as shown by quickly changing red litmus to blue, even when present in small amounts. The alkalis possess very marked power in neutralizing acids, and for this reason are said to be strong bases or strongly basic.

(2) *Alkaline earths.*—Calcium and magnesium belong to a group of elements often called *alkaline earths,* because their oxides and hydroxides have an alkaline reaction. While their power to neutralize acids is not so great as that of the alkalis, they are more strongly basic than are the bases of many other metals.

(3) *Carbonates with basic power.*—The compounds known as carbonates have the power to neutralize acids and are therefore basic. The alkali carbonates have an alkaline reaction, changing red litmus to blue. The most important carbonate in agriculture

is calcium carbonate, $CaCO_3$ (carbonate of lime), on account of its vital relations to soils and crops. The action of carbonates and acids is represented in the following equation, in which the neutral compound, calcium nitrate, is formed as the most important product.

$CaCO_3$	+	$2HNO_3$	=	$Ca(NO_3)_2$	+	H_2O	+	CO_2
Calcium carbonate		Nitric acid		Calcium nitrate		Water		Carbon dioxide

Acids *are compounds which have the power to neutralize bases and thereby form neutral salts.* Solutions of acids turn blue litmus red. For our purpose, acids consist of two chemical parts, (*a*) the element hydrogen (H) and (*b*) an acid-radical, consisting of one or more acid-forming elements. For example, hydrochloric acid, HCl, consists of (*a*) H and (*b*) Cl (acid-radical); nitric acid, HNO_3, composed of (*a*) H and (*b*) NO_3 (acid-radical); sulfuric acid, H_2SO_4, consists of (*a*) H_2 and (*b*) SO_4 (acid-radical); phosphoric acid, H_3PO_4 consists of (*a*) H_3 and (*b*) PO_4 (acid-radical).

These compounds are often spoken of as *free* acids. By this is meant an acid unneutralized, or the acid itself, uncombined with any other element or elements, in distinction from a neutralized acid, or an acid in combination as a salt.

In Table 3 (p. 24) is a list of the acid-forming elements, together with the formulas and names of those acids which possess special interest in our study of plant-food materials.

Salts.—We have already learned: (1) That a base consists of some basic element, generally a metal, in union with OH (forming an hydroxide) or with O (forming an oxide); (2) that acids are made up of (*a*) hydrogen (H) in union with (*b*) an acid-radical consisting of one or more acid-forming elements.

When we bring acids and bases together, we obtain two products as the result of the chemical change, (1) water (H_2O), and (2) *a compound consisting of the metal of the base combined with the acid-radical of the acid; compounds made up in this way are called salts,* as illustrated by the following equations:

	Bases			*Acids*			*Salts*		
(1)	Na	[OH	+	H]	Cl	=	NaCl	+	H_2O
	Sodium hydroxide			Hydrochloric acid			Sodium chloride		Water
(2)	Ca	$[(OH)_2$	+	$H_2]$	CO_3	=	$CaCO_3$	+	$2H_2O$
	Calcium hydroxide			Carbonic acid			Calcium carbonate		Water

(3)	Na	[OH	+	H]	NO_3	=	$NaNO_3$	+	H_2O
	Sodium hydroxide			Nitric acid			Sodium nitrate		Water
(4)	2K	[OH	+	H_2]	SO_4	=	K_2SO_4	+	$2H_2O$
	Potassium hydroxide			Sulfuric acid			Potassium sulfate		Water

In each equation there is formed, in addition to water, a *salt,* consisting of (*a*) the metal of the base combined with (*b*) the acid-radical of the acid. The hydrogen of the acid combines with the hydrogen and oxygen (HO) of the base to form water (H_2O); the metal of the base, in forming a salt, takes the place of the hydrogen of the acid.

Many of the substances with which we have to deal in connection with plant-foods belong to this class of chemical compounds, which we call *salts.*

Names and symbols of acids and salts.—Table 3 gives (1) the names of the acid-forming elements, (2) the names and symbols of the more common acids formed from them, and (3) the names of the salts corresponding to the acids. This table, together with Table 4 will be useful for future reference.

TABLE 3—ACID-FORMING ELEMENTS AND SOME OF THEIR ACIDS

Acid-forming elements	*Names and symbols of acids*		*Names of salts formed from acids*
Chlorine	Hydro-chlor-ic	HCl	Chlorides
Nitrogen	Nitr-ic	HNO_3	Nitrates
Sulfur	Hydro-sulfur-ic	H_2S	Sulfides
	Sulfur-ic	H_2SO_4	Sulfates
Carbon	Carbon-ic	H_2CO_3	Carbonates
Silicon	Silic-ic	H_2SiO_3	Silicates
Phosphorus	Phosphor-ic	H_3PO_4	Phosphates
Oxygen	....	....	....

(1) *Names of most acids* end in the syllable *ic,* added to the full or the shortened name of the acid-forming element in the acid. In the case of acids that contain no oxygen, like HCl and H_2S, the name is formed by placing before the name of the acid-forming element the syllable *hydro-,* and after the element the syllable *ic;* for example, hydro-chlor-ic acid, hydro-sulfur-ic acid.

In addition to these two simple forms of acids are many other kinds formed from these same elements, distinguished by special modifications of composition and name, but we confine our attention to the forms of most interest to us.

The element oxygen is present along with some other acid-forming element in most acids. The name oxygen, derived from two Greek words, means acid-producing. The presence or absence of oxygen in an acid is shown by the name of the acid, as previously explained.

(2) *Names of salts* derived from the oxygen-containing acids

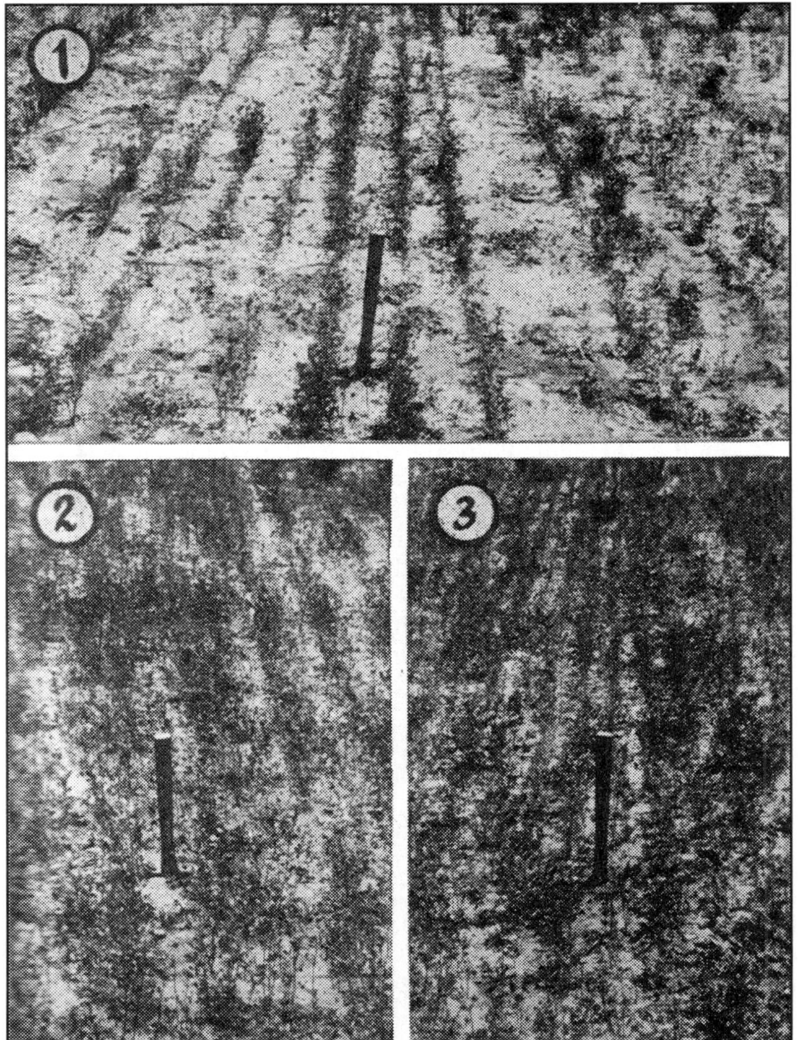

Alfalfa sown at fifteen pounds to the acre. 1. With six inch grain drill. Blanks due to too deep seeding and bunched feeding—4,450 pounds of hay. 2. Broadcast seeding harrowed in. Good stand if rain is sufficient—4,040 pounds of hay. 3. Sown with cultipacker grass seeder—4,375 pounds of hay.

in the list are designated by using first the name of the metal in the salt and adding to this the name of the acid, changing the last syllable from *-ic* to *-ate*. Salts derived from the acids that contain no oxygen end in the syllable *-ide* instead of *-ate;* the prefix *hydro-* in the name of the acid is dropped in naming the salt.

(3) *Symbols of salts.*—Every metal, or base-forming element, or group of elements (NH_4, for example) can form a salt with every acid, at least in theory. In the list (Table 4) are the symbols of some salts many of which are of interest in agriculture.

TABLE 4—SALTS FORMED BY METALS AND ACIDS

Salts of Sodium	*Salts of Potassium*	*Salts of Ammonium*	*Salts of Calcium*	*Salts of Magnesium*	*Names of Salts*
$NaCl$	KCl	NH_4Cl	$CaCl_2$	$MgCl_2$	Chloride
$NaNO_3$	KNO_3	NH_4NO_3	$Ca(NO_3)_2$	$Mg(NO_3)_2$	Nitrate
Na_2S	K_2S	$(NH_4)_2S$	CaS	MgS	Sulfide
Na_2SO_4	K_2SO_4	$(NH_4)_2SO_4$	$CaSO_4$	$MgSO_4$	Sulfate
Na_2CO_3	K_2CO_3	$(NH_4)_2CO_3$	$CaCO_3$	$MgCO_3$	Carbonate
Na_2SiO_3	K_2SiO_3	$(NH_4)_2SiO_3$	$CaSiO_3$	$MgSiO_3$	Silicate
Na_3PO_4	K_3PO_4	$(NH_4)_3PO_4$	$Ca_3(PO_4)_2$	$Mg_3(PO_4)_2$	Phosphate

In finding the name corresponding to the symbol of a salt, or in finding the symbol of a salt when the name is given, it may be stated that, as a general rule, the chemical name of a salt is intended to express the composition as closely as possible.

(*a*) To find the name of a salt that corresponds to any symbol in the table, take first the name of the metal at the head of the column in which the formula occurs and then add to it the name in the last column opposite the symbol. For example, to find the name of the salt whose symbol is Na_2SO_4, we find *sodium* at the head of the column and *sulfate* in the last column opposite; the name of the salt is sodium sulfate. Similarly, we find the name corresponding to K_3PO_4 to be potassium phosphate; $(NH_4)_2CO_3$, ammonium carbonate, etc.

(*b*) To find the symbol that corresponds to any given name of a chemical salt in the list, look in the column under the name of the metal contained in the salt and then go down the column until the symbol opposite the name in the last column that corresponds to the second part of the salt is reached. For example, the formula for sodium phosphate is in the column headed *sodium* in the line opposite the word *phosphate* of the last column; it is Na_3PO_4.

(*c*) Uniting or combining power of metals and acid-radicals as shown by acids and salts: One can readily write all the symbols of salts in the table by committing to memory the three following statements:

(1) In case of acids that contain one H, as HCl and HNO_3, *one acid-radical* (-Cl or -NO_3) combines *with one Na* or *K* or *NH_4*; while *two* (-Cl_2, -$(NO_3)_2$) are required to unite with one *Ca* or *Mg*.

(2) In case of acids that contain two H, as H_2S, H_2SO_4, H_2CO_3, etc., *one acid-radical* (-S or -SO_4 or -CO_3, etc.) combines with two of Na, K or NH_4 (*Na_2*, K_2 $(NH_4)_2$); while one such acid-radical requires only *one Ca* or *Mg*.

(3) In case of acids that contain three H, as H_3PO_4, *one acid-radical* (-PO_4) combines with *three* of *Na, K* or *NH_4* (Na_3, K_3, $(NH_4)_3$); while with *Ca* or *Mg two acid-radicals* ($(PO_4)_2$) combine with *three Ca* or *Mg* (Ca_3, Mg_3).

The combination of Ca and Mg with -PO_4 is not as easily understood as the others. For instance, we take Ca_3 and $(PO_4)_2$ in writing the symbol of calcium phosphate. The combining power of Ca is 2 in comparison with -PO_4 as 3, or one -PO_4 equals 1½ Ca; but since we cannot have a fraction connected with a symbol, such as Ca_1 .5 PO_4, we make the number even by multiplying both 1½ Ca and PO_4 by 2, which gives us $Ca_3(PO_4)_2$. Explained another way, Ca has a combining power equal to H_2; one Ca would not be sufficient to equal H_3 in H_3PO_4 and Ca_2 would equal H_4 or more than H_3, but taking H_3PO_4 twice, we have $H_6(PO_4)_2$, and H_6 exactly equals the combining power of Ca_3, giving Ca_3 $(PO_4)_2$. This point has been dwelt upon, because it is desirable to understand this and other compounds of phosphoric acid.

(4) *Acid salts.*—In the case of acids that have more than one H, it is possible to form two different salts with the same metal. For example, sulfuric acid can form two salts with potassium, one of which is contained in Table 4, viz., K_2SO_4, in which H_2 of H_2SO_4 has been replaced by K_2; the other potassium salt is formed by replacing one H of the acid by one K and leaving one H, making $KHSO_4$, which is known as potassium acid sulfate or potassium hydrogen sulfate. Such salts usually have an acid reaction.

In connection with our study of plant-foods, the most important case of acid salts is in connection with phosphoric acid, H_3PO_4. This can form three different salts with metals depending on whether the metal replaces the H_3 completely or only in part. For example, sodium can form three salts with phosphoric acid, because we can put the element Na one, two or three times in place of hydrogen (H_3), as follows:

(1)	$HHHPO_4$ or H_3PO_4	Phosphoric acid
(2)	$NaHHPO_4$, NaH_2PO_4,	one-sodium phosphate, or sodium acid phosphate
(3)	$NaNaHPO_4$, Na_2HPO_4,	two-sodium phosphate, or di-sodium hydrogen phosphate
(4)	$NaNaNaPO_4$, Na_3PO_4,	three-sodium phosphate, or tri-sodium phosphate

Calcium forms three different phosphates with phosphoric acid. The formation of these salts is of the same character as shown above in case of sodium, except that since one Ca equals H_2 and has double the combining power of Na, we use twice as much phosphoric acid, $2H_3PO_4$. The following arrangement illustrates the relations of these salts:

(1) $2H_3PO_4$ or	$H_6(PO_4)_2$,	Phosphoric acid
(2)	$CaH_4(PO_4)_2$,	one-calcium phosphate, or mono-calcium phosphate (soluble, p. 33)
(3)	$Ca_2H_2(PO_4)_2$,	two-calcium phosphate, or di-calcium phosphate (reverted, p. 33)
(4)	$Ca_3(PO_4)_2$,	three-calcium phosphate, or tri-calcium phosphate (insoluble, p. 33)

Complex and simple compounds.—Compounds are simple when they consist of few elements in simplest combination, while complex compounds contain a larger number of elements or in larger aggregations. For example, hydrochloric acid (HCl) is the simplest compound possible. It contains only two elements in the simple combination. Calcium phosphate, $Ca_3(PO_4)_2$ is a more complex compound. Some of the silicates are highly complex. They contain half a dozen different elements.

When we speak of a chemical compound decomposing or breaking down into simpler compounds, we mean that it forms several compounds, each with a smaller aggregation of elements. For example, ammonium carbonate, $(NH_4)_2CO_3$, decomposes, when heated, into NH_3 and CO_2 and H_2O, all of which are simpler compounds. When an organic substance like blood or meat or cottonseed-meal decomposes, the organic compounds, which are complex, form simpler compounds, generally the same as in case of ammonium carbonate or others equally simple. The terms simple and complex are, of course, relative; for example, HNO_3 is complex compared with HCl, but simple compared with $Ca_3(PO_4)_2$.

IV

PLANT-FOOD ELEMENTS AND THEIR COMPOUNDS

Sources of nitrogen.—Nitrogen occurs in nature chiefly in the following forms: (1) Atmospheric nitrogen, (2) ammonia, (3) in animal and vegetable matter, and (4) as nitrate. It has been a common custom to speak of nitrogen in its various compounds as ammonia, as, for example, "the ammonia in sodium nitrate." Such usage is chemically inaccurate and may be wholly misleading. The name, ammonia, should never be applied to nitrogen in any form except that of ammonia and ammonium compounds.

(1) *Atmospheric nitrogen.*—Nitrogen in the free or uncombined form constitutes about three-fourths of the air by weight, an amount equal to many millions of pounds for each acre on the surface of the earth. Nitrogen cannot be used directly from the air by most crops; but certain plants which belong to the legume family are able to use it indirectly through the medium of certain micro-organisms (p. 147).

(2) *Nitrogen in ammonia.*—Ammonia (NH_3) is a compound containing 14 parts by weight of nitrogen (82.35 per cent.) combined with 3 parts by weight of hydrogen. It is a colorless gas and dissolves easily in water, forming the familiar solution called spirits of hartshorn or ammonia water, which in its strong form contains 28 per cent. of ammonia. In water solution the composition is NH_4OH. Ammonia is present in the air in small amounts, being produced when vegetable or animal substances decompose (p. 137). It is often present in horse-stables and in fermenting manure heaps in amounts sufficient to produce its pungent odor.

The leaves of some plants have the power of absorbing ammonia directly from the air, so obtain nitrogen as food in this way, but only in small amounts. Some plants utilize nitrogen directly from ammonium compounds in the soil, but in general ammonia and its compounds must change into the form of nitrate in the soil before their nitrogen can be utilized by plants (p. 139).

(3) *Nitrogen in animal and vegetable matter, or organic nitrogen.*—Nitrogen in combination with hydrogen, carbon, oxygen, and sometimes sulfur and phosphorus, occurs in animals and plants in a

great variety of compounds, known under the general name of *proteins* (p. 14). In such combinations the nitrogen is called *organic* nitrogen. The nitrogen in slaughter-house by-products, such as blood, meat, tankage, etc., is organic; also in fish, bone, cotton-seed-meal, tobacco stems, green-crop manures, the solid portion of animal manures, etc. Before the nitrogen in such compounds can be used as food by plants, the substances, under proper conditions

Each element of plant food is necessary. *Left,* Tobacco plant grown on pure chemical plant food elements only. If any one of these elements is lacking, little or no growth may occur as shown in the illustration. In the other crocks phosphorus (6), nitrogen (5), potash (17) and all three of these elements (16) have been omitted.

of warmth, moisture, etc. (p. 138), through the action of micro-organisms (p. 139), must decay, form a large number of simpler compounds, be changed sooner or later into ammonia and finally into nitrate nitrogen.

(4) *Nitrogen as nitrate.*—Nitrogen combines with hydrogen and oxygen to form nitric acid (HNO_3), which combines with metals to form nitrates, such as sodium nitrate, $NaNO_3$, calcium nitrate, $Ca(NO_3)_2$ etc. We can conveniently speak of nitrogen in the form of nitric acid or nitrates as *nitrate nitrogen.* In the soil nitrate nitrogen is formed from organic nitrogen compounds and ammonia through a process called *nitrification* (p. 139). Most of the nitrogen used by plants as food is in the form of nitrate nitrogen; this is the most quickly available source of nitrogen in plant nutrition.

Compounds of nitrogen.—The most important compounds that contain nitrogen in available form and useful in crop growing are: (1) Nitrates of sodium, potassium, calcium and magnesium; (2) ammonium compounds, the sulfate, nitrate, phosphate and carbonate; (3) calcium cyanamid; (4) urea (p. 32).

(1) *Sodium nitrate,* $NaNO_3$ (known in its commercial, impure form as nitrate of soda, p. 165), contains 16.47 per cent. of nitrogen. It is easily soluble in water, one part by weight of the salt dissolving in an equal amount of water under ordinary conditions. It has for some years been one of the chief nitrogen-containing constituents used in many commercial fertilizers.

When sodium nitrate is treated with strong sulfuric acid, the nitric acid may be lost as a gas, as indicated by the following chemical reaction:

170		98		126		142
$2NaNO_3$	+	H_2SO_4	=	$2HNO_3$	+	Na_2SO_4
Sodium nitrate		Sulfuric acid		Nitric acid		Sodium Sulfate

This fact has a practical bearing, since sulfuric acid should never be mixed with a nitrate. It may happen when a poorly made phosphate, in which too much sulfuric acid has been used or the operation of mixing improperly managed, is mixed with sodium nitrate. In moist air sodium nitrate absorbs moisture and tends to become liquid.

(2) *Potassium nitrate,* KNO_3 (known in its commercial form as *niter* or *saltpeter,* p. 165), contains 13.85 per cent. of nitrogen, is quickly soluble and is available as plant-food for both nitrogen and potassium. One part by weight of the salt dissolves in about one and one-half parts of water.

(3) *Calcium nitrate,* $Ca(NO_3)_2$ (known in its commercial form as *lime nitrate,* p. 165), contains 17 per cent. of nitrogen. It has so strong an attraction for water that, when in contact with moist air, it tends to become liquid.

(4) *Ammonium sulfate,* $(NH_4)_2SO_4$ (called *sulfate of ammonia* in its commercial form, p. 166), contains 21.2 per cent. of nitrogen, which is equal to 25.75 per cent. of ammonia (NH_3). One pound of the salt dissolves in about one and one-third pounds of water. It does not absorb moisture from the air, as the nitrate of sodium or of calcium does.

(5) *Ammonium nitrate,* NH_4NO_3, contains 35 per cent. of nitrogen, one-half as ammonia (NH_3) nitrogen, and one-half as nitrate (NO_3) nitrogen. This combination gives it special value, but its price is so high it has been little used in general agriculture.

It is one of the most easily soluble compounds, one pound of water dissolving two pounds of the salt.

(6) *Ammonium carbonate,* $(NH_4)_2CO_3$, contains nearly 29.2 per cent. of nitrogen. It is easily soluble in water, one pound dissolving in three or four pounds of water. It is rarely, if ever, used in commercial fertilizers, but is of interest because it is found in the decomposition of animal and vegetable matter, especially of urine. This compound is one which breaks up or decomposes easily to form ammonia gas (NH_3), carbon dioxide gas (CO_2) and water (H_2O); this action takes place rapidly at temperatures above 100°F. When urine, especially of horses, ferments, ammonium carbonate is formed, and in a warm stable or manure pile, the ammonia gas escapes into the air, producing a more or less strong, characteristic ammonia smell.

(7) *Ammonium phosphates* (p. 36).

(8) *Calcium cyanamid,* $CaCN_2$ (commercially known as cyanamid, p. 167), contains 35 per cent. of nitrogen when pure.

(9) *Urea,* CON_2H_4, contains 46.6 per cent. of nitrogen when pure.

PHOSPHORUS

Source of phosphorus.—The original source of all phosphorus compounds is the earth's crust. As immediate sources of supply for plant-food uses we have (1) the soil, (2) the large phosphate deposits and (3) bones of animals.

Compounds of phosphorus.—In nature phosphorus is always found in combination with other elements and usually in the form of phosphates, which are combinations of phosphoric acid (H_3PO_4 or $-PO_4$) with metals (p. 26), especially calcium. The form in which phosphorus is generally found in animals and plants is calcium phosphate, with small amounts of magnesium, sodium and potassium phosphates. In the soil we have, in addition, iron and aluminum phosphates. With phosphoric acid, calcium forms several compounds, all of which are important as sources of plant-food phosphorus.

Use of terms, phosphoric acid and phosphates.—The use of the term, phosphoric acid, as commonly applied in connection with fertilizers, always means a compound whose composition is expressed by the symbol P_2O_5, which contains 43.7 per cent. of phosphorus. This usage is a survival of a custom established by chemists for expressing results of analysis. Accurately speaking, P_2O_5 is phosphoric anhydride; phosphoric acid is H_3PO_4. In expressing the composition of a fertilizer, however, "phosphoric acid" still commonly means P_2O_5.

When we say a compound or a fertilizer contains, for example

10 per cent. of phosphoric acid (P_2O_5), we mean the same thing as when we say it contains 4.4 per cent. of phosphorus (P). On many accounts it would be preferable to state the percentages as phosphorus (P) rather than as P_2O_5; but present usage has become so firmly established that many practical difficulties would be encountered in making the change. Such a change in connection with fertilizers has, however, been under discussion by agricultural chemists for years and will probably be brought about gradually in time.

The word "phosphate" is in popular use, often erroneously as a general term covering all kinds of commercial fertilizers, whether they contain phosphoric acid compounds only, or mixtures of these with nitrogen and potassium compounds, or even when commercial mixtures contain no phosphorus compounds at all.

TABLE 5—NAMES AND COMPOSITION OF CALCIUM PHOSPHATE COMPOUNDS

Names	*Chemical Symbols*	*Percentage of phosphorus (P)*	*Percentage of phosphoric acid (P_2O_5)*
(1) SOLUBLE CALCIUM PHOSPHATE Mono- (or 1-) calcium phosphate Calcium acid phosphate Acid calcium phosphate	$CaH_4(PO_4)_2$	26.5	60.7
(2) CITRATE-SOLUBLE OR REVERTED CALCIUM PHOSPHATE Di- (or 2-) calcium phosphate	$Ca_2H_2(PO_4)_2$, or $CaHPO_4$	22.8	52.2
(3) INSOLUBLE CALCIUM PHOSPHATE Tri- (or 3-) calcium phosphate Calcium phosphate Normal calcium phosphate Insoluble phosphate of lime Bone phosphate of lime, (B. P. L.)	$Ca_3(PO_4)_2$	20.0	45.8
(4) TETRA-CALCIUM PHOSPHATE Tetra-basic calcium phosphate Thomas phosphate powder Basic-slag phosphate	$(CaO_4)P_2O_3CaSiO_3$	17.0	39.0

Phosphoric acid compounds or phosphates.—The existence of four phosphate compounds of calcium is recognized. These noticeably differ in their composition in respect to the amount of calcium they contain, since calcium is present in the ratio of one, two, three and four parts *or combining weights* (p. 20). The fourth compound is different from the three others in respect to its source and is not so closely related to them as they are to each other. The one, two and three calcium phosphates together are known under an unusual variety of names, which may easily lead to confusion, but they can readily be kept distinct by knowing their individual characteristics with reference to one important property, viz., *their solubility in water.* For convenience they may be distinguished primarily by the following names: *Soluble, citrate-soluble* or *reverted* and *insoluble*—terms most commonly used in studying the composition of phosphatic materials. Table 5 shows the following facts with reference to each: (1) Various names, (2) chemical composition as expressed by formula, (3) percentage of phosphorus and (4) equivalent in the form of P_2O_5.

The special points to remember in connection with Table 5 are (1st) that the calcium in these compounds is in the relation of one, two, three and four; and (2) that the proportion of phosphoric acid or phosphorus is greatest in the compound that contains least calcium and decreases in the other compounds as the calcium increases.

Briefly stated, the insoluble can be converted into the reverted form by treatment with acid and into the soluble form by the use of twice as much acid. The soluble can be changed into the reverted form by various compounds.

In the commercial manufacture of "superphosphate" (p. 176), sulfuric acid is used to dissolve the insoluble phosphate as indicated by the following equations. The figures over each symbol give the amount of each compound used in the reaction. These equations do not give the full details of the chemical changes that take place—only the final results. Intermediate changes that occur will be considered in connection with the manufacture of superphosphates.

(*a*) Formation of reverted from insoluble phosphate:

310		98		272		136
$Ca_3(PO_4)_2$	+	H_2SO_4	=	$Ca_2H_2(PO_4)_2$	+	$CaSO_4$
Insoluble calcium phosphate		Sulfuric acid		Reverted calcium phosphate		Calcium sulfate (or gypsum)

(*b*) Formation of soluble from reverted phosphate:

272		98		234		136
$Ca_2H_2(PO_4)_2$	+	H_2SO_4	=	$CaH_4(PO_4)_2$	+	$CaSO_4$
Reverted calcium phosphate		Sulfuric acid		Soluble calcium phosphate		Calcium sulfate (or gypsum)

(*c*) Formation of soluble directly from insoluble without any reverted phosphate, showing the two preceding changes in one:

310		196		234		272
$Ca_3(PO_4)_2$	+	$2H_2SO_4$	=	$CaH_4(PO_4)_2$	+	$2CaSO_4$
Insoluble calcium phosphate		Sulfuric acid		Soluble calcium phosphate		Calcium sulfate (or gypsum)

In actual commercial operations, water is present and, instead of $CaSO_4$ (water-free calcium sulfate), there is formed $CaSO_4.2H_2O$, which is gypsum, or hydrated calcium sulfate.

(1) *Insoluble calcium phosphate* is contained in varying amounts in all agricultural soils, in animal and vegetable matter, especially in bones and seeds, in many mineral deposits and in water that has been in contact with soils. The commercial materials in which this compound occurs are considered on pages 173-185.

This form of calcium phosphate is called insoluble because it does not easily dissolve in water. One part by weight requires for solution 50,000 parts of pure water, more by the presence of some substances and less by others. Water that contains carbon dioxide (p. 47), as in the case of soil water, dissolves much more insoluble phosphate than does pure water. Some of the compounds used in fertilizers increase the solubility of this phosphate in water. Among them are sodium nitrate, ammonium sulfate, and potassium sulfate. Gypsum, iron and aluminum compounds make it less soluble in water. Substances which appear to have little influence are magnesium sulfate and chloride, sodium chloride (common salt) and potassium chloride (muriate).

(2) *Soluble calcium phosphate* is the principal constituent in the commercial product, "*superphosphate*" (acid phosphate) (p. 176), more extensively used in fertilizers than any other compound. It is a manufactured product. The special importance of this compound is due to its solubility in water, one part by weight dissolving completely in 100 parts of water under ordinary conditions, but it is easily changed by other compounds into the reverted form. On account of its ready solubility, it is the compound most extensively used by plants as the source of phosphorus.

(3) *Reverted calcium phosphate* receives its name from the fact that it can be formed from soluble calcium phosphate and is less soluble in water, so that, when soluble phosphate is changed into this form, it is said to "revert" or go back to an insoluble or less soluble condition. Reverted phosphate is, however, much more readily soluble in water containing carbon dioxide than in pure water. One part by weight of reverted phosphate requires for its solution about 7,500 parts of pure water, but only about 1,800 parts of water containing carbon dioxide. It is thus seen that this compound is fairly soluble in soil water. Reverted phosphate is known also as "*citrate-soluble*," which comes from the fact that, in ascertaining the amount of reverted phosphate by chemical analysis, it is dissolved in a solution of ammonium citrate.

Reverted phosphate is formed when the insoluble phosphate is treated with a limited amount of sulfuric acid and also from soluble calcium phosphate by such compounds as the insoluble phosphate, calcium carbonate, etc., as shown by the following reactions:

(*a*) Formation of reverted by insoluble phosphate:

234		310		544
$CaH_4(PO_4)_2$	+	$Ca_3(PO_4)_2$	=	$2Ca_2H_2(PO_4)_2$
Soluble calcium phosphate		Insoluble calcium phosphate		Reverted calcium phosphate

This action occurs in a superphosphate on standing, when the insoluble phosphate has not been completely dissolved by acid.

(*b*) Formation of reverted phosphate by calcium carbonate:

234		100		272		44		18
$CaH_4(PO_4)_2$	+	$CaCO_3$	=	$Ca_2H_2(PO_4)_2$	+	CO_2	+	H_2O
Soluble calcium phosphate		Calcium carbonate		Reverted calcium phosphate		Carbon dioxide		Water

In the presence of large amounts of calcium carbonate or of hydroxide (slaked lime), the soluble and reverted may be changed completely to insoluble calcium phosphate.

(*c*) Formation of insoluble from soluble phosphate by calcium carbonate:

234		200		310		88		36
$CaH_4(PO_4)_2$	+	$2CaCO_3$	=	$Ca_3(PO_4)_2$	+	$2CO_2$	+	$2H_2O$
Soluble calcium phosphate		Calcium carbonate		Insoluble calcium phosphate		Carbon dioxide		Water

(3) *Tetra calcium phosphate,* a manufactured by-product, is commonly known as basic-slag phosphate (p. 181).

(4) *Magnesium phosphates* are much like those of calcium in their general properties.

(5) *Ammonium phosphates.*—Phosphoric acid and ammonia combine to form three phosphates.

(*a*)	H_3PO_4	+	NH_3	=	$(NH_4)H_2PO_4$
	Phosphoric acid		Ammonia		Mono-ammonium phosphate
(*b*)	H_3PO_4	+	$2NH_3$	=	$(NH_4)_2HPO_4$
					Di-ammonium phosphate
(*c*)	H_3PO_4	+	$3NH_3$	=	$(NH_4)_3PO_4$
					Tri-ammonium phosphate

These three ammonium phosphates contain nitrogen and phosphorus in the percentage given below, with the equivalent percentage of ammonia and phosphoric acid. They are readily soluble in water and are valuable constituents of some concentrated fertilizers (p. 354).

	Nitrogen (*N*) *Per cent.*	*Ammonia* (*NH_3*) *Per cent.*	*Phosphorus* (*P*) *Per cent.*	*Phosphoric acid* (*P_2O_5*) *Per cent.*
Mono-ammonium phosphate ...	12.2	14.7	26.9	61.7
Di-ammonium phosphate	21.2	25.4	23.5	54.1
Tri-ammonium phosphate	28.2	34.0	20.8	47.9

(6) *Three potassium phosphates* are formed by the combination of phosphoric acid and potassium hydroxide, KOH (caustic potash); namely, mono-potassium phosphate (KH_2PO_4), di-potassium phosphate (K_2HPO_4), and tri-potassium phosphate (K_3PO_4). The percentages of potassium and phosphorus, with the equivalent percentages of potash and phosphoric acid are given below. These three potassium phosphates are easily soluble in water and are useful in making concentrated fertilizers.

	Potassium (*K*) *Per cent.*	*Potash* (*K_2O*) *Per cent.*	*Phosphorus* (*P*) *Per cent.*	*Phosphoric acid* (*P_2O_5*) *Per cent.*
Mono-potassium phosphate	28.7	34.4	22.8	52.4
Di-potassium phosphate	44.9	54.1	17.8	41.0
Tri-potassium phosphate	55.2	66.2	14.6	33.6

(7) *Iron phosphate,* $FePO_4$, and *aluminum phosphate,* $AlPO_4$, are noted for their insolubility in water. When iron or aluminum compounds, such as oxides, are present in large amounts in soils, they may act upon and convert any soluble calcium phosphate applied into the insoluble phosphate of iron or aluminum, practically causing the complete loss of the soluble phosphate for immediate crop uses. This loss can be prevented by keeping in the soil an abundance of calcium carbonate (p. 40). These insoluble phosphates are changed by calcium carbonate into insoluble calcium phosphate, which can become more readily available. Some of these phosphate compounds are among our most important commercial products (p. 173).

POTASSIUM

Experiments have shown that when potassium compounds are lacking in the soil plants suffer severely, though they may not die. Potassium forms a larger part than any other mineral element in the ash of plants.

Source of potassium.—The element potassium is never found in nature uncombined, only in compounds. It is a constituent of many minerals. The partial decomposition and solution of the rocks that contain these minerals give rise to the presence of potassium compounds everywhere in the soil. These are taken up by plants and used as food. When vegetable material is burned, the potassium forms a part of the unburned ash, in which it is present as potassium carbonate, K_2CO_3. When wood-ashes are treated with water and leached, the potassium carbonate along with some other compounds is dissolved, forming "lye"; and this evaporated to dryness leaves impure potassium carbonate, which was years ago an article of commerce known as *pot-ash* on account of being made in iron pots. In this way came the word, *potassium,* the Latin form of which is Kalium. K, the first letter of the Latin word, is therefore used as the chemical symbol of potassium.

The word *potash* is almost universally used in agricultural literature in referring to potassium compounds. In the analysis of these compounds, it became the custom to give the results of analysis not as potassium (K) but as potassium oxide (K_2O), which was called potash. This old custom has unfortunately persisted; so when we analyze a fertilizer for potassium, we do not often state the result as potassium (K) but as potash (K_2O). In the interest of accuracy of statement, it would be desirable if we could abolish the use of the word potash in analytical and agricultural chemistry and use only potassium. It is wholly incorrect to say that potassium chloride, for example, contains 50 per cent. of potash (K_2O), since it contains no oxygen at all. Of course, under such circumstances it simply means that the amount of potassium (K) in the compound is equivalent to 50 per cent. of K_2O. Another expression, *actual* potash, is also used in fertilizer analysis to express K_2O. In stating amounts, we shall commonly make use of both forms K_2O and K.

Compounds of potassium.—Some of the chief characteristics of a few potassium compounds are of interest as valuable plant-foods. Their commercial forms, as used in making fertilizers, include carbonate, chloride, sulfate, nitrate and silicates.

(1) *Potassium carbonate,* K_2CO_3 (known also as carbonate of potash), contains 56.6 per cent. of potassium, which is equivalent to 68 per cent. of potash (K_2O). It is strongly alkaline (p. 21), easily soluble in water, two parts by weight of water dissolving one part of carbonate. The only form of carbonate once used in the feeding of plants was that contained in wood-ashes (p. 189).

(2) *Potassium chloride,* KCl (known commonly in commercial forms as muriate of potash), contains 52.5 per cent. of potassium, which is equivalent to 63.2 per cent. of K_2O. It is easily soluble in water, three parts by weight of water dissolving one part of the

Grades used in studying effect of soil reaction on production of scabby potatoes. Highest yields of marketable potatoes were on soils with a pH reaction between four and eight tenths and five and four tenths—just sufficiently acid to guard against serious losses from scab. The use of liming materials to reduce acidity may easily result in the production of scabby potatoes, unless proper precautions are taken to guard against losses due to scab.

chloride, closely resembles common salt in appearance and taste, is a valuable source of plant-food (p. 188) and is more extensively used in fertilizers than any other potassium compound.

(3) *Potassium sulfate,* K_2SO_4 (known commercially as sulfate of potash), contains 45 per cent. of potassium, which is equivalent to 54 per cent. of K_2O. While it is easily soluble in water, it is less so than carbonate or chloride, one part by weight of sulfate requiring eight to ten parts of water for solution. This compound has special value as a source of potassium in plant nutrition, since it must be used with some crops which are injured by the chloride (p. 188).

(4) *Potassium nitrate,* KNO_3 (known commercially as nitrate of potash, niter, and saltpeter), contains 38.6 per cent. of potassium, which is equivalent to 46.6 per cent. of KaO. It is about as easily soluble in water as the chloride. This compound possesses additional value in plant nutrition on account of its being in the nitrate form (p. 165).

(5) *Potassium phosphates.*—(p. 180.)

(6) *Silicates.*—Potassium is present in soils in several mineral combinations with oxygen, silicon, aluminum and other metals.

CALCIUM

Calcium is essential to the growth of plants. Its compounds are commonly called *lime* compounds, as the result of a former usage by chemists. The term *lime* should properly be used only when speaking of the compound calcium oxide, CaO, so we shall generally use the word calcium in speaking of so-called lime compounds.

Source.—Calcium, like the other mineral elements, has its source in the crust of the earth, where it occurs in a variety of compounds, especially carbonate and various silicates.

Compounds of calcium most useful in agriculture are carbonate, phosphates, nitrate, and sulfate. They are more fully discussed in connection with indirect fertilizers (p. 192).

(1) *Calcium carbonate,* $CaCO_3$ (known in commerce as *carbonate of lime, ground limestone,* etc., p. 195), contains 40 per cent. of calcium, which is equivalent to 56 per cent. of lime (CaO). The carbonate is the chief constituent of limestone, marble, chalk, shell-marl, coral, shells, etc., (p. 195). Calcium carbonate is only slightly soluble in pure water, but in water that contains all the carbon dioxide it can hold at ordinary temperatures, it is soluble to the extent of one part by weight in 1,000 parts of water, forming calcium bicarbonate, $CaH_2(CO_3)_2$. Calcium carbonate has basic (p. 20) properties, neutralizing acids and forming salts, as illus-

trated by the following reaction which takes place in the soil when nitric acid is formed (p. 139):

$$\begin{array}{ccccccccc} 100 & & 126 & & 164 & & 44 & & 18 \\ CaCO_3 & + & 2HNO_3 & = & Ca(NO_3)_2 & + & CO_2 & + & H_2O \\ \text{Calcium} & & \text{Nitric} & & \text{Calcium} & & \text{Carbon} & & \text{Water} \\ \text{carbonate} & & \text{acid} & & \text{nitrate} & & \text{dioxide} & & \end{array}$$

(2) *Calcium oxide,* CaO (commercially known as *quicklime, burnt lime, caustic lime, stone-lime, lump-lime, building-lime,* etc., p. 193), contains 71.4 per cent. of calcium. It is made by burning calcium carbonate, the carbon dioxide being driven off as indicated by the following equation:

$$\begin{array}{ccccccc} 100 & & & & 56 & & 44 \\ CaCO_3 & + & \text{heat} & = & CaO & + & CO_2 \\ \text{Calcium} & & & & \text{Calcium} & & \text{Carbon} \\ \text{carbonate} & & & & \text{oxide} & & \text{dioxide} \end{array}$$

This shows that when we burn 100 pounds of pure calcium carbonate, we obtain 56 pounds of calcium oxide, 44 pounds of carbon dioxide gas going into the air. Calcium oxide has basic properties (p. 22), that neutralize acids and form salts, as illustrated by the following reaction which occurs in the manufacture of calcium (lime) nitrate (p. 165):

$$\begin{array}{ccccccc} 56 & & 126 & & 164 & & 18 \\ CaO & + & 2HNO_3 & = & Ca(NO_3)_2 & + & H_2O \\ \text{Calcium} & & \text{Nitric} & & \text{Calcium} & & \text{Water} \\ \text{oxide} & & \text{acid} & & \text{nitrate} & & \end{array}$$

(3) *Calcium hydroxide,* $Ca(OH)_2$ (commonly called *slaked lime, hydrate lime,* etc., p. 194), contains 54 per cent. of calcium which is equal to 75.7 per cent. of CaO. One part by weight of calcium hydroxide is soluble in 800 to 1,000 parts of water, the solution being known as *lime-water.* It is formed when calcium oxide is treated with water—when it undergoes the change known as *slaking* according to the following chemical reaction:

$$\begin{array}{ccccc} 56 & & 18 & & 74 \\ CaO & + & H_2O & = & Ca(OH)_2 \\ \text{Calcium} & & \text{Water} & & \text{Calcium hydroxide} \\ \text{oxide} & & & & \text{(slaked lime)} \\ \text{(quick lime)} & & & & \end{array}$$

Slaked lime may therefore be properly regarded as quicklime (CaO) diluted with about one-third of its weight of water, except that the water is not merely mixed, but is chemically combined with the lime.

Air-slaked lime is calcium oxide that has absorbed moisture from the air and changed into hydroxide. The operation is slow, depending largely on the amount of moisture in the air. As the slaked lime is formed it slowly absorbs carbon dioxide from the air, forming carbonate, as shown by the following reaction:

74		44		100		18
$Ca(OH)_2$	+	CO_2	=	$CaCO_3$	+	H_2O
Calcium hydroxide or slaked lime		Carbon dioxide		Calcium carbonate		Water

Air-slaked lime is, therefore, generally a mixture of hydroxide and carbonate; but in case of long exposure the change to carbonate becomes complete. It is thus seen to be of uncertain composition, since it may vary from nearly pure hydroxide to nearly pure carbonate.

Calcium hydroxide is an alkaline base (p. 22) which changes red litmus to blue. It is therefore useful in neutralizing acids.

(4) *Calcium phosphates* (p. 33).

(5) *Calcium nitrate* (p. 31 and p. 165).

(6) *Calcium sulfate,* $CaSO_4.2H_2O$ (known in commerce under various names, as *gypsum, land-plaster, sulfate of lime,* etc.), contains about 23 per cent. of calcium. When this is heated to about 120° C. (248° F.), three-fourths of the water is driven off and the resulting product forms a powder, called *plaster of Paris.* This takes up water readily and hardens or "sets." Gypsum has been extensively used in agriculture. When used in stables mixed with manure, it combines with ammonium carbonate, to form ammonium sulfate and calcium carbonate, as shown in the following reaction:

136		96		132		100
$CaSO_4$	+	$(NH_4)_2CO_3$	=	$(NH_4)_2SO_4$	+	$CaCO_3$
Calcium sulfate		Ammonium carbonate		Ammonium sulfate		Calcium carbonate

Calcium sulfate in the form of gypsum ($CaSO_4.2H_2O$) is soluble, one part by weight in about 400 parts of water.

MAGNESIUM

In chemical behavior, magnesium and its compounds resemble calcium in many ways but are not commonly used in agriculture for fertilizing purposes.

Source.—Magnesium compounds are present in varying amounts in all agricultural soils. One source of commercial supply is the German potash mines (p. 186).

Compounds of magnesium.—The most familiar compounds of magnesium are oxide, hydroxide, carbonate, chloride, nitrate, phosphates, sulfate and silicates.

(1) *Magnesium carbonate,* $MgCO_3$ often occurs mixed with calcium carbonate in so-called dolomitic limestone.

(2) *Magnesium oxide,* MgO (commercially known as *calcined magnesia*) is best known as a constituent of burnt lime prepared from magnesian limestone, in which it occurs in amounts that vary approximately from 10 per cent. up to 40 per cent. or more. Its presence in large amounts is objectionable for many purposes, since it slakes much more slowly than calcium oxide.

(3) *Magnesium hydroxide,* $Mg(OH)_2$, is formed by treating magnesium oxide with water. This compound, unlike calcium hydroxide, is only slightly soluble in water. It is present in slaked magnesian lime.

(4) *Magnesium sulfate,* $MgSO_4$, known commonly as *Epsom salts,* is contained in some of the German potash salts (p. 186), in which likewise magnesium chloride, $MgCl_2$, is found.

(5) *Magnesium silicates.*—The minerals, serpentine and talc, are hydrated silicates of magnesium, varying in their percentage of water. Magnesium occurs also in some of the complex silicates.

SULFUR

Sulfur compounds are always found in plants and are essential. They are not always present in soils in sufficient abundance to meet agricultural demands.

Source.—Sulfur is found in a great many minerals in combination with other elements. Under the usual conditions of soil formation, all sulfur compounds are sooner or later changed into sulfates. These are the only forms in which agricultural plants utilize sulfur.

Compounds of sulfur found in soils, some of which are present in fertilizers in combination with other essential constituents, are the following: (1) Potassium sulfate (p. 40), (2) calcium sulfate (p. 42), (3) sodium sulfate (p. 45), (4) magnesium sulfate (above), (5) ammonium sulfate (p. 31). The sulfur in these compounds varies from 18 to 26 per cent.

Sodium sulfate, Na_2SO_4 (known also as Glauber's salt and sulfate of soda) is found in soils of dry regions, being a prominent constituent of "white alkali" and "alkali dust"; it is an undersirable component, since it is a plant poison when present beyond certain limits; it is, however, less poisonous to plants than common salt.

Ferrous or iron sulfate, $FeSO_4$, (commonly called green-vitriol or copperas), is sometimes found in considerable amounts in soils that have been under water, then drained and brought under cultivation. It is a plant poison when present beyond certain amounts. Such soils bears poor crops until, by treatment generously with calcium carbonate and thorough cultivation, the iron sulfate is changed into harmless forms of combination.

IRON

Though used in small amounts, iron compounds are essential to the growth of plants. This element does not, however, have general interest for us in connection with the use of fertilizers, because most agricultural soils have an abundance and need no additions, so far as we know at present. An exception occurs in various Hawaiian pineapple lands. They contain iron but in such insoluble forms that fertilization with small amounts of soluble iron salts is necessary.

Source.—Numerous minerals contain iron and furnish the supply now in our soils. Iron compounds impart characteristic colors to soils (p. 83).

Compounds of iron.—The form in which iron is most commonly found in soils is in combination with oxygen and hydrogen either as *iron (ferric) oxide,* Fe_2O_3, as *hydroxide,* $Fe(OH)_3$ or some similar form of combination, known as hydrated oxides, of which there are many forms. These compounds are reddish brown or yellow and resemble iron rust in composition. Whatever kind of compound iron in the soil may originally exist in, it is sooner or later changed into one of the forms mentioned. Iron (ferrous) sulfate has already been mentioned in connection with sulfates. The iron compounds commonly present are only slightly soluble in water, but sufficiently so to furnish all that is needed by plants unless the soil contains an abundance of calcium carbonate. Iron compounds in solution are changed by soluble calcium phosphate into the insoluble iron phosphate.

CHLORINE

Source.—Chlorine is always found in soils and plants, but only in the form of compounds. Large amounts are present in sea water and in special deposits in the earth.

Compounds of chlorine.—In combination with hydrogen, chlorine forms hydrochloric (or muriatic) acid (HCl), and in combination with metals it forms salts called chlorides (p. 26), the most familiar of which are (1) potassium chloride (p. 188), (2) sodium chloride, (3) magnesium chloride.

Sodium chloride, NaCl, *common salt,* contains 60.7 per cent. of chlorine and 39.3 per cent. of sodium. Its ready solubility in water is a familiar fact of everyday experience. *Agricultural salt* is an impure form of sodium chloride whose agricultural importance is not great. When used at all, it is as an indirect fertilizer (p. 192). Salt is generally a constituent of the alkali soils of dry regions.

SODIUM

Source.—Inasmuch as sodium generally occurs in soils as sodium chloride, its origin and relations are essentially the same as those of chlorine.

Compounds of sodium.—The most abundant compound of sodium is the *chloride*. *Sodium sulfate* is the chief constituent of the "white alkali" of arid regions. *Sodium carbonate* is a constituent of the "black alkali" spots in soils of arid regions, the dark color being due to the action of the alkali on the organic matter of the soil. The most important agricultural compound of sodium is the nitrate (p. 212), the value of which is due to the nitrogen.

SILICON

Source.—Next to oxygen silicon is the most abundant element in the earth's crust and, therefore, in soils. It is a constituent of many of the most common rocks and is of more interest agriculturally for its relations to soil texture than to plant nutrition directly.

Compounds.—Silicon is found in various compounds; for instance (1) silicon dioxide and (2) silicates.

(1) *Silicon dioxide*, SiO_2, commonly known as *silica,* is found in rocks and soils as *quartz* and is the chief constituent of sand. It contains about 47 per cent. of silicon.

(2) *Silicates.*—Some compounds in this class are complex, containing several different metals in combination with some form of silicic acid. The simplest forms are represented by potassium and sodium silicates (K_2SiO_3, Na_2SiO_3), which dissolve in water and are then known as water-glass. Common feldspar contains potassium or sodium or calcium in combination with aluminum and silicic acid, as, for example, $KAlSi_3O_8$, $NaAlSi_3O_8$, etc. Pure clay or kaolin is aluminum silicate combined with water; it is a hydrated aluminum silicate ($H_2Al_2\ (SiO_4)_2.\ H_2O$).

HYDROGEN

Hydrogen is present in plants in much smaller amounts than is carbon or oxygen, constituting only 6 or 7 per cent. of the dry matter of plants.

Source.—Hydrogen is rarely found in nature in other than combined form. We can conveniently classify its compounds as (1) water hydrogen, (2) mineral hydrogen, (3) organic hydrogen.

(1) *Water,* the most abundant hydrogen compound, contains one-ninth of its weight as hydrogen. This is the source of hydrogen in plants.

(2) *Mineral sources.*—Many compounds in the crust of the earth contain hydrogen. Minerals that contain water in some form of combination are called *hydrated.*

(3) *Organic hydrogen.*—All plant and animal tissues contain hydrogen in many different organic compounds.

Among the important uses of hydrogen is the synthetic production of ammonia (p. 163).

OXYGEN

Oxygen, which occurs naturally in both combined and uncombined forms is the most abundant of all chemical elements, forming a large part of earth, air and water.

Source of oxygen.—For convenience, we can consider its forms under four divisions: (1) Atmospheric oxygen, (2) oxygen in minerals, (3) oxygen in water and (4) oxygen in organic matter.

(1) *Atmospheric oxygen.*—In the air oxygen is found free or uncombined, being simply mixed with the other constituents. It constitutes about one-fifth of our entire atmosphere, which means that in the air resting upon one acre of the earth's surface there are over 10,000 tons of oxygen.

(2) *Oxygen in minerals.*—Nearly one-half of the crust of the earth consists of oxygen in various combinations with many other elements.

(3) *Oxygen in water.*—Eight-ninths of water by weight consists of oxygen.

(4) *Oxygen in organic matter.*—Over one-third of all animal and vegetable substances consists of oxygen.

CARBON

Source.—The source of supply of carbon as plant-food is the carbon dioxide of the air. Carbon may properly be called, in a chemical sense, the central element of all animal and vegetable substances, for it is a necessary constituent of every living organism. Carbon constitutes about 45 per cent. of the dry matter of plants, an amount greater than any other element.

Compounds of carbon.—While the number of carbon compounds is innumerable, we need consider only one.

Carbon dioxide, CO_2, commonly known as *carbonic acid gas,* is a compound that contains 27.3 per cent. of carbon and 72.7 per cent of oxygen. Under ordinary conditions it is present in the air to the extent of about 3 parts in 10,000. Though present in so small proportions, we have in the atmosphere resting upon one acre more than 28 tons of this gas, containing over 7½ tons of carbon, an amount sufficient to supply many times the needs of any crop. To furnish the carbon in a tree containing 10,000 pounds of dry matter would require about 15,000,000 cubic yards of air. There is, however, no danger that the carbon dioxide of the air will ever be appreciably reduced in amount, because the supply is being constantly replenished in various ways, as by the burning of fuel, the breath of animals and all natural processes of decay of organic matter. Thus, one ton of good coal produces, in burning, over 3 tons of carbon dioxide gas. The people in a town of 10,000 inhabitants daily breathe out not less than 5 tons of this gas, an amount that would furnish enough carbon to grow several acres of any large-yielding crop. Carbon dioxide is an important constituent of the air contained in the soil.

V

ACTION OF NUTRIENT CONSTITUENTS IN PLANT NUTRITION

It is a matter of practical interest and importance to know in what way nutrient elements are of value in the growth of plants. We know that in the animal body certain food constituents perform definite duties peculiar to each. In a similar way, each constituent of plant-food performs some function or accomplishes some task peculiar to itself. We may regard the plant as a factory in which the labor is carefully and systematically distributed; each plant-food element may be regarded as a specialist and as having specific duties to perform in the field of its specialty. These specialists are mutually interdependent and must all work together in order to produce best results. If one element is overabundant or deficient, the harmony of action is disturbed and the output is likely to be abnormal in amount or quality.

We shall find it interesting also to notice how the elements are distributed in plants in a way to indicate the location and, to some extent, the character of their activities.

While the organic or air-derived elements (carbon, hydrogen, nitrogen, oxygen) contribute most of the material of which plants are made, we shall see that the mineral elements have functions in plant nutrition so important as to make their presence absolutely essential and that no one element can take the place of any other except, perhaps, to a limited extent in some special cases under unusual conditions.

NITROGEN

In the study of plant nutrition, especially as applied to crop growing, probably no element has received so much attention as nitrogen. It is present in the dry matter of plants to the average extent of only 1.5 per cent., and yet its compounds are of the highest value in agriculture as well as in commercial industries. Its agricultural importance is indicated by the following facts: (1) This element is an essential constituent of plant nutrients; (2) the amounts ordinarily available in nature for immediate use are small; (3) it is one of the elements of the soil most heavily used by crops;

(4) of all forms of plant-foods in fertilizers, it is and always has been the most expensive; (5) it is the most elusive of plant-food constituents, since it changes sooner or later into some form easily lost to soil and plant.

Action of nitrogen in plants.—The nitrogen-containing compounds in plants are of peculiar importance in relation to the life and growth, since in the form of proteins (p. 14), they constitute an essential part of the important substance known as *protoplasm* (p. 122), which is the living part of plant-cells. On account of this relation no other substance, when properly applied, can so quickly show its beneficial effect upon plant growth.

Its influence upon plant growth is shown by several striking effects, more or less closely associated, but which for convenience we will consider under separate heads, as follows: (1) Growth of foliage, (2) flowering process, (3) retarding maturity of plants, (4) color of plants, (5) regulation of plant growth, (6) quality of crops, (7) health of plants, and (8) composition of plant.

(1) *Effect of nitrogen on growth of foliage.*—An abundance of nitrate promotes growth, as shown by extreme luxuriance of stem and leaf. This effect is of the highest importance, because as a necessary preliminary for complete development of plants, we must have a good growth of leaf and stem, since the leaves especially constitute the working laboratory of the plant. In the case of many crops succulent foliage is the product for which the crop is grown, for example, celery, lettuce, cabbage, cauliflower, grasses, etc. Lack of nitrogen results in a general decreased or stunted growth. In the case of a grain crop, like wheat or barley, the stems will be short and the leaves small; the weight of straw in comparison with the weight of grain will be less than in case of a crop grown under normal conditions of nitrogen supply.

(2) *Effect of nitrogen upon the flowering process.*—The growth of flowers is retarded by an abundance of nitrogenous plant-food. Ordinarily, most plants at a certain stage of growth cease to produce new branches and leaves and no longer increase those already formed, but commence to produce flowers and fruits for the development of seed. If a plant is provided with all the nitrate nitrogen it can use just at the time its flower-buds begin to develop, other conditions being kept normal, the process of flower formation may be checked, while the activity of growth may be transferred back to, and renewed in, stems and leaves, which take on new vigor, multiply and show remarkable luxuriance. Should flowers be developed under such conditions, they are likely to be sterile, and to produce either no seed or only unproductive ones.

(3) *Effect of nitrogen retarding the maturing of plants.*—From what has preceded, it can be inferred that the effect of furnishing

crops large amounts of available nitrogen is to retard the maturing process, since the vegetative activity is then carried on in the leaves and stems at the expense of flower and fruit. A crop grown on a soil excessively rich in nitrogen always tends to ripen slowly; it shows overgrowth of leaves and stems, which, in the case of a grain crop, often results in lodging badly before maturity, due to the formation of long-jointed, soft, weak stems. Delayed maturity may subject a crop to danger from frost.

(4) *Effect of nitrogen on color of plants.*—The presence of abundant nitrogen in plants is usually accompanied by deep green foliage, which is regarded as a sign of vegetative activity and health. The substance that makes plants green is called *chlorophyl* (p. 122), which is present in the living, active cells of plants. Nitrogen in combination with other elements is a prominent constituent of chlorophyl. Protoplasm and chlorophyl are intimately associated and their united action is responsible for combining nutrient elements into the compounds manufactured in the plant. Therefore, it can be readily seen why nitrogen is closely connected with the production of green color in plants and why this is a sign of vegetative vigor and activity.

(5) *Effect of nitrogen in the regulation of general plant-growth.*—As a result of the influence of nitrogen on plant foliage, we may say the amount of nitrogen regulates the general growth of the whole plant and, therefore, *the amount of nutrient constituents that a plant can use.* For example, it does not matter how large amounts of available phosphorus and potassium compounds may be in a soil; the amount of these constituents used by a crop is more or less regulated by the amount of nitrogen within reach of the crop. This is one of the reasons why nitrogen is so highly important in plant feeding.

(6) *Effect of nitrogen on quality of crops.*—The quality of crops is affected by the amount of nitrogen used. Large applications cause softness of tissues, which may be desirable or undesirable, according to the kind of crop. In the case of crops in which decided succulence, tenderness and crispness are desired, such as lettuce, celery, asparagus, etc., relatively large amounts of nitrogen are needed. Cabbages and similar vegetables, given large amounts of nitrate grow rapidly and have delicious, tender crispness, but their keeping quality is impaired. In cases of crops for immediate consumption one can, therefore, use larger amounts of nitrogen than when the crop is to be stored long. The quality of grain is seriously affected by too much nitrogen. Wheat, for example, is lighter, weighs less per bushel. Some fruits and vegetables show poorer shipping quality when grown with an excess of nitrogen.

(7) *Effect of nitrogen on the resisting power of plants.*—Excessive nitrogen is likely to make crops less resistant to the attack of fungous diseases. This is a matter of common experience. Nursery stock grown on land fertilized exclusively for years with large amounts of farm manure is susceptible to plant diseases and, in some cases, also to attack of some insects. Rust is more abundant on wheat crops too heavily fed with nitrogen. Greenhouse crops are grown on soils rich in nitrogen and much trouble is experienced with plant diseases. Shrubbery and young trees which have been pushed too rapidly in the growth of new wood by excessive feeding with nitrogen, are less resistant to the effects of severely cold weather.

(8) *Effect of nitrogen on the composition of plants.*—While the changes in composition of plants due to use of increasing amounts of nitrogen are not large, the fact of such a tendency is well established, showing that both straw and grain of plants grown with large amounts of nitrogen contain somewhat more of this element than do plants grown with less.

Distribution of nitrogen in plants.—In the growing plant, the nitrogen is present in larger proportions, first in the leaves and finally in the seeds; it is found as a living constituent of the protoplasm (p. 122) and as a reserve food in the cells. Nitrogen is prominent in the *living* parts of plants; in the inactive or formerly living portions, it appears as a remnant. These statements are illustrated in Table 6.

PHOSPHORUS

Phosphorus compounds (p. 84), found everywhere in soils, are of great value to plants. All plants used as food contain small amounts of phosphorus compounds, which they obtain from the soil. The phosphorus present in bones in the form of calcium phosphate is taken by animals as vegetable foods. The phosphorus compounds in the foods consumed by animals are largely passed into the excrements. Soils often become deficient in available phosphorus compounds, especially in regions where grain crops are grown continuously for long periods. Phosphorus compounds are present in seeds in larger amounts than in any other portion of plants.

Action of phosphorus in plants.—In connection with crop growth, phosphorus compounds produce several important effects of practical interest, as follows: (1) Effect on germination of seed, (2) effect on root development, (3) effect on early ripening, (4) effect on relation of grain and straw, (5) effect on nitrogen in grain, and (6) relation to protoplasm.

(1) *Effect of phosphorus on germination of seed.*—Available phosphorus compounds favor rapid development of the young seedling. A familiar practice with progressive farmers is to give a crop, corn for example, a generous application of soluble phosphate at planting time to insure prompt growth of seed and to establish the plant in the soil under conditions favorable to continued development. Superphosphate has marked favorable effect on wheat sown in fall.

TABLE 6—POUNDS OF NITROGEN IN 1,000 POUNDS OF DRY MATTER

Variety and part of plant	*Lbs. of nitrogen*	*Variety and part of plant*	*Lbs. of nitrogen*
APPLES		GARDEN PEAS	
Flesh of apple	3.8	Seeds	44.5
Skin	4.7	Pods	18.7
Core	10.4	Leaves	38.5
Leaves (old)	8.1	Vines	12.2
New wood	9.7	BEETS	
RASPBERRIES		Roots	31.0
Berries	12.1	Tops	30.1
Leaves	23.3	Young tops and roots	41.3
New wood	7.6	WHEAT	
Old wood	4.3	Grain	23.0
ASPARAGUS TOPS		Straw	6.0
May 5	23.1	PEAS	
Aug. 18	13.2	Seeds	41.8
Berries	24.5	Straw	16.6

(2) *Effect of phosphorus on root development.*—The application of available phosphate to a crop increases the development of roots. The effect on seed germination, stated in the preceding paragraph, is due to the stimulation of the growth of roots, thus giving the young plant a favorable start. The use of available phosphate is, therefore, of special value in case of root crops, especially on heavy soils which tend to limit root development. Lack of phosphorus results in poor development of roots, as well as of leaves and stems in cereals.

(3) *Effect of phosphorus on early ripening of crops.*—Phosphates tend to favor the early maturing of crops. The formation of grain begins sooner when soluble phosphate is freely used. Other conditions being favorable, it has been shown there may be fully a week's difference in the ripening of field crops due to the generous application of soluble calcium phosphate. This maturing effect is due to the close relation phosphorus has to seed production. Plants do not come to maturity and do not produce seeds, unless supplied with available phosphorus compounds. It is not surprising, therefore that such compounds are always found in larger amounts in seeds than in any other part of plants.

The ripening effect of phosphorus is the reverse of that of available nitrogen, which, when applied abundantly, tends to prolong the season of growth and delay the maturing process.

(4) *Effects of phosphorus on relative production of grain and straw.*—Available phosphates, generously used, increase the proportion of grain to straw, and tend to strengthen the latter—the reverse of the action of nitrogen.

(5) *Effect of phosphorus on nitrogen content of grain.*—When available phosphorus compounds are liberally supplied to grain crops, the grain produced contains a smaller percentage of nitrogen than under ordinary conditions; reverse of the effect of nitrogen.

(6) *Relation to protoplasm.*—Phosphorus is an essential constituent of protoplasm and therefore is connected with cell activities. It is essential to grain formation, is associated with cell division, and takes part in the change of starch into sugars.

Distribution of phosphorus in plants is largely and intimately associated with seed production, as suggested by a recapitulation of the statements already made: (1) Stimulation of seed germination and more rapid development of young seedlings; (2) tendency of generous application of soluble phosphates to favor early maturing of crops; (3) the necessity of abundance of soluble phosphorus compounds to produce good yields of grain and to promote increase of grain in relation to straw; (4) the effect of soluble phosphates upon the composition of the grain, reducing the percentage of nitrogen. These facts are still further emphasized when we examine the plants with reference to the phosphorus content of their various parts.

TABLE 7—POUNDS OF PHOSPHORUS IN 1,000 POUNDS OF DRY MATTER

Variety and part of plant		*Variety and part of plant*	
APPLES		GARDEN PEAS	
Flesh of apple	0.6	Seeds	4.8
Skin	0.7	Pods	1.5
Core	1.3	Leaves	2.2
Leaves (old)	0.5	Vines	1.1
New wood	1.3	BEETS	
RASPBERRIES		Roots	2.7
Berries	2.5	Tops	3.2
Leaves	2.0	Young roots and tops	4.4
New wood	1.0	WHEAT	
Old wood	0.4	Grain	4.1
ASPARAGUS TOPS		Straw	1.2
May 5	3.2	OATS	
Aug. 18	1.4	Grain	2.8
Berries	3.3	Straw	0.9
PEAS		CORN	
Seeds	4.4	Kernels	2.8
Straw	1.9	Leaves and stalks	1.9

POTASSIUM

Action of potassium in plants.—Potassium compounds are essential to several important activities in plant life. Generally speaking, they are most abundant in young and growing parts where vegetative activity is greatest and least abundant in the older parts that have ceased to grow. The effects of potassium upon plant growth and products are shown in several ways: (1) Formation of carbohydrates, (2) formation and transference of starch, (3) growth of stems and leaves, (4) fleshy fruits, (5) plant-cells, (6) relation to plant acids, (7) prolonging the growing period, (8) plant resistance to disease, and (9) leguminous crops.

(1) *Influence of potassium upon formation of carbohydrates.*—One of the main forms of activity in the cells of leaves is the formation of carbohydrates. In order that a plant may produce starch, sugar, cellulose and other carbohydrates, potassium compounds are necessary. In some way, not yet clearly understood, potassium enables the plant to combine the carbon, hydrogen and oxygen furnished by carbon dioxide and water. Thus it influences and makes possible the process of assimilation by which these elementary constituents are built into carbohydrate products.

(2)*Effect of potassium on formation and transference of starch.*—Potassium aids in doing important work also in the transference of starch from one part of a plant to another. As first produced in the leaf the starch is insoluble, but it is changed into soluble compounds within the plant-cells. Then it can pass through the cell-walls gradually and later be carried into the fruit or the seed where it accumulates and changes back to its usual insoluble condition. Potassium, as well as calcium, in association with other elements, is essential to render assistance in some way in the performance of these important duties.

(3) *Effect of potassium on growth of stems and leaves.*—Potassium compounds are important in plant nutrition because they have a marked influence upon the development of leaves and of woody parts of stems. When they are deficient the stems are likely to be weak and brittle, and the leaves to be not normal in color, showing a tendency to die early from the margin toward the center. The close relation of this plant-food constituent to the development of leaves and stems is suggested by the fact that larger amounts of potassium compounds are found in these portions of plants than in any other part. This fact is consistent also with the function potassium compounds perform in the manufacture of carbohydrates.

(4) *Effect of potassium upon fleshy fruits.*—It is commonly

believed that potassium compounds are requisite to a normal development of the fleshy portions of fruits. This is supported by analysis of several varieties of grapes shown in Table 8.

TABLE 8—POUNDS OF POTASSIUM IN 1,000 POUNDS OF DRY MATTER

Variety	*Berries*	*Leaves*	*New wood*
Delaware	17.4	8.3	5.3
Concord	14.0	13.5	5.5
Catawba	13.5	9.1	5.0
Diamond	11.2	8.9	7.2

(5) *Effect of potassium on turgor of plant-cells.*—In order that plant-cells may satisfactorily do their work of manufacturing plant compounds from food constituents, it is necessary that they be somewhat swollen and turgid (p. 125). Potassium compounds are believed to be the agents mainly associated with the maintenance of this condition.

(6) *Relation of potassium to plant acids.*—Potassium acid salts are present in numerous sour plant juices. In these cases, potassium is combined as an acid salt (p. 14) with such organic acids as tartaric, oxalic, citric, etc. For example, the acidity of grapes is due to the acid salt, potassium acid tartrate ($KHC_4H_4O_6$); the sourness of sorrel comes from acid potassium oxalate (KHC_2O_4).

(7) *Effect on seed.*—Potassium compounds are influential in relation to seed formation, increasing the weight of individual grains in the case of cereals and producing heavy, plump kernels.

(8) *Effect of potassium in prolonging period of growth.*—A relative excess of potassium compounds supplied to a crop tends to prolong the period of growth in stems and leaves and thus delay the maturing of the crop. In this respect it resembles the effect of nitrogen.

(9) *Effect of potassium on resistance to disease.*—Potassium compounds enable plants to withstand more effectively attacks of fungous diseases. For example, in the absence of sufficient potassium, wheat is prone to rust. Grasses on soils poor in potassium compounds are subject to various fungous diseases. In general, crops which do not receive their full supply of potassium are more susceptible to disease, and, in this condition, the resisting power is weakened still more if the plant receives large amounts of available nitrogen.

(10) *Effect of potassium on leguminous crops.*—Potassium compounds produce a pronounced, favorable effect on the growth of the leguminous crops, clovers, alfalfa, beans, peas, etc. When they are applied to soils deficient in this plant-food element, as sandy and gravelly lands, the effects upon the growth of legumi-

nous crops are usually marked. In what way potassium does this work is not definitely known, but it has been suggested that potassium stimulates the formation of carbohydrates in the nodules and thereby makes them better fitted to support the nitrifying bacteria (p. 146).

Distribution of potassium in plants.—The data in Table 9 illustrate the relations of potassium to the various parts of plants.

TABLE 9—POUNDS OF POTASSIUM IN 1,000 POUNDS OF DRY MATTER

Variety and part of plant		*Variety and part of plant*	
APPLES		GARDEN PEAS	
Flesh of apple	6.2	Seeds	14.3
Skin	8.5	Pods	14.1
Core	10.0	Leaves	28.4
Leaves (old)	3.5	Vines	30.0
New wood	4.5	BEETS	
GRAPES		Roots	52.0
Berries	13.0	Tops	41.2
Leaves	10.0	Young tops and roots	60.0
New wood	4.5	WHEAT	
RASPBERRIES		Grain	5.4
Fruit	13.0	Straw	4.8
Leaves	9.0	OATS	
New wood	4.1	Grain	4.0
Old wood	3.0	Straw	9.3
PEAS		CORN	
Seeds	9.5	Kernels	3.2
Straw	10.3	Leaves and stalks	16.0

CALCIUM

Some of the practical relations of calcium compounds in the growing of crops have been long appreciated in practice, though the exact reasons for the results obtained were not understood, and especially so far as the plant itself was concerned. We are here interested, not in the action of calcium compounds on the soil, (p. 200), but in the plant as a source of nutrition.

Action of calcium on plants.—Some of the functions of calcium in plant nutrition are fairly well known, at least by inference, though the evidence may be, perhaps, not yet regarded as actually established beyond doubt. They are: (1) Effect on cell-walls, (2) effect on transportation of starch, (3) effect on root-hairs, and (4) relation to acids.

(1) *Effect of calcium on cell-walls.*—One of the probable, important functions of calcium compounds is to aid in the growth and solidification or strengthening of cell-walls.

(2) *Effect of calcium on transportation of starch.*—Calcium, as

well as potassium appears to be associated with the transference of starch in plants (p. 54). This function was formerly thought to depend chiefly upon the presence of potassium compounds, but the exact relations of different elements to the changes cannot be regarded as known with a satisfactory degree of fullness or certainty.

When land needs lime the power spreader economizes time, labor and material besides reducing the disagreeable handling.

(3) *Effect of calcium upon the development of root-hairs.*—It has been noticed that in the absence of calcium compounds root-hairs (p. 123) do not develop as profusely as they do when such compounds are abundant.

(4) *Relation of calcium compounds to acids.*—Calcium is found in plants in combination with various acids, especially oxalic. It forms calcium oxalate, which can often be seen as small crystals. It is inferred that in thus uniting to form oxalate, which is not soluble in water, the calcium is utilized to remove the oxalic acid from the solution and thus neutralize the injurious effects it might have when present as an uncombined acid. Calcium is also found in plants as carbonate and phosphate and sometimes as sulfate.

Distribution of calcium in plants.—In a study made by the author of the composition of the fruit, leaves and new wood in several varieties of fruit trees and bushes, it was shown that the new wood uses more than three times the amount of calcium used by the fruit, while the leaves use over fifteen times as much as the fruit. Calcium is thus seen to be present in largest amounts where there is the greatest vegetative activity. It is found most abundantly in those tissues that have attained their growth, and in which the work of food-making in the chlorophyl-containing cells of leaves is going on. It is less abundant in those organs of the plant in which the manufactured foods are stored, and in dead parts such as old wood.

It is seen that while nitrogen, phosphorus and potassium are absorbed rapidly in the early stages of plant growth but decrease during later growth, the process is different with calcium. As plant growth advances, the absorption of calcium continues. While much of the nitrogen and phosphorus, present before the leaf falls, is transferred back to the stems or to the seed, calcium remains in the leaf. These statements are illustrated in Table 10.

TABLE 10—POUNDS OF CALCIUM IN 1,000 POUNDS OF DRY MATTER

Variety and part of plant		*Variety and part of plant*	
APPLES		BEETS	
Fruit	0.3	Roots	1.6
Leaves	22.0	Leaves	21.4
New wood	1.8	TURNIPS	
PEACHES		Roots	6.1
Fruit pulp	0.6	Leaves	27.3
Stones	0.4	WHEAT	
Leaves	34.8	Grain	0.5
New wood	1.9	Straw	2.1
PEARS		OATS	
Fruit	0.6	Grain	0.1
Leaves	20.3	Straw	3.0
New wood	11.4	CORN	
PLUMS		Kernels	0.3
Fruit pulp	0.7	Leaves and stems	4.3
Stones	1.0	PEAS	
Stems	18.1	Seeds	1.0
Leaves	30.3	Straw	15.4
New wood	17.6		

MAGNESIUM

Magnesium was formerly regarded as of secondary importance in relation to plant nutrition, but increased knowledge of what it does has modified this opinion. It is absolutely indispensable to all plants. While magnesium and calcium have points of close chemical resemblance, neither one can take the place of the other in their relations to plant life.

Action of magnesium in plants.—The work of magnesium in plants is less understood even than that of calcium. Four points for us to consider are: (1) Relation to seeds, fruit, etc., (2) relation to chlorophyl, (3) relation to oil formation, (4) effect as poison.

(1) *Relation of magnesium to seeds and other parts of plants.*—Magnesium compounds accumulate in seeds, fruits, flowers, roots and tubers to a larger extent than does calcium, but less in stems, leaves and wood, while the reverse is true of calcium. Hence, magnesium compounds appear to be especially important in the formation of seeds.

(2) *Relation of magnesium to chlorophyl.*—Magnesium is an essential constituent of chlorophyl (p. 122) and, therefore, is of use in the formation of starch. Lack of magnesium results in the production of plants without the green color characteristic of the presence of chlorophyl.

On small areas, such as home gardens, a home made spreader will be found useful for lime distribution.

(3) *Magnesium appears to be essential* for the formation of fat or oil in plants. Seeds rich in oil contain a larger proportion of magnesium than do those rich in starch.

(4) *Effect of magnesium compounds as poisons.*—Magnesium compounds, especially the chloride, may act as poisons and injure plants in the absence of calcium compounds. Solutions, even of great dilution, injure plant roots under such conditions. It is held by some scientists that magnesium compounds are injurious to plants when the ratio of magnesium to calcium goes beyond a certain limit, but the exact ratio is a matter of dispute and appears to vary according to many conditions.

Distribution of magnesium in plants.—The concentration of magnesium compounds in comparison with calcium, appears to be greater in seeds, fruits, flowers, roots and tubers, but the reverse in leaves, stems, straw and new wood. These relations are as indicated by Table 11, which should be compared with Table 10.

TABLE 11—POUNDS OF MAGNESIUM IN 1,000 POUNDS OF DRY MATTER

Variety and part of plant		*Variety and part of plant*	
APPLES		BEETS	
Fruit	0.6	Roots	1.8
Leaves	6.3	Leaves	10.2
New wood	0.[illegible]	TURNIPS	
PEACHES		Roots	1.8
Fruit pulp	1.0	Leaves	2.8
Stones	0.1	WHEAT	
Leaves	8.8	Grain	1.6
New wood	0.3	Straw	0.8
PEARS		OATS	
Fruit	0.6	Grain	1.6
Leaves	4.2	Straw	1.3
New wood	1.8	CORN	
PLUMS		Kernels	1.3
Fruit pulp	0.9	Leaves and stems	1.8
Stones	0.9	PEAS	
Stems	4.3	Seeds	1.3
Leaves	8.0	Straw	2.6
New wood	3.2		

SULFUR

It is well established that in the entire absence of sulfur compounds (sulfates) plants do not grow. Sulfur is used by crops in about the same amounts as phosphorus.

Action of sulfur in plants.—While much remains to be learned about the relations of sulfur to plant nutrition, a few facts are available that bear on the subject, for instance: (1) Relation to proteins and (2) relation to certain oils.

(1) *Relation of sulfur to proteins.*—Sulfur is always a constituent of protoplasm and also of plant proteins. Experiments show that crops rich in protein may be increased by the use of sulfates.

(2) *Relation of sulfur to odorous oils.*—Sulfur is a constituent of mustard oil, which imparts characteristic flavors to cruciferous plants, such as mustard, horseradish, turnip, cabbage, etc. It is also contained in garlic oil, the highly odoriferous compound characteristic of onions, garlic, leeks, etc.

Distribution of sulfur in plants.—Sulfur compounds occur much more largely in leaves and stems than in other parts of plants, as shown in Table 12.

IRON

Action of iron in plants.—Unless green plants receive enough iron compounds, no chlorophyl, that is green coloring matter, (p. 122) can be formed, consequently no normal growth. Although no iron compound appears to be a constituent of protoplasm or of any coloring-matter in chlorophyl, yet the presence of an iron compound is necessary for the production of green color in leaves, stems, etc. Under some conditions plants fail to assimilate iron and remain white or nearly so, when they are said to be suffering from *chlorosis*, which is regarded as the result of a diseased condition. Sometimes a shoot develops so rapidly that iron compounds do not reach the growing parts quickly enough, and then the newly formed leaves are white instead of green.

TABLE 12—POUNDS OF SULFUR IN 1,000 POUNDS OF DRY MATTER

Variety and part of plant		*Variety and part of plant*	
WHEAT		CORN	
Grain	0.170	Kernels	0.170
Straw	0.119	Leaves and stalks	0.126
OATS		CABBAGE LEAVES	1.82
Grain	0.189	HORSERADISH	2.00
Straw	0.195	TURNIPS	
		Roots	0.82

Distribution of iron in plants.—Plants take iron from the soil in smaller amounts than any other necessary constituent. From the data in Table 13 it is seen that iron is present generally in leaves in larger amounts than elsewhere in plants.

TABLE 13—POUNDS OF IRON IN 1,000 POUNDS OF DRY MATTER

Variety and part of plant		*Variety and part of plant*	
WHEAT		PEAS	
Grain	0.2	Seeds	0.14
Straw	0.2	Straw	0.6
OATS		BEETS	
Grain	0.3	Roots	0.4
Straw	0.6	Tops	1.5
CORN		TURNIPS	
Kernels	0.1	Roots	0.4
Leaves and stalks	0.8	Tops	1.3

SODIUM

Sodium appears as a regular, though small, constituent of plants.

Action of sodium in plants.—The weight of evidence appears to show that sodium is not essential to plant growth. The functions of potassium in plant nutrition cannot be performed by sodium, although it has been shown that, in the presence of an abundance of sodium salts, some plants use less potassium than otherwise, especially when potassium salts are not abundant.

Sodium compounds have been observed to increase crops, especially when there is lack of potash in the soil.

Distribution of sodium in plants.—Sodium is found mostly in the lower parts of plants, little being present in seeds, as illustrated in Table 14.

TABLE 14—POUNDS OF SODIUM IN 1,000 POUNDS OF DRY MATTER

Variety and part of plant		*Variety and part of plant*	
WHEAT		PEAS	
Grain	0.3	Seeds	0.2
Straw	0.5	Straw	1.6
OATS		BEETS	
Grain	0.4	Roots	9.2
Straw	1.8	Leaves	22.4
CORN		TURNIPS	
Kernels	0.2	Roots	6.0
Stems and leaves	0.5	Leaves	8.3

CHLORINE

All plants contain some chlorine, generally as sodium chloride. The necessity of chlorine in plant nutrition may be regarded as still an unsettled question, although the weight of evidence appears to be against the need of this element in normal plant growth.

Action of chlorine in plants.—Chlorine compounds are needed, if at all, in only small amounts. Beyond a limited amount, sodium chloride (salt) acts as a poison to plants, the chlorine and the sodium both acting injuriously. The ability of plants to withstand the injurious action of salt is shown by strand and marine plants, which contain much more salt than inland plants and, indeed, an amount that would kill inland plants. Some experiments indicate that a small amount of sodium chloride may be helpful to some crops.

Distribution of chlorine in plants.—Chlorides accumulate chiefly in the lower portions of plants, apparently as accidental constituents. Table 15 illustrates the general location of chlorine in plants.

SILICON

Action in plants.—The present general belief is that silicon performs no essential function in plant life, its constant presence being due simply to the fact that its compounds are in the soil solution and are taken into the plant as accidental constituents. Silicon compounds are present in such relatively large amounts, especially in the straw of cereals, it was taken for granted that they must perform some essential function. The function assigned to

TABLE 15—POUNDS OF CHLORINE IN 1,000 POUNDS OF DRY MATTER

Variety and part of plant	
OATS	
Grain	0.3
Straw	3.1
CORN	
Kernels	0.1
Leaves and stalks	0.7
PEAS	
Seeds	0.4
Straw	2.9
BEETS	
Roots	7.5
Leaves	25.0
TURNIPS	
Roots	4.0
Leaves	11.8

them was that of contributing to the stiffness of straw, but it has been shown that plants fully mature, with normal erectness in stems, when no silicon is present. It has been suggested by experimental work that soluble silicates may cause an increased assimilation of phosphoric acid by the plant, but there is no evidence that silicon can take the place of phosphorus in plant nutrition.

Distribution of silicon in plants.—Silicon is found in largest amounts in the older portions of plants, the leaves, stalks, straw, chaff, etc., especially when vegetative activity has diminished in those parts, as illustrated in Table 16.

TABLE 16—POUNDS OF SILICON IN 1,000 POUNDS OF DRY MATTER

Variety and part of plant	
WHEAT	
Grain	0.2
Straw	14.5
OATS	
Grain	4.9
Straw	13.4
CORN	
Kernels	0.1
Leaves and stalks	6.1
PEAS	
Seeds	0.1
Straw	1.4
BEETS	
Roots	0.6
Leaves	2.3
TURNIPS	
Roots	0.6
Leaves	1.8

CARBON

The element carbon may properly be called, in a chemical sense, the central element of all animal and vegetable substances, a necessary constituent of every living organism. It constitutes about 45 per cent. of the dry matter of plants, an amount greater than any other element in plants.

Action of carbon in plants.—Carbon is a constituent of practically all organic compounds among which are starch, cellulose, proteins, sugars, oils, acids, etc. Hence, its chief function is to furnish the large amount of carbon found in plant compounds. Plants obtain their carbon chiefly by taking in carbon dioxide through their leaves directly from the surrounding air. In the cells of the leaves, the carbon dioxide, by means of energy furnished by the sun's rays, is separated into its two elements, its carbon uniting with other elements to form the various compounds mentioned, most of the oxygen being returned to the air uncombined. The process of starch formation from CO_2 and H_2O is represented by the equation: $6CO_2+5O_2=C_6H_{10}O_5+6O_2$. This process is called *photosynthesis* because it depends upon light.

Effect of fresh and rotted manure. 7. One-quarter clay, three-quarters sand with fresh manure. 8. Same soil mixture with well rotted manure. 9. Sand with fresh manure. 10. Sand with well rotted manure.

In an indirect way, carbon dioxide performs another service for plants, since the carbon dioxide usually contained in soil water in considerable amounts has a pronounced dissolving action upon some of the mineral plant-food constituents of the soil, especially potassium and phosphorus compounds.

Distribution of carbon in plants.—Carbon is distributed as various compounds throughout all parts of plants; unlike the mineral elements, it is not concentrated in any one place. It is true, however, that some of the specific carbon compounds are locally concentrated, as, for example, sugar and starch in grain, fruit, roots, tubers, etc.

OXYGEN

The element oxygen constitutes about 42 per cent. of the dry matter of plants and is, therefore, second only to carbon in respect to its abundance in vegetable matter. If, however, we consider the green plant and take into consideration the oxygen of the water in the plant, oxygen constitutes a far larger proportion of growing plants than any other single element; indeed, it amounts to more than all the other constituents put together. Oxygen is present as an important constituent of all important plant compounds.

Action of oxygen in plants.—Oxygen serves at least two functions in plants. (1) It supplies the oxygen needed to make the various organic compounds. This oxygen comes largely, if not wholly, from water and not from atmospheric oxygen. (2) Free oxygen supports oxidation processes on which most plants depend for energy. It comes mainly from the atmospheric supply, but during photosynthesis is produced also in the green parts of the plant. Green plants do not thrive in the absence of oxygen. Seeds require free oxygen for germination. In the process of flowering, the absorption and chemical action of oxygen in the blossom are so marked in some cases as to develop a measurable increase of temperature. Oxygen is needed in the soil for the activity of roots and of micro-organisms as well as for effecting important chemical changes in the soil constituents.

Distribution of oxygen in plants.—Oxygen is distributed all through the plant in the various compounds of which it forms a part.

HYDROGEN

Action of hydrogen in plants.—The chief function of hydrogen in plants is to furnish the supply needed for the various combinations formed in the plant. The source of supply of hydrogen in plant nutrition is water.

As a constituent of water, hydrogen along with oxygen performs many important duties in connection with growing plants, so its action will be considered under the subject of water (p. 101).

Distribution of hydrogen.—Like the two other air-derived elements, carbon and oxygen, hydrogen is not concentrated in any particular part, but is distributed through the whole plant.

Up to this point we have learned that there are ten chemical elements essential to plant growth; we have studied in a preliminary way such compounds of these elements as are to have a special interest for us in connection with plant-feeding; we have seen that each plant-food constituent performs a special kind of work and produces specific effects in relation to plant growth; and finally we have observed that all the constituent elements must work together, that while some are used relatively in large, and others in small amounts, yet all are necessary and no one can take the place of another.

PART II

FUNCTIONS AND PHYSICAL PROPERTIES OF SOILS

VI

FUNCTIONS AND PHYSICAL PROPERTIES OF SOILS

General relations of soils and plants.—The soil performs two distinct functions in relation to plants: (1) It furnishes a dwelling place, in which, under conditions more or less favorable, plants begin and continue growth from seed to maturity. (2) It is intimately associated with the nutrition furnishing directly the mineral constituents (p. 15) plants use and serving as a medium for conveying to them a portion of the air-derived constituents (p. 12).

Soil and subsoil.—When we speak of the soil in connection with agriculture, we mean the superficial layer of earth in which plant roots can grow and from which they largely obtain their inorganic food supply. It is made up of disintegrated, powdered and partially decomposed rocks, mixed with varying amounts of organic matter formed mostly by decay of vegetable substances.

We distinguish by the name of *subsoil* the layer of earth below the soil proper. While the change from soil to subsoil is gradual, certain properties of the two are distinct; for instance: (1) Owing to the prevalence of root growth in the upper layers, the soil contains more decaying organic matter (p. 96) than the subsoil, this tends to make it darker in color. (2) Owing to the greater amount of organic matter, micro-organisms are largely confined to the soil, diminishing rapidly in the subsoil and finally disappearing altogether (p. 130). (3) The *texture* (p. 71) of soil and subsoil usually differs, the water carrying the finest particles into the subsoil and leaving the coarser particles in the upper layer. (4) The soil is usually richer than the subsoil in available plant-food, since conditions in the upper layer are more favorable for the conversion of unavailable into available forms.

Formation of soil.—While soil formation is a matter of general interest and often of much importance in connection with the plant-food supply in soils, we do not need to give more than a brief outline to its consideration. The principal part of the soil was once solid rock, which as the first step toward the formation of soil was broken up and powdered. The conversion of rocks into soil has been accomplished by means of various agencies, physical and chemical, and biochemical. The chief agents, *acting mechanically,* in rock disintegration are: (1) Heat, (2) cold, (3) flowing water, (4) moving ice (glaciers), (5) wave action, (6) winds, (7) vegetable growths, and (8) animals. The more prominent *chemical* and *biochemical* agencies are: (1) Oxygen of air, (2) the dissolving effect of water, especially when containing carbon dioxide (p. 47), (3) the decomposition of vegetable and animal matter by micro-organisms (p. 130), and (4) other forms of micro-organic action. Some of these agencies are more or less intimately associated.

That the soil has come from rocks is conclusively shown by the fact that we are able often to identify minute crystalline fragments of the characteristic minerals originally present in the rock from which the soils have come. The extent to which a soil consists of undecomposed, powdered rock-minerals and of products of rock decomposition depends upon the kind of rock and the intensity of action of the decomposing agents. In the case of some resistant crystalline rocks, the original rock constituents may not be removed by chemical decomposition more than one-half, while, in extreme cases, 90 per cent. of the rock may be changed. In the case of limestone rocks, the change of original material is always much greater, depending largely upon the purity of the limestone. Not only are the constituents of rocks decomposed, but the compounds into which they are changed by decomposition are frequently soluble and are carried from the surface into the lower portions of the soil.

A soil for agricultural purposes depends for its value upon (1) the original material from which it has been derived, (2) the state of fineness to which it has been reduced, and (3) the extent of loss it has undergone through chemical decomposition and leaching of essential plant-food constituents. In the processes of soil decomposition, there is a large proportional loss of compounds that contain calcium, magnesium and potassium.

PHYSICAL PROPERTIES OF SOILS

Under this head, we study those soil factors which contribute to make a suitable dwelling place for plants or, more particularly, for plant roots. It is, therefore, pertinent to inquire: (1) What important work do roots perform for plants? (2) What are the soil conditions essential for doing their work?

Function of plant roots.—The important things that roots do for plants in relation to the soil are:

(1) *Mechanical support.*—Roots serve to hold plants firmly in

place. They penetrate the soil downwards and sideways, and brace plants firmly in upright positions.

(2) *Absorbents of food materials.*—From the soil, roots absorb water containing dissolved nutrient materials (p. 124). Roots are the feeding organs of plants in relation to inorganic plant-food.

Soil conditions essential for work of roots.—In order that plant roots may do to the best advantage the work for which they are designed, the soil must offer favorable physical conditions, among which are the following:

(1) *Firmness and mellowness.*—For the most favorable condition of root growth, a soil should be firm in order to support the plant in its place, and mellow in order to enable the delicate rootlets to penetrate easily in their search for water and dissolve plant-food.

(2) *Moisture.*—Roots are the organs by which the large amounts of water required for plant growth are taken up. Hence the soil must be sufficiently moist to furnish the needed water.

(3) *Warmth.*—Root cells do not perform their work below certain temperatures and therefore require a soil of some degree of warmth.

(4) *Ventilation.*—Roots must breathe; they must therefore be supplied with atmospheric oxygen. This means that the soil must be well ventilated by being kept in a somewhat porous or open condition, as opposed to too great compactness.

From the preceding statements it is obvious that the physical properties which agricultural soils should possess are (1) firmness, (2) mellowness, (3) ability to hold enough moisture, not too much, (4) the power of absorbing and retaining heat, and (5) a porous, open condition that permits sufficient circulation of air, but not too much. These physical properties are dependent upon the mechanical conditions which, in turn, are governed by the fineness and arrangement of the particles of which the soil is made up.

Soil texture.—The physical properties of soils are, in large measure, primarily dependent upon the size of the individual particles. In speaking of the size of the individual particles, we use the word *texture*. The fineness of division required in the particles of a good agricultural soil is not easily appreciated. According to reliable estimates, a single ounce of good soil contains many *thousand millions* of separate particles. If the surfaces of all the particles in one cubic foot of soil could be spread out in one continuous surface, they would cover an area of one to three acres, varying according to the fineness of the particles; so it must be kept in mind that this great spread of surface in each cubic foot of such soil can be made accessible to the root systems of plants. It should be mentioned in passing that soil particles vary greatly in shape;

some are rounded and smooth; others are irregular, angular, jagged or somewhat rounded or smoothed fragments.

The mineral or inorganic constituents of the soil vary in size from large stones and coarse sand to finest dust. The number of particles of any one size varies greatly in diverse soils. The relative amounts of coarse and fine particles can be used as a basis for classifying soil material into textural groups, since the physical properties are largely affected by the proportions of different-sized particles.

Classification of soil materials according to texture.—The inorganic materials to be considered in the physical make-up of soils may be divided into the following groups or classes, according to fineness or coarseness of texture: (1) Coarse part, (2) sand, (3) silt and (4) clay. These textural groups are known as "soil separates."

(1) *The coarse part* includes all material larger than one millimeter (one twenty-fifth of an inch) in diameter and is divided into two parts, (1) *gravel* and *stone,* all material larger than two millimeters (two twenty-fifths of an inch) in diameter, and (2) *fine gravel,* all material smaller than two millimeters and larger than one millimeter in diameter.

(2) *Sand,* as applied to the texture of soils, includes those particles whose size is between one millimeter and one-twentieth of a millimeter (one twenty-fifth and one five-hundredth of an inch) in diameter. This group is subdivided into four subgroups or grades: (*a*) coarse, (*b*) medium, (*c*) fine and (*d*) very fine sand, which may be regarded as roughly corresponding in size, respectively, to the grains in coarse, medium and fine granulated sugar and fine table salt.

Sand, in the ordinary chemical sense, consists, when pure, of grains of quartz or silicon dioxide (SiO_2), also called silica (p.45); the quartz particles come from the original rock material of the earth. They are more or less rounded as the result of constant rubbing and wearing under the action of moving water and are generally made up in large part of quartz or silica because of the hardness of the original rock from which the particles come. Such material resists chemical and physical action; the remnants due to such action are left largely as more or less coarse sand particles. But sand, as applied to soil texture, usually contains also worn fragments of resistant minerals other than silica. Generally speaking, the finer the sand particles, the larger is likely to be the amount of mineral fragments other than silica, and therefore, of plant-food material. Sands are generally more or less colored by admixture of iron compounds (p. 83). In the case of sands on seashores bordered by coral rock, the grains consist largely of calcium carbonate.

Whatever their origin or the composition of the constituent

grains, sands are distinguished by one physical peculiarity; namely, the particles in a mass lack coherence when dry and easily fall apart.

(3) *Silt* includes soil particles that vary between one-twentieth and one two-hundredth of a millimeter (one five-hundredth and one five-thousandth of an inch) in diameter. In its physical properties, it is intermediate between sand and clay. When wet, silt feels fine, smooth and sticky like clay. If a soil that contains a large amount of silt and clay is shaken up with water and then allowed to stand, the silt settles slowly, while the clay remains suspended through the water a much longer time. A good illustration of silt is seen in the fine deposit in ditches and drains in which water moves slowly.

Deep cracks characteristic of soil in bad physical condition. Water passes rapidly into the air.

(4) *Clay,* as applied to soil texture, refers only to the size of the particles without reference to chemical composition. It includes all those whose size is less than one two-hundredth of a millimeter (one five-thousandth of an inch) in diameter.

Considered chemically, pure clay is a compound that contains aluminum, silicon, hydrogen and oxygen, known as kaolinite or hydrated aluminum silicate ($Al_2Si_2H_4O_9$), which results largely from the decomposition of feldspar. Particles that have the clay texture generally contain silicate or real clay particles in large proportions, because the original silicate minerals from which the particles come are not sufficiently resistant to withstand the forces that grind them into the fineness of clay texture; but they also contain, in variable though not large proportions, small grains of silica or quartz and fragments of broken-down minerals other than silicates. The reddish color commonly associated with many clays is due to the presence of iron compounds.

Clay particles in mass possess characteristic properties in marked contrast to those of the sand texture; the larger the proportion of particles of clay texture in a soil, the stronger will be the manifestation of the characteristics peculiar to clay texture; namely, puddling, plasticity, shrinking, suspension and flocculation.

(*a*) Puddling. When wet clay is worked or stirred, its constituent particles unite and become impervious to water. This process is called *puddling*.

(*b*) Plasticity. Clay is plastic in moist condition; it can be molded into any desired shape, which it retains on drying, and does not, like sand, fall apart into separate particles.

(*c*) Shrinking. When wet clay dries, it shrinks considerably and becomes so hard and tough that the mass can be broken or made fine only with difficulty. Owing to this property, the surface of wet clay soils splits into numerous characteristic cracks on drying.

(*d*) Suspension in water. When clay is stirred up with water, the finer particles remain long in suspension and the water may remain turbid for days or even weeks before all the particles completely settle and leave the water clear. This property is well illustrated by the water of some turbid rivers.

(*e*) Flocculation is discussed on page 78.

The characteristic properties of clay particles are due not so much, if at all, to chemical composition, as to fineness of division and to the resulting colloidal condition of the particles themselves. On being powdered, a mass consisting of pure clay (kaolinite) shows the physical properties of clay only when the size of the particles is reduced beyond a certain degree of fineness; large particles in mass possess no plasticity and on drying fall apart easily into powder, behaving much like quartz particles of the same size. On the other hand, when reduced to particles as small as those of clay, pure quartz shows all the characteristics of clay particles of the same size. It is generally true that in the finest particles of soils kaolinite particles predominate, because the material from which this is formed by chemical decomposition naturally undergoes extensive powdering by the action of water and weathering. In the case of sand particles, the grains are likely to be made up of quartz, because this material is so hard as to resist the agencies that produce fine particles in the case of less resistant material. Grains of quartz or silica are therefore more abundant in the coarser parts of soils. While we can distinguish chemically between pure sand (quartz) and pure clay (kaolinite), the chemical distinction does not apply sharply to the relations to texture of the materials, sand and clay, because particles of clay texture generally contain some quartz.

Classification of agricultural soils on the basis of texture.—The physical properties of soils vary according to the varying proportions of the diverse-sized soil particles they contain. While there is an indefinite number, soils for our purpose, can be considered under three general types or classes, (1) sandy, (2) clay and (3) loamy.

(1) *Sandy soils.*—In soils characterized as sandy, the amount of silt and clay particles is relatively small (usually less than 20 per cent.), the larger portion consisting of sand particles of various grades of fineness. Soils made up wholly of sand have little or no agricultural value. Soils with so high a percentage of sand particles as to be distinctly sandy have the following characteristics: (*a*) They permit water to pass downward too rapidly; (*b*) they admit air too freely, causing too rapid destruction of organic matter; (*c*) they absorb heat too rapidly, which results in excessively rapid evaporation of soil moisture, marked heating of the soil, and, in consequence, a premature ripening and even burning or drying of crops; (*d*) their absorbent and retaining powers with reference to soluble compounds are slight, so plant-food leaches rapidly when rainfall is abundant.

With increase of slit, clay and so-called colloidal organic matter, these properties are favorably modified. The texture of sandy soils can be changed by application of clay soil, but rarely is this procedure economically practicable. Little can be done to make a sandy soil into a good crop-bearing medium.

It is common to speak of sandy soils as light, because the particles move freely; they are loose and easy to work, offering little resistance.

(2) *Clay soils.*—In their physical properties, clay soils that contain large proportions (30 per cent. or more) of fine-textured particles are the reverse of sandy soils, as indicated by the following statements: (*a*) They hold water tenaciously and have a tendency to be deficient in ease of drainage; (*b*) they are much less accessible to air, and on this account organic matter decomposes slowly; (*c*) they are commonly known as cold soils; (*d*) clay soils are often called heavy because the particles stick together; they easily become compact, are hard to work and do not allow plant roots to penetrate easily; (*e*) after rains, clay soils quickly bake and harden on the surface so that when worked later the soil breaks into hard lumps; (*f*) when worked in too wet a condition, clay soils become "puddled," that is, the crumb-structure (p. 76) is destroyed and the individual clay particles are brought into the closest possible con-

tact. This has the effect of making the soil practically impervious to water and air, thus interfering seriously with drainage and ventilation.

A clay soil may be made less compact by mixing with it sand that is not too coarse. This also renders it more porous and admits air and water more easily. The adverse physical properties of clay soils are also favorably modified by increasing the amount of organic matter (p. 96).

(3) *Loamy* soils, intermediate in texture between sand and clay, contain a mixture of the various textural groups in such proportions that the properties of no one group predominate. In loamy soils the particles of sand, silt and clay textures are present in such relative amounts that the excessive porosity of the sand texture and the undesirable compactness of clay are favorably modified. Loams usually possess a desirable structure (p. 76), but are more or less sticky when wet. It is obvious there must be many varieties of loamy soils, their differences being caused by variation in the proportions of their various textural soil particles. Thus, we have sandy loams, silt loams, clay loams, gravelly loams, stony loams, etc.

(*a*) Sandy loams consist of somewhat more than one-half sand with less than one-half of silt and clay. They are among the lightest soils adapted to crop growing and are usually found useful for such crops as corn, cotton, small fruits and vegetables.

(*b*) Silt loams consist largely of silt with relatively small proportions of clay and sand. They are intermediate in physical properties between sand and clay, being more permeable to water than clay and less so than sandy soils. The subsoils of many of the western prairies belong to this type. While difficult to till, when properly drained, they are well adapted for general farm crops.

(*c*) Clay loams contain larger proportions of clay than silt loams do. As a rule they are somewhat difficult to manage, since they are inclined to be stiff and sticky, but they are usually regarded as being best adapted of all soils for general farming. When properly managed, they retain their crop-producing powers for many years.

(*d*) Gravelly loams consist of coarse-grained particles which vary in proportion from one-fourth to two-thirds, mixed with sandy, silt or clay loam. Fruit and forest trees do well on some kinds of gravelly loam.

(*e*) Stony loams consist of sandy, silty or clay loam mixed with stones, the latter varying in proportion from one-fourth to three-fourths. The stones interfere with ordinary methods of tillage so such soils are best utilized for some kinds of fruit and forest trees.

Soil structure.—By structure of soils we mean the manner in which the soil particles are arranged or grouped with reference to one another. The particles may be separated for the most part, each being free from the surrounding particles as in the case of sand; or they may perhaps be held together by colloidal material to form a group in which they remain united under ordinary conditions. When soil particles thus adhere, in larger or smaller grains, the structure is known as the *granular* structure or the *crumb* structure. Granules are of all sizes; they may become so enlarged as to interfere with favorable conditions for managing the soil in crop growing. They are then called clods. The tendency to form a crumb-like structure is strong in clay soils and practically absent in sandy soils.

The farmer usually recognizes structure in a practical way by the manner in which the soil behaves under the plow, whether it plows hard or easy. When a spading-fork is pushed into a soil that is just moist enough to work well and a mass of earth is removed and turned over the granular, crumb-like structure readily reveals itself. A good loam falls apart easily into the characteristic soil crumbs. A clay loam may need some strokes of the fork to make it crumble and the resulting crumbs will be larger than in loam less clayey. Sandy soils and sandy loams fall apart so readily that a forkful may fall apart and run between the tines. A clay soil is lumpy and falls apart only by vigorous strokes and then not into fine crumbs.

Factors that influence soil structure.—It is of practical importance to know what conditions are favorable to the crumb or granular structure and what to the single-grain structure; and it is desirable especially to study those factors which a farmer can make use of in modifying and controlling the soil structure. Control of structure is desirable in connection with loams, silts and clays; the modification of coarser-textured sandy soils is difficult to change appreciably by practicable means. The following conditions are among those that affect soil structure: (1) Variation of moisture, (2) freezing, (3) cultivation, (4) organic matter, (5) growth of roots, (6) soluble salts, (7) animal life, (8) hard rains.

(1) *Variation of moisture.*—It has been observed that the alternate drying and wetting of a fine-textured soil is favorable to the production of the crumb structure. When other conditions are uniform, the greater the number of times a soil changes from one condition to the other the smaller will the size of the granules become. Continued wetness or continued dryness does not favor the crumb structure.

(2) *Freezing.*—In the freezing of wet soils, the water forms crystals in the larger, open spaces between the soil particles, and the soil mass is broken up or separated, resulting in a condition that favors granular structure.

(3) *Cultivation.*—Manipulation of the soil by plowing or other forms of tillage, when the moisture content is not too great or too small, favors the formation of the granular structure. On the other hand, tillage when the soil is too wet or too dry is unfavorable.

(4) *Presence of organic matter.*—The presence of decomposing organic matter in soils favorably affects the soil structure (p. 76).

(5) *Growth of roots.*—It is a matter of common observation that crops differ in their effects upon soil structure. Plants with fine roots, like grass, wheat, etc., leave the soil in better working condition or tilth than coarse-rooted plants do.

(6) *Soluble salts.*—Certain compounds like calcium hydroxide (slaked lime), calcium sulfate (gypsum), calcium carbonate, mineral acids, common salt, etc., have the power of causing fine particles to unite and form aggregations. For example, if fine clay is mixed with water to form a turbid mixture and left to itself, it may remain turbid several days before the finest particles settle to the bottom and leave the water clear. If, however, we put into the mixture a little calcium oxide, which at once forms hydroxide (p. 41), or add hydroxide directly, the fine particles in suspension will be quickly affected. First, the particles will come together and form little flakes or floccules which soon settle, leaving the liquid clear. This process is known as *flocculation.* Calcium carbonate being less soluble than the hydroxide has much less flocculating power. This action is one of practical importance in connection with fine-textured soils, because it favors the granular structure. It is one of the important effects of liming (p. 192) clay soils.

Some substances, on the other hand, produce just the opposite effect, known as *deflocculation;* they break up the crumb structure, and cause the grains to separate, or they prevent the formation of the crumb structure. Among such deflocculating compounds are sodium carbonate and potassium carbonate which are found in many alkali soils. Long-continued use of nitrate of soda (p. 291) on clay soils may result in the formation of sodium carbonate in amounts sufficient to cause deflocculation.

(7) *Animal life.*—The soil crumbs formed by the action of earthworms are familiar. Ants and animals with burrowing habits also influence soil structure.

(8) *Hard rains.*—The effect of driving, heavy rains is readily seen on the surface of a cultivated clay soil. The granular structure is destroyed, and when the surface dries, a crust is formed which affects seriously the properties of the soil in relation to the growth of plant roots. These unfavorable effects are prevented by any kind of covering that protects the soil, such as mulches, sod, etc.

VII

THE INORGANIC CONSTITUENTS OF SOILS

ALL agricultural soils contain two general classes of constituents, (1) mineral or inorganic and (2) organic. The inorganic part comes from the rocks of the earth's crust; the organic portion has its origin in vegetable and animal matter.

Composition of earth's crust.—We can obtain an approximate idea of the general composition of soils by noticing the average composition of the earth's superficial crust, the material which enters most largely into soils; this is approximately as given in Table 17.

TABLE 17—CONSTITUENTS OF EARTH'S CRUST

Element	*Per cent.*	*Element*	*Per cent.*
Oxygen	47.0	Magnesium	2.5
Silicon	28.0	Sodium	2.5
Aluminum	8.0	Potassium	2.5
Iron	4.5	Sulfur, phosphorus and other constituents	1.75
Calcium	3.5		

It is seen that oxygen, silicon and aluminum, the materials most abundant,—over 83 per cent. of the rock constituents from which soils are formed—do not contribute any material to the nourishment of plants.

Mineral compounds in soils.—When rocks are converted into soil, those portions that undergo chemical change generally form certain definite chemical compounds, then cease to change further to any marked extent under the conditions present. The arrangement in Table 18 shows in general the various kinds of chemical compounds usually present in soils.

Some of these compounds are present only under exceptional or abnormal conditions; others are always present in all agricultural soils. The points that deserve particular attention are the special compounds, the amounts present in soils, the ways in which they may be removed and their action on other soil constituents or on the soil as a whole.

(1) *Calcium.*—(*a*) Compounds. Calcium is present in all good agricultural soils as carbonate ($CaCO_3$) in larger proportion than in any other form. It is also generally present, but to a much less extent, as phosphate, nitrate, sulfate and in some of the complex silicates. The carbonate may be regarded as the commonly available form or source of plant-food calcium; in most silicates calcium is of little importance as an immediate source of plant-food, although pure calcium silicate ($CaSiO_3$) is regarded as furnishing available calcium.

TABLE 18—CHEMICAL COMPOUNDS IN SOILS

Name of element	*Forms of combination*
Calcium	Carbonate, sulfate, silicate, phosphate, nitrate
Potassium	Carbonate, sulfate, silicate, phosphate, nitrate, chloride
Sodium	Carbonate, sulfate, silicate, phosphate, nitrate, chloride
Magnesium	Carbonate, sulfate, silicate, phosphate, nitrate, chloride
Iron	Oxide, hydroxide, hydrated oxides, phosphate, silicate
Aluminum	Oxide, hydroxide, hydrated oxides, phosphate, silicate
Manganese	Oxide, hydroxide, hydrated oxides, phosphate, silicate

(*b*) Amount in soils. Calcium carbonate ranges in variable amounts from a fraction of 1 per cent. to over 50 per cent. in some limestone soils. Soils are not capable of raising good crops when the calcium carbonate drops below 0.2 per cent. In most cases an amount in excess of 1 per cent. (equal to 30,000 pounds an acre in the upper foot of soil) is considered desirable. Coarse sands and soils with high percentages of organic matter usually contain insignificant amounts of calcium carbonate; clay soils also are commonly deficient. The subsoil may often contain less than the soil. *Marly soils* are mixtures of clay and finely divided calcium carbonate. *Shell-marl* is sometimes found in deposits that consist of nearly pure calcium carbonate, generally mixed with varying amounts of impurities. In alkali soils calcium compounds are present in considerable amounts, especially the sulfate. The chloride may be present also.

(*c*) Removal. Calcium carbonate in soils is constantly undergoing solution by the carbon dioxide present in soil water, also by acids formed in the decomposition of organic matter (p. 89). In this dissolved form it may escape from the soil in the drainage water (p. 118). Calcium is also removed by crop and weed in rather large amounts, varying in the case of diverse kinds of plants. On account of its important relations to soils and to plant growth, it is essential that the amount present be not allowed to diminish much below 1 per cent. This involves the occasional addition of calcium compounds to cultivated soils (p. 192).

(*d*) Relation to texture. Calcium carbonate in soils may exist in particles of any size, varying from the texture of clay to that of sand. The finer the particles, the more quickly they undergo solution in water containing carbon dioxide.

Sweet corn leaves curl during dry weather when the supply of water in the soil is insufficient to supply the loss by transpiration.

(*e*) Action in soil. Calcium carbonate acts in several ways of great value in agricultural soils. (1) It neutralizes soil acids. In the decay of vegetable matter, the acids formed (p. 94) accumulate in the absence of calcium carbonate and may injure the crop-producing power of the soil. When calcium carbonate is present in abundance, the calcium unites with the acids to form neutral, harmless salts. Calcium carbonate is the most effective acid neutralizer we find in soils. (2) It changes insoluble aluminum and iron phosphates into the more available form of tri-calcium phosphate (p. 33). In many soils that contain insufficient amounts of calcium carbonate, the soluble phosphate applied in fertilizers is largely changed into unavailable iron and aluminum phosphates and so is practically lost to the immediate use of crops. (3) Dissolved in water that contains carbon dioxide it may react with insoluble potassium compounds to form soluble potassium carbonate. (4) The action of soluble calcium compounds in flocculation of fine clay particles has been noticed (p. 78). (5) After calcium carbonate is changed into soluble carbonate in the soil water, it is often redeposited under specific conditions in solid form and acts as a cement to hold together the finer particles of silt and clay, thus changing the structure of the soil particles favorably.

(2) *Potassium.*—(*a*) Compounds. Potash is present in soils in both soluble and insoluble compounds. The soluble forms include sulfate, chloride (muriate), nitrate, and probably phosphate and carbonate. The insoluble forms are generally silicate compounds present in such minerals as potassium or orthoclase feldspar (about 14 per cent. of potassium), leucite (about 18 per cent. of potassium) and several others (p. 38). In ordinary soil analysis, the percentage of potassium given is the acid-soluble, which roughly represents the amount available, but which is usually only a small part of the total potassium present.

(*b*) Amount in soils. Potassium is present in soils in amounts ranging from a few hundredths of 1 per cent. to 3 per cent. or more, only a small proportion of which is generally available. A good agricultural soil does not generally contain less than 0.2 per cent. of acid-soluble potassium (equal to about 6,000 pounds an acre in the upper foot). Potassium is generally abundant in clay soils and deficient in muck, peat and sandy soils. In alkali soils it is often abundant, running in some cases as high as 17 per cent. in the upper 12 inches, though it is usually between 1 and 7 per cent.; in these soils it exists in the forms of sulfate, carbonate and chloride.

(*c*) Removal. Potassium compounds are removed from soils by drainage water (p. 113) in amounts somewhat less than by means of crops (p. 111).

(*d*) Relation to texture. Generally speaking, potassium is found in largest amounts in the finer soil particles.

(3) *Sodium* is present in the soil in combinations much like those of potassium, generally in smaller amounts. Its compounds are prominent, and often predominant, constituents of alkali soils. The compounds present in such soils are one or more of the following: Sulfate, carbonate, bicarbonate, chloride, nitrate and phosphate (p. 80). In the case of excessive use of sodium nitrate in the absence of acid-forming compounds, amounts of sodium carbonate may be formed to such an extent as to injure seriously the

crumb structure of soils (p. 77). The amount of sodium does not appear to vary much in diverse-textured particles.

(4) *Magnesium* is usually present in soils in much the same forms of combination as calcium, but generally in much smaller amounts. Soluble magnesium compounds are believed to injure plants when present in amounts nearly equal to calcium. The sulfate and chloride of magnesium, often prominent constituents of alkali soils, are poisonous to plants in even comparatively small amounts, especially if the amount of calcium compounds is relatively deficient. Magnesium compounds in soils act like the corresponding calcium compounds in neutralizing acids. Magnesium is removed from the soil in much less amounts than calcium, either by drainage water or crops. The amount of magnesium is greatest in the finest particles, decreasing with those of coarser texture.

(5) *Iron* is more or less prominently present in soils in the form of hydroxide, $Fe(OH)_3$, and oxide, Fe_2O_3 and in hydrated oxides, which are intermediate in composition between hydroxides and oxides—forms familiar under the general term iron rust. These compounds are red, yellow or brown and form the main coloring matter of soils. They are generally products of decomposition of complex iron-containing silicates that undergo chemical change when air has abundant access to moist soils. In some cases, as in clay subsoils, in soils rich in iron compounds, in muck-beds, etc., where air is largely excluded, iron compounds poisonous to plants are formed and, if present in sufficient amounts, make a soil sterile; for example, sulfur compounds of iron known as sulfides may be changed into the injurious compounds, iron sulfate (p. 44) and free sulfuric acid. This condition can be usually prevented or overcome by generous use of calcium carbonate or quicklime (p. 192). The amount of iron compounds present in soils varies greatly, but the cases are rare in which a soil does not contain enough iron to supply the needs of crops indefinitely. The amount of iron removed from soils by drainage water or crops is generally less than that of any other soil constituent. Iron compounds may be useful in forming insoluble iron phosphate and thus preventing the risk of loss of phosphorus by drainage. Iron compounds, like those of calcium, may act as cementing material to hold together smaller particles of silt or clay. The amount of iron, greatest in the finest soil particles, diminishes as the particles increase in size.

(6) *Aluminum* is present in soils chiefly as silicate, hydrated oxide, and phosphate. It is especially prominent in soils derived from clay (kaolinite); and present in largest amounts in the particles of finest texture. Any value which aluminum may possess in crop growing is confined to its action on other soil constituents.

(7) *Silicon* is present in soils in much larger amounts than any

of the preceding constituents. With oxygen it makes up about 75 per cent. of the mineral compounds present in soils. Silicon occurs as the characteristic constituent of all silicates. It is also present in soils to a variable extent as silicon dioxide (SiO_2), commonly called silica, familiar as the mineral, quartz. White quartz sand consists of silica and in all sandy soils silica is a prominent constituent.

(8) *Phosphorus* is present in soils in smaller amounts than any of the elements thus far discussed. A soil that contains 0.25 per cent. of total phosphorus (equal to 7,500 pounds an acre in the upper foot of soil) is rich in this constituent; the usual amounts in agricultural soils lie between 0.03 and 0.11 per cent., only a part of this being available. In the finer particles of soils, phosphorus compounds are present in larger amounts than in coarser particles.

Phosphorus is present in soils mainly as phosphates of calcium, iron, aluminum and magnesium. Iron and aluminum phosphates (p. 37) are so insoluble as to be practically unavailable plant-food; however, in these forms, they are not easily leached from soils. Insoluble or tri-calcium phosphate (p. 33) is a more readily available form. On soils that contain large amounts of iron and aluminum compounds, the soluble phosphoric acid in fertilizers is changed into a form of iron and aluminum phosphate unless there is considerable calcium carbonate in the soil. Phosphates are removed largely from soils by crops and, to some extent, by drainage water, though in much smaller amounts than any other important plant-food constituent.

(9) *Sulfur* is present in agricultural soils as sulfate. In special cases, which may be regarded as abnormal, it may exist temporarily as sulfide. It is found in relatively small amounts in most soils but in proportions large in comparison with the amounts used by crops. It is a prominent constituent of alkali soils in combination as sulfate of potassium, sodium, magnesium and calcium sulfate. Sulfates are found in drainage water in amounts greater than are used by crops and in some soils may become deficient. However, when commercial fertilizers are applied in the form of potassium sulfate, or acid phosphate, which may be about half calcium sulfate (p. 196), considerable amounts of sulfate are added to the soil.

AMOUNTS OF PLANT-FOOD CONSTITUENTS IN SOILS

In studying the results of chemical analysis of soils, the following analysis of a soil known to be good for crop production and therefore recognized as fertile will serve as an illustration.

Nitrogen (N)..... 0.20 per cent.		
Phosphorus (P)... 0.07 "	(equal to 0.16 per cent.	phosphoric acid, P_2O_5)
Potassium (K).... 0.35 "	(" 0.42 "	potash, K_2O)
Calcium (Ca)..... 0.60 "	(" 1.50 "	calcium carbonate, $CaCO_3$)
Sulfur (S)........ 0.06 "	(" 0.18 "	sulfate, SO_4)

An examination of these figures suggests the following statements:

(1) Soil constituents that plants use for food are present in soils only in relatively small proportions. The great bulk consists of such inert materials as compounds of silica, aluminum, etc.

(2) The combined amounts of nitrogen, phosphorus and potassium, even in fertile soils, are usually less than 1 per cent.

(3) Although the figures do not furnish the information, it is true that, of the nutrient compounds in soils, only small amounts are in forms immediately available.

TABLE 19—PROPORTIONS OF TOTAL PLANT-FOOD CONSTITUENTS IN AGRICULTURAL SOILS

Constituent	*Per cent.*	*Pounds in one acre to depth of 9 inches*
Nitrogen	0.10 to 0.30	2,500 to 7,500
Phosphorus (P)......	0.03 to 0.11 (= 0.07 to 0.25 phosphoric acid, P_2O_5)	750 to 2,750 (1,750 to 6,250, P_2O_5)
Potassium (K).......	0.8 to 1.60 (1.00 to 2.00 potash K_2O)	20,000 to 40,000 (25,000 to 50,000, K_2O)
Calcium (Ca)........	0.20 to 1.50 (0.50 to 3.75 calcium carbonate, $CaCO_3$)	5,000 to 37,500 (12,500 to 93,750, $CaCO_3$)
Sulfur (S)...........	0.02 to 0.20 (0.06 to 0.60 sulfate SO_4)	500 to 5,000 (1,500 to 15,000, SO_4 ion)

Variation of plant-food constituents in agricultural soils.— In examining the results of analysis of large numbers of crop-producing soils, it will be found that the proportion of each plant-food constituent varies widely in diverse soils. It is desirable to state the results of soil analysis not only in the form of percentages but also as pounds per acre. In analyzing soils, samples are taken to a specified depth, which varies, but is usually 6, 9 or 12 inches. The samples taken to a depth of 6 or 7 inches represent the portion usually plowed, in which the roots mostly grow. At depths of 9 or 12 inches portions of soil are included in which plant-roots penetrate more or less and the analysis of which gives an idea of the amount of plant-food near enough at hand to be utilized easily. In stating the results of soil analysis, it is necessary, therefore, to state the depth to which the sample of soil is taken; this is particularly important in stating the results as pounds per acre. It will suffice for our purpose to give results based on a depth of 9 inches, and to assume that the soil in one acre to that depth weighs, on an average, 2,500,000 pounds. In soils of good crop-producing power, the amounts of constituents usually lie between the limits indicated in Table 19.

In extreme cases, figures run below and much above these. Thus, potassium may in some cases run above 100,000 pounds an acre and calcium still higher, while phosphorus may exceed 10,000 pounds, and nitrogen in muck and peat soils may run above 50,000.

VIII

THE ORGANIC CONSTITUENTS OF SOILS

In addition to inorganic, rock-derived materials, soils contain variable amounts of organic matter, which has its origin for the most part in the former generations of plants, including micro-organisms, that have grown in the soil and undergone decomposition; to a less extent, soil organic matter comes also from animal remains. In the case of agricultural soils, the upper layer is continually receiving supplies of organic matter in the form of leaves, plant residues left in the soil (roots, stubble, etc.), green-crop manures (pp. 256-265), farm manures (pp. 235-255), dead cells of micro-organisms and various kinds of vegetable and animal waste materials. This organic matter, whatever its original source or character, is found in soils in all conditions or stages of decomposition, varying from the fresh material to that which has undergone more or less extensive chemical changes, and which appears in the forms known as leaf-mold, peat, muck, humus, vegetable mold, etc., according to the nature and extent of the changes it has undergone. Sooner or later, decomposition so alters vegetable or animal matter in soils as to form a dark-colored mass, in which all trace of any characteristic tissue structure in the original material disappears. The decomposed material is more or less completely incorporated as an integral part of the soil mass; in this condition it is called *humus*.

The presence of decomposing and decomposed organic matter in agricultural soils is essential, generally speaking, as an aid in furnishing conditions which enable crops to use applied fertilizers most efficiently. The organic matter of soils may affect the production of crops in more ways than any other soil constituent, because (1) it is a source of plant-food, (2) it helps to make available some of the insoluble mineral plant-food, and (3) it influences favorably the physical conditions of soils. (4) It may also, on the other hand, contain compounds which under some conditions are poisonous to plants.

On account of the extreme importance of organic soil constituents, we shall study (1) the changes that organic matter undergoes in soils, (2) the conditions that influence such changes, (3) the relations of the decomposition products to other soil constituents, (4) the specific action in relation to the use of fertilizers.

Composition of organic matter.—The composition of organic matter, whether fresh or in various stages of decomposition, is

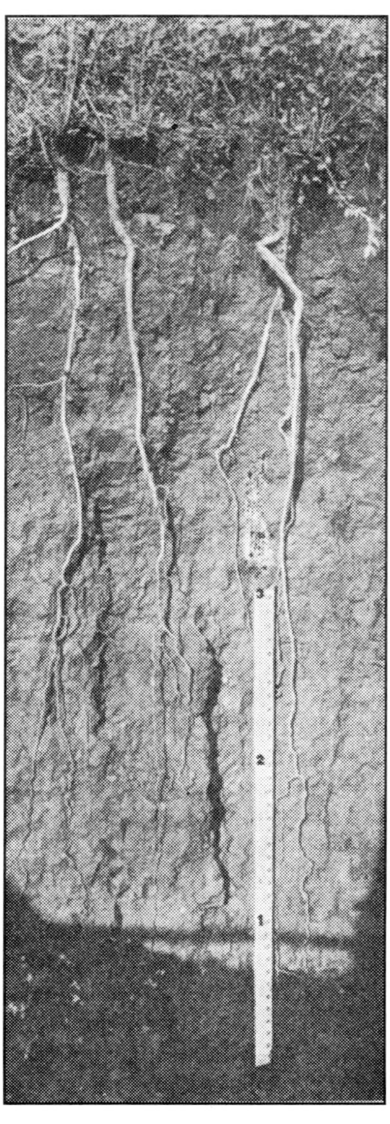

Alfalfa roots often reach depths of more than six feet.

extremely variable, since it is a complex mixture of numerous compounds, the kinds and amounts of which constantly vary. In general, the decomposition products of organic matter are governed by the source and character of the original material and by the conditions that determine the kind and extent of decomposition, such as moisture, air supply, temperature, kind of micro-organisms and mineral constituents, factors which, in turn, are affected by tillage, drainage, fertilization, irrigation, inoculation, etc.

Without going into too great detail, it is desirable that some few fundamental facts regarding the composition of fresh organic matter be stated and then the general changes traced through which the compounds of the fresh material go during the processes of decomposition.

Fresh organic matter is made up mostly of three general classes of compounds: (1) *Nitrogen-free organic compounds* (carbon, hydrogen and oxygen), which include (*a*) carbohydrates, sugars, starches, etc., (*b*) celluloses, ligno-cellulose or lignin (p. 14), (*c*) fats or oils and waxes, and (*d*) organic acids. (2) *Nitrogen-containing compounds,* mostly proteins (contain nitrogen, usually sulfur, and, in some cases, phosphorus, in addition to carbon, hydrogen and oxygen). (3) *Inorganic or mineral constituents* need not be considered as they are present in only small amounts. These three classes of compounds in fresh organic matter undergo decomposition in soils and furnish variable, complex mixtures, which may contain some or all of the following products of decomposition, according to the kind and extent of change.

(1) *Nitrogen-free compounds,* the largest part of soil organic matter, consist largely of various indefinite, little-understood, dark-colored substances, more or less acid in character, rich in carbon (so-called humic-acid compounds) and, in smaller amounts, of certain definite organic acids (acetic, butyric, etc.). They come in largest amount from the celluloses, starch and sugar of fresh organic matter.

(2) *Nitrogen compounds* are present in decomposing and decomposed organic matter in much smaller amounts than nitrogen-free products. They include (*a*) dark-colored, insoluble nitrogen-containing carbon compounds, (*b*) ammonia, and (*c*) nitric acid.

(3) *Inorganic compounds* include the mineral constituents or salts, mainly as calcium, magnesium and potassium compounds, such as phosphate, carbonate, sulfate, etc., always present in small amounts in fresh organic matter.

Decomposition of organic matter.—The process of decomposition by which fresh organic matter is changed into different forms of combination is a biochemical process primarily brought about by micro-organisms, chief among which are bacteria (pp. 130-153), fungi and actinomyces (p. 154). Decomposition is of two general kinds, one takes place in the presence of the oxygen of the air (aërobic, p. 136) and the other with exclusion of air (anaërobic, p. 137). The composition and amount of the products of decomposition in the soil differ greatly in the two processes.

(1) *Decomposition in presence of air.*—With abundance of air (oxygen), organic matter (chiefly carbon, hydrogen, nitrogen and oxygen) decomposes into simpler compounds, and practically all passes away, when the process is complete, in the form of gaseous compounds (carbon dioxide, ammonia, free nitrogen, water, etc.), without leaving any organic residue. This is illustrated when leaves or twigs fall upon the ground. Without darkening much, such organic materials, under favorable conditions of moisture and temperature gradually undergo a process akin to burning or union with oxygen and in time disappear, leaving about the same constituents that would be found in the ashes produced by rapidly burning the same material. This process can be easily observed in its various stages in forests in the case of fallen trees; in the course of time, under the action of fungi, bacteria, etc., the solid wood softens, becomes "rotten," loses all signs of structure of grain or fiber, and sinks down into a shapeless heap; the material may remain nearly white or may become reddish or brownish. The pile slowly flattens out and in time all traces of organic matter disappear.

Other conditions being uniform, this destructive form of decomposition takes place to a greater or less extent in soils according to the amount of air present. It is a fact of practical experience that vegetable or animal matter disappears much more rapidly in light, porous, well-drained, well-ventilated soils than in compact or over-moist soils where the air supply is generally limited. Other conditions being the same, organic matter disappears more rapidly at higher temperatures, as illustrated in case of a warm, dry climate.

(2) *Decomposition in absence of air.*—Away from air (oxygen), as when leaves or twigs fall into a swamp or a pond and are buried under water, the organic matter decomposes slowly and only partially, leaving a comparatively large amount of incompletely decomposed organic material, since the loss of organic compounds through change into gaseous products is comparatively small. In connection with this process, the following facts are of interest:

(*a*) Visible effects of decomposition. Signs of change in the original organic material are made visibly evident, first, by change in the color of the substance through various shades of brown to black, and, second, by the loss of any characteristic shape or tissue structure that the material may have had originally, resulting in the formation of a finely granular or powdery mass when dry, or a gelatinous mass in colloidal condition when wet.

(*b*) Increase of carbon and resistance to change. In the conversion of vegetable material into the dark-colored compounds that constitute a large part of well-decomposed organic matter, hydrogen and oxygen pass from the organic matter in the form of gases in a much larger proportion than does carbon; consequently, the carbon accumulates and the decomposed organic matter is richer in carbon and poorer in hydrogen and oxygen than the original organic material. It is also a fact of interest that these dark-colored, high-carbon compounds show increasing resistance to further decomposition, the resistance becoming more pronounced as the percentage of carbon in the substance increases. On this account the dark-colored material long remains in the soil. This resistance to decomposition finds an explanation in the composition of the material. The micro-organisms first use up the least resistant compounds of the fresh organic matter. In the remaining more resistant portions, it is increasingly difficult for the organisms to obtain food so their activity diminishes. In time, when only the most resistant compounds are left, decomposition practically ceases from lack of food supply.

(*c*) Increase of nitrogen and insolubility. Proteins, in absence of air, are decomposed with greater difficulty and less completeness than starch, sugar, some forms of cellulose, etc. Nitrogen, therefore, tends to accumulate, so the decomposed organic matter contains a larger percentage of nitrogen than the original material. The nitrogen compounds slowly change into dark-colored substances of variable composition and the nitrogen is left in forms of combination not easy to decompose further. More or less soluble nitrogen is made insoluble by being taken into the body substance of the micro-organisms themselves. In this anaërobic process of decomposition, nitrogen compounds tend to become less available as plant-food.

Decomposition of organic matter in soils.—(1) *Compounds undergoing change.*—Of the many compounds involved in the decomposition of organic matter in soils, some are changed quickly, others slowly. The first compounds acted on include starches, sugars and proteins soluble in water, after which come the normal celluloses (p. 13) and insoluble proteins. The compounds that more strongly resist decomposition are the ligno-cellulose or lignin (woody-fiber), oils or fats, waxes and resins. The residues from these compounds therefore constitute the material from which humus mostly comes.

(2) *Relation of carbon to nitrogen.*—When organic matter de-

composes in soils, carbon and nitrogen tend to accumulate in the residue, but the accumulation of nitrogen is greater than that of carbon, compared with the amount in the original undecomposed material. Thus, in fresh plants carbon is present in amounts twenty-five to forty times that of organic nitrogen, while in the decomposed organic matter of the soil carbon is only ten to twelve times that of nitrogen. This decrease in the proportion of carbon to nitrogen is due to micro-organic decomposition of the carbon compounds while the nitrogen is worked into proteins as part of their bodies.

A heavy duty tractor may replace a dozen mule team and handle a six foot disc plow cutting five inches deep, thus speeding up soil preparation.

Intermediate decomposition products between fresh organic matter and the products of advanced decomposition, the decomposing material, as a whole, acquires characteristics, which represent the kind and extent of change, or the stage, in the process of decomposition. The mass of organic material in these intermediate stages is known under various names, such as leaf-mold, peat, muck, etc. The character of the products is largely controlled in each case by the conditions of air supply, temperature, moisture, etc., since these conditions determine what particular set of decomposing agents predominates. The main characteristics of some of these intermediate decomposition products of organic matter are:

(1) *Leaf-mold* or *mild humus* is the dark-colored, organic material formed by the incomplete decomposition of leaves, twigs, etc.,

on the surface of well-drained alkaline or neutral forest soils. Of a similar nature is the brown, powdery material often found inside dead, hollow trees. Such materials represent a stage of decomposition, in the presence of air, intermediate between the fresh organic matter and the gaseous products formed by complete action. Leaf-mold and similar materials are as indicated different from humus in two ways: (1) Leaf-mold is either neutral or alkaline, while humus is acid; (2) it is easily changed in air by micro-organisms into simple compounds, while humus is only slowly changed.

(2) *Peat* consists of plant remains in the early stages of decomposition under water and therefore away from air. The original structure of the plant tissues is more or less plainly visible. Peat is generally formed in extensive deposits, often of considerable depth. Decomposition is usually delayed or prevented by exclusion of air through saturation with water, but takes place when the mass dries out somewhat; consequently there are usually alternating periods of active decomposition and suspension of such changes. The partially decomposed material may be prevented from undergoing further change by the accumulation of acid decomposition products which poison the micro-organisms. The alternate saturation and drying of the mass furnish conditions which delay or advance decomposition. The plant-food material in peat is only slowly available. Two general classes of peat are recognized, (1) grass peat from swamp-grass, etc., and (2) moss-peat from sphagnum moss.

(3) *Muck,* known also as meadow-peat, swamp-muck, etc., consists of plant remains that have undergone more complete decomposition that peat, though the line of division between muck and peat is not sharp. It more nearly approximates the condition of the dark constituents of humus. Muck is brown or black and shows much less of the original structure of plant tissues than does peat. Its plant-food constituents are not quickly available, but are more so than those of peat. Drainage, liming (p. 192), application of phosphorus and potassium compounds, and tillage make peat or muck soils productive.

Active and inactive organic matter.—In soils organic matter is said to be active when its less resistant portions are changing its own nutrient constituents from insoluble to soluble forms, and furnishing products that become available compounds (p. 6). When the more active stages of decomposition have passed, there remains a portion which so strongly resists further change it may require years for complete decomposition. This portion is known as inactive or slowly available organic matter. While not a source of quickly available plant-food, it is useful on account of its effect upon the physical properties of the soil (p. 97). The application of fresh organic matter to soils is generally attended with much more rapid and effective results than the use of such inactive materials as peat and muck.

Acids in decomposing organic matter.—In the decomposition of organic matter, acid compounds deserve special attention on account of their important, practical relations. For convenience they may be considered under the following heads: (1) Acids in plant juices, (2) organic acids from carbohydrates, fats, waxes, resins, etc., (3) miscellaneous organic acid compounds, (4) carbon dioxide, (5) nitric acid.

(1) *Acids in plant juices.*—Plant juices are generally acid owing to the presence of the organic acids or acid salts they contain. These are commonly called vegetable acids. They are present in the fresh organic matter and, although they are decomposed sooner or later, they probably persist long enough to produce some solvent effect in the soil.

(2) *Organic acids from carbohydrates, fats, waxes, resins, etc.*—In those portions of soils well aerated and warm, sugar, starch, part of the cellulose, etc. are converted by micro-organisms rapidly into definite organic acids. It is a familiar experience that carbohydrate materials sour easily and oils become rancid. For example, acetic acid, familiar in vinegar, ensilage, fermenting pea-vines, fruit-refuse about canning works, etc., results from carbohydrate decomposition. Many of the acids formed in this way are of definite, well-known composition. They or their salts may be decomposed into carbon dioxide and water in the presence of abundant air and warmth, and, may, therefore, not be present in soil organic matter that has reached an advanced stage of decomposition under these conditions.

(3) *Miscellaneous organic acid compounds.*—When the organic matter of the soil is in an advanced stage of decomposition organic compounds more or less acid in character occur in a variable number. These compounds, as well as others of different chemical character, are found in that part of the organic matter called "humic acid." This substance was supposed to be made up of a few well-defined compounds, but numerous other organic compounds have been shown to have acid properties. When these acid compounds are treated with any acid-neutralizing compound such as calcium carbonate (limestone) or calcium hydroxide (slaked lime), the acids are neutralized and form compounds insoluble in water.

(4) *Carbon dioxide gas* is always formed (p. 47) by the decomposition of organic matter in soils and is largely taken up by the soil water. Soils in which considerable amounts of organic matter are undergoing decomposition are rich in this gas, the soil

air often containing two or three hundred times as much carbon dioxide as does ordinary air.

(5) *Nitric acid* is produced as the final product of the bacterial change (p. 132) of the protein compounds present in fresh organic matter.

Poisonous compounds or toxins formed from organic matter.—Research chemists in the Bureau of Soils, United States Department of Agriculture, have isolated from the organic matter of soils numerous decomposition products. Among them are some that appear to be useful plant-foods, while some act as poisons, or are toxic, to plants grown in water solutions that contain these compounds in small amounts. Some of them are definite organic acids. So far as indications point at present, organic toxic compounds in the soil are rendered partially or completely harmless to plants by good drainage, aeration, tillage, crop rotation, liming and fertilizers, conditions that have been recognized as essential to good farming.

Action of organic matter on soil constituents.—Under agricultural conditions, vegetable and animal matter is distributed through the soil and comes into intimate contact with many of the soil particles. These particles to some extent are modified in composition during the process of decomposition of the organic matter, especially by the acid compounds present at various stages. The chemical changes produced in soil constituents by these acids may be summarized as follows:

(1) *Potassium compounds* in some insoluble forms are changed into soluble compounds, such as potassium carbonate, potassium nitrate, and combinations of potassium with organic compounds.

(2) *Phosphorus compounds,* especially insoluble calcium phosphate, are changed into soluble form. Probably even the insoluble iron and aluminum phosphates may be dissolved to some extent.

(3) *Organic acid compounds* combine with calcium, magnesium, etc., to form insoluble compounds, as already stated; this has the effect of preventing rapid loss of the decomposed organic matter by leaching.

(4) *Nitric acid* combines with calcium, potassium, sodium, etc., to form nitrates.

(5) *Carbonates.*—The carbon dioxide of the soil, which comes largely from the decomposition of organic matter dissolves insoluble soil constituents and makes them available. Important among these are calcium carbonate and phosphate, and compounds of potassium, magnesium, etc.

(6) *Neutralization of acids.*—In combining with acid compounds in organic matter, the basic inorganic constituents of the soil usually form neutral salts; the free acids originally in fresh

organic matter and those produced during decomposition are thus neutralized and the soil is prevented from becoming acid.

From the foregoing statements, it is readily seen that *the available supply of plant-food in soils is closely associated with the presence and decomposition of organic matter.*

Relation of soil character to decomposition of organic matter.—The rapidity with which organic matter is decomposed and the length of time the products remain in soils are points closely associated with soil conditions. In warm, light, porous soils, which easily admit abundance of air, the fresh organic matter is more rapidly decomposed and with greater loss than in compact, cold, moist soils, which are not well ventilated. In the case of compact soils, decomposition is slow and continues long. On account of less free admission of air, organic matter is more abundant in pastures and meadows than in cultivated soils.

Physical characteristics of decomposition products.—Owing to its colloidal condition so-called decomposed organic matter (humus) in soils has certain characteristic physical properties which give it peculiar value in relation to crop production. Those of special interest are (1) absorbing power, (2) expansive nature, (3) lightness, (4) cementing power.

(1) *Absorbing power.*—Decomposed organic matter acts much like fine charcoal or a sponge in absorbing and retaining water, gases and solutions. For example, a cubic foot of soil rich in such material can hold twice as much water as the same amount of sand.

(2) *Expansive power.*—Decomposed organic matter shrinks when dried and expands similarly when it takes up water.

(3) *Lightness.*—A given volume of the organic portion of the soil averages about half as heavy as an equal volume of the mineral portion.

(4) *Cementing power.*—Though much less sticky than clay, decomposed organic matter is enough so to possess moderate cementing power; it holds together the coarse particles of light sandy soils, and also separates the crumbs in clay soils, thus favoring granular structure and promoting lightness.

Amount of organic matter in soils.—The amount of organic matter in soils varies widely. It is generally larger in the soil than in the subsoil, decreasing with depth; it is usually greater in clay than in sandy soils, and in meadows or pastures than in cultivated soils; it is generally greater in moist than in dry soils. In good soils where rainfall is fairly abundant, it averages about 2 per cent., varying from below 1 to over 5 per cent. In subsoil the average is about 0.80 per cent. In deep acid peat or muck soils the amount may rise as high as 85 per cent. In soils of dry climates it is usually below 1 per cent. In connection with such soils it is an interesting fact that, while the amount of humus is small, the percentage of nitrogen in the humus is high, while the nitrogen is largely retained, since the carbon in the organic matter is largely destroyed by the action of micro-organisms under the conditions of high temperature and low moisture.

Action of organic matter in soils.—Organic matter is of extreme importance in connection with crop growing on account of the useful effects it produces in soils. Some of these are purely physical, some chemical, and some biological. These are summarized as follows:

(1) *Water-holding power.*—Organic matter in the form of humus increases the power of a soil to take up and retain water and soluble materials in the water. On account of its colloidal condition, it also absorbs gases (ammonia, carbon dioxide, etc.) more abundantly than does any other soil constituent. Soils with a generous supply of decomposed organic matter resist the effects of drouth better than others.

(2) *Structure.*—Because of its colloidal properties organic matter as humus greatly improves the mechanical condition of soils. It promotes crumb-like structure, making heavy soils less sticky and easier to work, enabling them to receive, distribute and hold water more effectively, improving ventilation, and making it easier for roots to penetrate. It serves to bind sandy soils, diminish excessive porosity and ventilation, enables them to take up and retain moisture, prevent abnormal leaching of plant-food, helps roots to get a firmer hold, and prevents or lessens drifting by winds and erosion by rains.

(3) *Warmth.*—The warmth of a soil is largely dependent upon the color. Other conditions being uniform, the darker a soil, the more readily it absorbs the sun's heat. Soils well supplied with decomposed organic matter are darker and therefore warmer and more uniform in temperature than others. This is of special advantage when an early start is desired in a crop, as in early market gardening.

(4) *Availability of mineral nutrients.*—The action of decomposition of organic matter which produces acid compounds that dissolve mineral constituents of the soil, is of special value in relation to potassium and phosphorus compounds, the available forms of which are mostly associated with the organic matter.

(5) *Relation to soil nitrogen.*—Most of the soil's nitrogen comes from organic material. Therefore, organic material constitutes the main, natural source of the soil's nitrogen supply. The nitrogen decomposed in organic matter is held in slowly available form, and gradually undergoes conversion into nitrate under normal soil conditions.

(6) *Relation to micro-organisms.*—Organic matter furnishes the food essential to the growth of micro-organisms. Other conditions being uniform, the greater its amount in soils, the greater the work of micro-organisms. Its effect upon the moisture and temperature of soils makes the conditions favorable for the growth of micro-organisms. Material like farm manure not only furnishes food for bacteria, but adds new supplies of micro-organisms to those already in the soil. Such materials as green-crop manures and ordinary organic fertilizing materials (cottonseed-meal, tobacco waste, tankage, etc.) are valuable more for the food they furnish bacteria already in the soil than for the numbers of bacteria they add.

Loss of organic matter in soils.—It is well known in farm experience that soils lose their organic matter. The effects of this loss are shown in various ways, as by increased compactness, decreased power to receive and hold moisture, greater difficulty of working and in diminished productivity. Generally speaking, any condition favorable to increased action of the destructive, aërobic micro-organisms (p. 136) hastens the disappearance of soil organic matter, especially when such conditions are accompanied by a failure to furnish fresh supplies of such material. Among the specific conditions that hasten such losses are the following:

(1) *Abundance of air.*—In very porous soils where air penetrates freely, decomposition of organic matter by micro-organisms is rapid, especially when accompanied with high temperature and moderate moisture. When green-crop manures are plowed under there is often danger that they may not be sufficiently covered with soil or properly compacted, especially when the growth of organic material is large; under such conditions micro-organisms may decompose it too completely.

(2) *Moderate amount of soil moisture.*—Too rapid action is prevented in a wet soil on account of the exclusion of air. The presence of only a moderate amount of moisture favors action of micro-organisms.

(3) *High temperature.*—Temperatures above 75° F. favor rapid action of decomposition organisms. Organic matter is used up more rapidly in warm climates than in cool. In temperate climates fresh organic matter is most rapidly transformed into humus during summer.

(4) *Too much cultivation* has the effect of making the conditions of air-supply and moisture favorable for rapid decomposition of organic matter.

(5) *Supply of micro-organisms.*—When a soil contains abundant micro-organisms that decompose organic matter, action goes on more rapidly than when the supply is small. It is well known that green-crop manures, which add only small numbers of bacteria to

soils, undergo decomposition more rapidly if two or three tons of partially fermented farm manure are spread on the surface before plowing under. The manure furnishes large numbers of the desired kinds of bacteria.

(6) *Abundance of calcium carbonate.*—In a soil well supplied with any basic compounds such as calcium carbonate or the hydroxide (slaked lime), the acids formed by decomposition are neutralized as fast as formed and the decomposing organic matter is kept neutral. Under these conditions, no free acids accumulate to poison the decomposition organisms. These continue their work unimpeded. Organic matter is decomposed more rapidly in such soils than in soils that contain little or no basic or neutralizing material. There acids accumulate quickly and retard or stop decomposition.

(7) *Lack of organic matter.*—It is obvious that in a cultivated soil if no fresh supply is furnished the organic matter must be exhausted in time. Failure to supply organic matter by the regular routine of crop-growing is due, first to removing the entire crops, leaving little or no organic material in the soil, and, second, to such kinds of crop-rotation as do not supply organic matter.

(8) *Summer-fallowing,* allowing a field to lie bare for an entire crop season, furnishes conditions of temperature and moisture which favor rapid destruction of organic matter with conversion of organic nitrogen into nitrate, and danger of loss by leaching before the soil is occupied by a crop.

Accumulation of organic matter in soils.—Reverse conditions prevent loss of organic matter in soils and favor increase. Generally speaking, those conditions favor retention of organic matter in soils which prevent or delay decomposition. Stated more in detail, the following conditions promote accumulation of organic matter in soils:

(1) *Limited air supply.*—Absence of air in slow decomposition has already been discussed (p. 137).

(2) *Low temperature.*—Micro-organisms that decompose organic matter work more slowly as the temperature drops below 75° F.

(3) *Excessive supply of water* has the effect of excluding air and keeping temperature low, thus preventing or delaying decomposition.

(4) *Absence of cultivation* has the effect of diminishing air supply in the upper layer.

(5) *Absence of micro-organisms* that cause decomposition necessarily prevents changes in organic matter.

(6) *Lack of calcium carbonate* or other basic compounds permits the accumulation of acids which poison the micro-organisms and delay or prevent regular decomposition.

(7) *Addition of organic materials.*—The supply of organic matter in soils may be kept up by addition to soil of (*a*) green-crop manures, (*b*) crop residues (stubble, roots, etc.), (*c*) farm manure and waste materials, (*d*) organic matter in commercial forms, such as dried blood, cottonseed-meal, etc., (*e*) application of muck, peat, etc., (*f*) animal manure dropped in pastures.

Keeping up supply of organic matter in soils.—In considering the conditions under which organic matter is used up or is maintained in cultivated soils, it is obvious that the supply will not take care of itself, but must be given special attention. Speaking in general terms, there must be, in the first place, a sufficient supply of organic material; then, the conditions of decomposition, especially as to supply of air and moisture, must be so regulated that on the one hand, the process will not go on too rapidly and result in needless loss of organic material, or, on the other hand, proceed so slowly as to waste time in properly influencing soil and crops. In addition, care must be taken by use of calcium carbonate to prevent an accumulation of acid products of decomposition to a degree that will unfavorably affect crops.

Relation of humus to use of fertilizers.—*Many thousand tons of commercial fertilizers have been and are being applied without evidence of any appreciable effect upon yield or quality of crops, simply because used upon soils deficient in organic matter* The reason is readily understood when we bear in mind the fact that absence or deficiency of soil organic matter produces a set of physical, chemical and biological conditions, under which plants cannot grow to best advantage. Such a condition, for example, as compactness of soil, can be best remedied, not by application of concentrated fertilizers, but by a generous and wise supply of organic matter. In general, the use of commercial fertilizers on soils deficient in organic matter aggravates the abnormal conditions, does harm more often than benefit, and results in direct financial loss.

IX

RELATIONS OF WATER TO SOILS AND CROPS

GENERALLY speaking, the moisture supply is a large, dominating factor in plant growth. When other conditions are favorable, the efficiency of fertilizers in crop growing is dependent upon the water supply furnished by soils. Plants utilize food material only when it is in solution in water. It is well-known that the effect of an applied fertilizer may be wholly lost in a dry season, although it is also true that a well-fed crop withstands drouth better than one poorly fed. We shall consider the subject under the following divisions: 1, Functions of water in plant growth. 2, Distribution of water in soil. 3, Loss of soil water. 4, Control of soil water.

FUNCTIONS OF WATER IN PLANT GROWTH

Water performs the following functions in relation to plant growth: (1) It dissolves plant nutrients in the soil; (2) acts as a carrier of dissolved nutrients from soil to plants; (3) furnishes the hydrogen and oxygen used in large amounts by plants; (4) keeps the cells of plants turgid, an essential for plant nutrition; (5) and influences the temperature of plants by evaporation from leaves.

Solvent action of soil water.—The mineral constituents of the soil are somewhat soluble even in pure water, still more in water which contains carbon doxide and oxygen in solution, as in case of rain water in soils rich in decomposing organic matter. Various compounds in soils differ greatly in regard to solubility in soil water, but all in some degree continuously undergo solution, changing their plant-food constituents from unavailable to available forms.

Action of water as a carrier.—Water takes dissolved mineral plant-food in through the roots (p. 124) and through the plant from one part to another; carries through the plant the gaseous food constituents taken from the air; transfers within the plant the various soluble products formed by the plant, carrying them from one part to another for storage or assimilation (p. 124), as development requires, much as the blood in the animal body carries to every tissue and organ the nutriment needed.

Water as a source of plant-building furnishes materials which are used in large amounts. It is used either directly as water or it is decomposed into its two constituents, hydrogen and oxygen, which are then used to form new compounds (cellulose, starch, sugar, oils, proteins, etc.). Hydrogen makes up 6 to 7 per cent. of plant dry matter, all primarily from water. Oxygen present in the dry matter of plants is in considerably larger proportions than hydrogen, practically all supplied by water. If we consider plants in their fresh state, the proportion of water they contain as such is large, much larger than all other constituents together. The amount of unchanged water in a fresh crop at full growth averages about 80 per cent. of the weight. In general the limits of water in growing plants may be placed between 60 and 95 per cent.

Action of water on plant-cells.—Plant-cells maintain the conditions of highest activity only when well distended or swollen; this condition is technically called *turgor* (p. 125). Absence of a sufficient supply of moisture is indicated by wilting, which is caused by the shrinking of the cells, due to lack of water. The swollen condition of plant-cells gives firmness to the soft parts of plants, enabling small, soft stems to carry branches, leaves, flowers and fruit of considerable weight.

Large amounts of water required for plant growth.—Crops require large amounts of water, (1st) because plants average about 80 per cent. of water in weight and (2d) because large amounts of water pass through the plant and transpire from the leaves into the air (p. 105). Most of the water that enters a plant through the roots is finally transpired as water vapor into the air from the surface of leaves; this process is known as *transpiration.* The amount of water thus transpired is immensely greater than that retained in the plant. In hot, dry weather some plants exhale their own weight of water during 24 hours. The amount of moisture transpired varies with (1) the kind of crop, according to whether it has a large or a small leaf surface; (2) character of the climate; (3) size of crop; (4) amount of moisture supply and other conditions. For each pound of dry matter in a mature crop, the amount of water transpired during the growing season may vary from 200 to 900 pounds. The usual variations for common crops are between 300 and 500 pounds in the northern United States. Expressed in another way, the total amount of water transpired during growth by an average crop on one acre varies from 400 to 1,000 tons, which is equivalent to a mass of water covering the ground to a depth of 3.5 to 9 inches. When comparison is made at any given time during the season, soils that remain uncropped are found to contain more moisture than soils that bear crops.

DISTRIBUTION OF WATER IN SOILS

Water exists in soils in three conditions: (1) Free, (2) capillary, and (3) hygroscopic.

Free water (known also under the names of gravitational, hydrostatic, ground and bottom water) is readily apparent to the eye and the touch. It saturates the soil and fills the spaces between the soil particles. It is evident in surface soil after rain, when it can be easily recognized by sight as water. Under the influence of gravity, it moves downward into the subsoil until it reaches a fixed

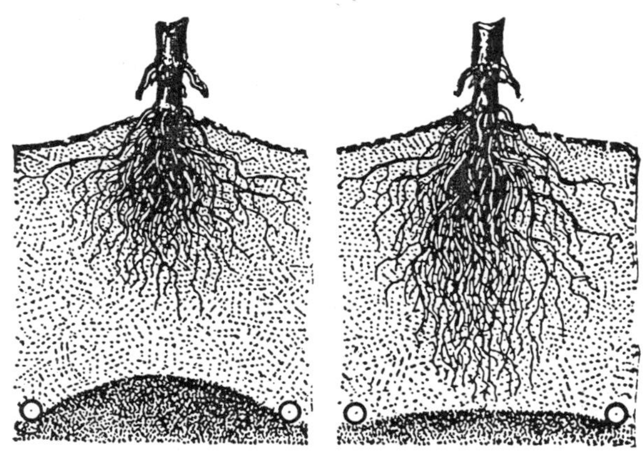

Good drainage encourages the roots to strike downward and when drouth comes the plants do not suffer.

level, where the pore-spaces are already full of water. This level varies greatly in diverse locations; in some, it may be many feet; in others, within a few inches of the surface, in some cases it is at or above the surface. This level of free water in soils is usually known as the *water-table.* The upper surface of free water can be ascertained by digging a hole in the ground until water stands in it. The level at which water stands in wells is the water-table. This free water is the source of springs and ordinary wells. In swamps and standing bodies of water, the water-table is at or above the surface of the ground.

Free or ground water is of use to plants when it is below the layer of root growth, but near enough to the surface to be pumped up by capillary attraction to the upper layers of soil. When so near the surface as to saturate any portion of the soil penetrated by crop roots it may be regarded as unavailable, since it may injuriously affect growth of most crops, because (1) it shuts off needed air supply; (2) makes the soil cold; (3) prevents formation of nitrate nitrogen (p. 144), while favoring decomposition and loss of any nitrate present; (4) promotes the formation of compounds poisonous to plants; (5) and retards decomposition of organic matter and produces other conditions unfavorable to crop growth. In such cases, the water-level must be lowered by a proper system of drainage (p. 108). Some plants, such as cranberry and rice, grow naturally when their roots are under water.

Capillary water surrounds soil particles in the form of a thin film and is held by colloidal matter to some extent. This furnishes the principal source of available water supply for crops; it is the most effective form for dissolving and holding the soluble plant nutrients. The capillary form of water moves from moist portions of soils into those less moist. The usual tendency, therefore, under ordinary conditions is for capillary water to move from lower to upper layers of soil. The distance that capillary water can move depends upon the arrangement of the soil particles and the resulting air-space. The stirring of soils of close structure, like clays for example, usually increases the air-space, the extent to which capillary water can move, and the water-holding power of the soil. In coarse, sandy soils, where the air-space is large, stirring has the opposite effect, since it decreases the extent of capillary movement and the water-holding power.

The amount of capillary water in soils depends upon (*a*) the size of the soil particles, (*b*) the structure, and (*c*) the amount of colloidal material. The finer the soil particles, the larger is the amount of capillary water. Loose, open structure, as in sands, is unfavorable to water-holding power; the close structure of clay soils may have the capacity for capillary water increased by promoting crumb-structure to a certain extent.

Hygroscopic water is that held intimately by the colloidal material of the soil. It is not moved by capillary attraction or by gravity since it is not in the liquid state. It is the moisture that remains when soil is dried in air. It forms the smallest proportion of the three forms of soil water. Plants do not appear to be able to make use of hygroscopic water.

LOSS OF SOIL WATER

Water in soils is lost (1) by percolation, (2) by evaporation, and (3) by growing plants.

Percolation.—When a soil is saturated with water, that is, when all the spaces between the soil particles are completely filled, the water tends to go down under the attraction of gravity until it reaches the water-table. This downward movement is called percolation. Its amount depends upon several conditions: (*a*) Texture and structure of soil, (*b*) amount of rainfall, and (*c*) the presence or absence of growing plants on the surface. In soils of fine texture, such as clay, percolation is slow, the pore spaces being small and often stopped up with colloidal material. In soils of coarse texture, like sand, the downward movement of water is much quicker. The larger the spaces between soil particles, the more rapid will percolation be. Soils covered with growing plants retain in the upper layer more water than bare soils. The amount of percolation is greater in case of heavy rainfalls than in light. Where rainfall is fairly abundant, loss of water by percolation may amount to one-half of the rainfall.

Evaporation.—A large amount of water is lost to soils by direct evaporation from the surface of the soil. It is influenced by (*a*) temperature, (*b*) humidity, (*c*) winds, etc. A combination of dry air, high temperature and strong winds affords the most favorable conditions for surface evaporation. Loose, open soils lose water by evaporation to a greater extent than do soils of close structure.

Transpiration.—The amount of water transpired by plants (p. 105) is essentially evaporation from the surface of the portion of plants above the soil, more particularly the leaves. Little can be done by soil treatment to influence this loss. When the amount of water transpired is greater than that supplied by the soil, plants wilt.

Water-holding power of soils.—Soils vary greatly in their capacity to hold water, depending upon: (1) the general character of soil and subsoil, (2) the amount of humus. Soils take up and hold water largely in proportion to the colloidal material they contain. Fine-textured soils and subsoils like clay take up water slowly and hold it tenaciously, while coarse-textured soils like sand admit water freely and let go of it easily. Compact soils receive and hold the water less satisfactorily than soils that possess a crumb-like structure.

CONTROL OF SOIL MOISTURE

In crop growing, one of the most important operations is so to control soil moisture that crops will be sufficiently supplied with water as they need it. This object can be accomplished within certain limits, depending upon the special existing conditions, such as amount of rainfall, character of the soil, and kind of crop. Control of soil moisture is brought about by decreasing the amount of water lost, increasing the water-holding power of soils, removing superfluous soil water, and direct addition of water in the following ways: (1) Tillage, (2) mulching, (3) organic matter, (4) fertilizers, (5) drainage, (6) irrigation. Our treatment of these methods must necessarily be in the nature of a summary. In their application, farmers in each case must be guided by the special existing conditions, such as local conditions of soil, climate, season, crop, and general character or system of farming.

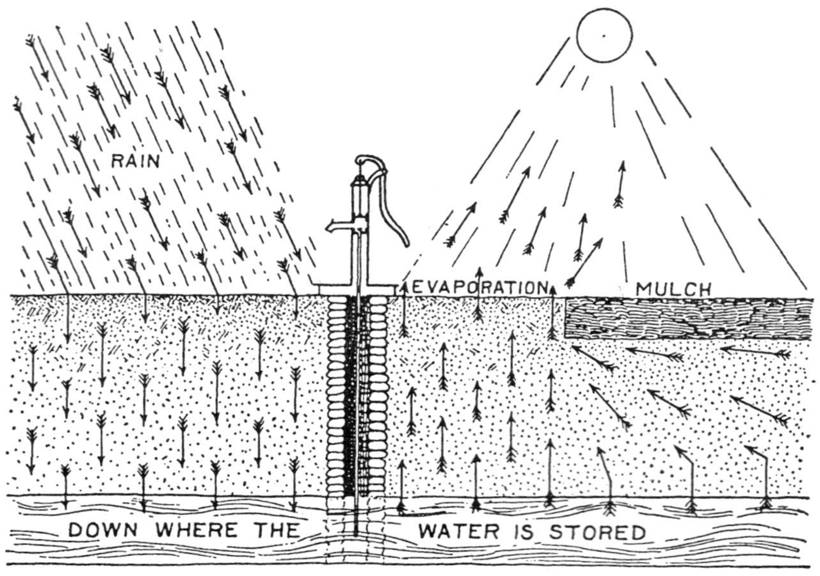

Rain enters the soil by gravity, circulates by capullarity and unless prevented by a mulch (right) passes into the air by evaporation.

Tillage.—Under this general term are included the various operations of soil handling or manipulating requisite for crop production, such as plowing, harrowing, rolling, cultivating, hoeing, etc.

(1) *Plowing.*—Only depth and time need be considered here. In general, depth of plowing is regulated (*a*) by season, (*b*) character of soil, (*c*) length of time soil has been cropped and (*d*) kind of crop. As a rule, clay soils should be plowed deeper than sandy soils. The longer a soil has been cropped, the deeper should be the plowing. Fall plowing is usually deeper than spring plowing. Deep plowing is attended with more effective results than shallow plowing in soils well supplied with organic matter, especially in case of light, sandy soils, which should not be plowed deeply unless so supplied. One advantage of late fall plowing, especially in re-

gions where fall and winter rainfall is not abundant, is the saving of moisture by preventing loss through evaporation and by enabling the soil to take in more water from winter rain and snow than if unplowed. Early spring plowing, accompanied by cultivation, is effective in holding soil moisture. The amount of moisture in a soil, especially in heavier soils, must be considered in connection with plowing. If plowed when too moist, such soils become puddled (p. 74), and when they dry, are hard and lumpy. This condition injures the water-holding power of the soil.

(2) *Harrowing and rolling* carry further and complete the pulverizing effect of the plow; rolling, in addition, compacts the surface, retarding percolation, though increasing evaporation.

(3) *Cultivation.*—The operations included under this head affect moisture conditions, especially in control of weeds, which interfere with crop growth, especially if they are prevalent when crops are getting started. They compete for water as well as for plant-food. Cultivation, at the same time, helps to keep the soil in good physical condition in relation to granulation and aeration.

Mulching.—Any material put on the surface of the soil to prevent evaporation or to suppress weeds is a mulch. Originally, the term was applied to such materials as straw, leaves, farm manure, etc., later it included the earth-mulch. Its present use extends to other materials. The formation of a loose, superficial layer of earth (soil-mulch) by cultivation has been a universal practice for the destruction of weeds and the conservation of soil moisture. It is doubtful if such a mulch, under usual conditions, lessens evaporation to any marked extent. Stirring of the surface by cultivation brings more moist soil into contact with the air and tends to increase evaporation.

The latest kind of mulch is a special, impervious paper, devised and first used in Hawaii to control weeds in raising sugar-cane and later to the growing of pineapples. Marked increase of crops has resulted from the use of the paper mulch. To what extent a similar mulch can be satisfactorily utilized with other crops under other climatic conditions has not yet been ascertained, but preliminary experimental work by the Bureau of Plant Industry, United States Department of Agriculture, has given encouraging results not only in relation to conservation of moisture by weed control but also in increasing germination, hastening maturity and producing crops superior in size, quality and cleanliness.

Organic matter.—The value of decomposing organic matter in soils as a material for holding moisture has been considered on page 96.

Fertilizers.—Fertilizers that contain soluble compounds, such as sodium nitrate, potassium chloride, common salt, etc., may exert a beneficial influence upon soil moisture in connection with crop growing, because water that contains dissolved salts more readily attracts the subsoil water to the surface and tends to keep the top layer more moist than does pure water. On the other hand, certain solutions like extracts of manure or organic substances that contain oil have the opposite effect; under such conditions, soils lift less water from the subsoil than when oil-free solutions are present and dry at the surface more rapidly; in such cases shallow-rooted crops have the appearance of being burned, owing to lack of moisture. This is noticeable when heavy applications of fresh farm manure are followed by dry, hot weather.

Drainage contractors use excavators which dig trenches rapidly and accurately at usually less cost than the work can be done by hand.

Drainage has for its object the removal of free water from the layer of soil occupied by roots; the removal is effected by providing open passages through which the water may pass. Drainage is necessary in soils such as swamps filled with standing water all the while, and also in soils saturated with water at intermittent periods. *Successful growing of crops primarily depends largely* upon good drainage. The application of fertilizers to a soil that needs drainage results in so much waste of time and money. The subject can be discussed here only superficially. Some special treatise on drainage must be consulted by anyone who desires full knowledge of drainage in farm practice. The most that we can attempt to do here is to emphasize the importance of drainage and point out some of the beneficial effects.

(1) *Improvement of soil structure.*—Wet soils, particularly clay soils, generally lack crumb structure. Drainage promotes its forma-

tion which is followed by results that affect other soil conditions favorable to plant growth, such as ventilation, warmth, etc. The soil is also made more firm by drainage.

(2) *Air supply increased.*—This results through removal of water from the pores of soils and also from increase of pore-space by granulation of soil particles.

(3) *Temperature favorably affected.*—Soils have a higher average temperature after drainage than before. Well-drained soils become warm more quickly in spring, permit earlier seeding and remain warm later in fall than do undrained ones; therefore, the effect of drainage is to increase the length of the growing season. Wet soils are late and their growing season is likely to be relatively short. The increased warmth promotes both chemical changes and activity of micro-organisms in soils.

(4) *Growth of desirable micro-organisms favored.*—The growth of micro-organisms that require air (p. 136), among them the organisms that promote decomposition of organic matter and those that produce nitrate nitrogen (p. 139), are favored by drainage. In wet soils that lack ventilation, denitrification is likely to take place (p. 137).

(5) *Effect upon supply of food.*—Drainage promotes those biochemical changes which result in changing unavailable into available plant-food.

(6) *Removal of injurious compounds.*—Drainage, along with heavy irrigation, is the most reliable method of removing from alkali soils the salts accumulated in amounts so large as to be injurious to plants.

(7) *Increase of available water supply.*—Drainage, while removing an excess of water, nevertheless increases the amount of water that plants can use, especially in case of fine-textured soils. Soils saturated only part of the time undergo wide variations in moisture, being at one time excessively wet and at another excessively dry. When such soils are drained, crops do not suffer from deficient supply of moisture in dry spells, since their change in structure resulting from drainage enables them to retain moisture longer. Although the total amount of water in a soil is diminished by drainage, the supply, as a whole, available for plants is increased for the growing season.

(8) *Effect on root growth.*—In well-drained soil, roots grow deeper than in undrained soil. This results in their coming in contact with an increased area of soil, and enables them to obtain more mineral food and moisture and therefore to withstand drouth for a longer time.

There are other advantages that good drainage brings, but those mentioned are sufficient to show its extreme importance. Stated in practical terms, good drainage results in surer and larger crops than poor or no drainage, while it enables one to control to better advantage the times of seeding and harvesting as well as all the operations of tillage.

In the case of soils that do not grow good crops, one of the first *inquiries to make in ascertaining the cause is in regard to drainage.* On a badly drained soil, efforts to raise crops meet with poor reward, so, in the interests of profitable crops, this condition must be made right before one can expect to reap the reward of tillage, fertilizers, etc.

Irrigation consists in the direct application of water to soils. It finds its most extensive applications necessarily where rainfall is less than 10 to 20 inches a year. However, in some special cases where crops of high value are concerned or where special market conditions prevail, irrigation has proved profitable when the annual rainfall is 40 inches or more. From the few experiments made it is doubtful if irrigation will find extensive applications outside of dry climates, except in a limited way under special conditions. The general subject cannot be treated satisfactorily here, so the reader interested in it is referred to special works.

The relation of soil moisture to special crops.—Soil moisture is of practical importance especially in the control it exerts in determining locally what crops can best be grown. Generally speaking, the amount in a soil often affects the time of sowing or planting, ripening, the yield, and in some cases, the quality of a crop. For example, early market-garden crops are best grown on light soils that hold comparatively small amounts of water; later crops of the same products are grown on loams with two or three times the water-holding capacity. In such cases, earliness is at the expense of yield, the later crops produce more heavily. In respect to influence on quality, it is well known that potatoes grown on light soils are, in general, superior in cooking quality to those grown on heavy soils.

X

LOSSES AND GAINS OF PLANT-FOODS IN SOILS

KNOWLEDGE of ways in which amounts of plant-foods in soils are decreased and increased, and appreciation of the extent to which various forms of plant-food are liable to loss, are essential to an intelligent and economical use of fertilizers in crop growing.

LOSSES OF PLANT-FOODS

The fact that plant-foods are removed from soils has been long established, but the precise ways in which, and the extent to which, such losses take place have not been clearly understood nor fully appreciated until recent years. Plant-food constituents are lost from soils: (1) By growth and removal of crops, (2) by leaching, (3) by mechanical agencies, (4) by conversion into gaseous form and escape into air.

Removal of plant-foods by crops.—The largest amounts of plant-food constituents are removed by crops grown on and taken away from the land. Various crops use the same plant-food constituents in variable amounts, a fact that must be considered in determining what to grow to advantage when the crops are to be sold. The amount of plant-food consumed is an essential factor in estimating the cost of growing a crop for market. It is not, however, of such importance where the crop is to be utilized on the farm to feed animals and a large proportion of its plant-food constituents returned to the soil.

Table 20 serves to illustrate the approximate amounts of the more important plant-food constituents used by a few typical crops. Data regarding other crops are given at the end of each chapter in Part Four. The figures given in this and other chapters are based upon the best data available, but it must be kept in mind that they are only approximations to what may be regarded as fair conditions of yield. They cannot be applied too closely to any individual case. It has seemed preferable to base our data upon moderate rather than upon maximum or minimum conditions. From the data given, one can easily estimate the amounts of plant-food constituents removed by a given crop when the yield is known.

For example, our estimate of the plant-foods removed by a wheat crop is based on a yield of 25 bushels of grain an acre. If the yield in a particular case is 12½ bushels, the amount of plant-foods removed will be one-half that given in the table; if the yield in another particular case is 50 bushels, the plant-food used by the crop will be twice that given in the table.

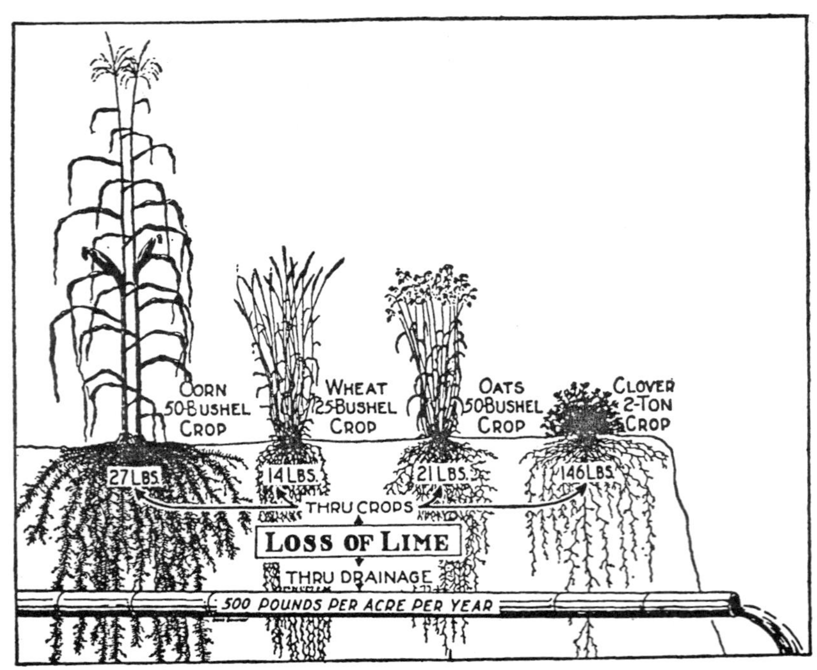

Average crops of corn, oats, wheat and red clover remove from the soil an amount of calcium that would take 208 pounds of limestone to replace. Where these crops are grown in rotation, this means an average loss of fifty-two pounds an acre a year. Losses that result from rains vary, but easily amount to 500 pounds or more an acre a year.

In those cases where unused portions of a crop go back directly to the soil, as with the leaves of fruit-trees, small fruits, the leaves and stems of many vegetables, etc., we have given results in most cases for only the portion of the crop actually removed from the soil, although in some cases we have included also information for the portions of the crop not removed.

When a crop is sold as taken from the field, the amount and composition of the crop are a direct measure of the loss of plant-food. When a crop is fed to animals on the farm and the manure put back on the soil on which the crop was grown, then only a small part of the plant-food constituents is actually lost.

Loss of plant-food by leaching.—Plant-food constituents in soils dissolve in soil water to a greater or less extent. When rainfall is in excess of what the soil can hold, the portions dissolved in water pass away in the drainage. While a small amount of dis-

TABLE 20—PLANT-FOOD CONSTITUENTS IN A CROP GROWN ON ONE ACRE

Kind of crop	*Portion of crop*	*Yield*	*Nitrogen lbs.*	*Phosphoric acid (P_2O_5) lbs.*	*Potash (K_2O) lbs.*
CORN	Grain	25 bu.	23.2	9.1 (4. P)	5.5 (4.6K)
	Stalks, etc.	1,500 lbs.	15.0	4.5 (2. P)	21.0 (17.4K)
	Cobs	250 "	1.0	0.2 (0.1P)	1.1 (0.9K)
			39.2	13.8 (6.1P)	27.6 (22.9K)
WHEAT	Grain	25 bu.	30.0	12.8 (5.6P)	6.0 (5. K)
	Straw	2,500 lbs.	12.5	3.8 (1.7P)	15.0 (12.5K)
			42.5	16.6 (7.3P)	21.0 (17.5K)
OATS	Grain	25 bu.	16.0	6.5 (2.9P)	4.8 (4 K)
	Straw	1,250 lbs.	8.0	2.5 (1.1P)	15.6 (13 K)
			24.0	9.0 (4.0P)	20.4 (17 K)
APPLES .... (30 trees per acre)	Fruit	300 bu.	6.0	3.0 (1.3P)	15.0 (12.5K)
	Leaves	1,000 lbs.	10.0	1.5 (0.7P)	3.5 (2.9K)
	New wood	100 "	0.5	0.2 (0.1P)	0.3 (0.2K)
			16.5	4.7 (2.1P)	18.8 (15.6K)
CABBAGE		10 tons	60.0	20.0 (8.8P)	80.0 (66.4K)
BEANS	Seeds	25 bu.	60.0	18.0 (8. P)	19.5 (16.2K)
	Straw	2,000 lbs.	28.0	6.0 (2.6P)	38.0 (31.5K)
			88.0	24.0 (10.6P)	57.5 (47.7K)
PEACHES .... (100 trees per acre)	Fruit	400 bu.	22.2	11.0 (4.8P)	45.5 (37.7K)
	Leaves	5,300 lbs.	47.7	8.0 (3.5P)	42.0 (35. K)
	New wood	1,500 "	8.6	2.0 (0.9P)	2.5 (2. K)
			78.5	21.0 (9.2P)	90.0 (74.7K)
TURNIPS	Roots	10 tons	50.0	20.0 (8.8P)	90.0 (74.7K)

Phosphorus, P. Potassium, K.

solved material is carried away in the surface drainage of sloping or hilly land, most of the water that falls on cultivated land is carried first into the subsoil with its dissolved material and then, to a greater or less extent, passes from the soil permanently in the form of drainage water. The dissolved soil constituents in drainage water sooner or later find their way to the sea. It has been estimated that about 5,000,000,000 tons of mineral matter *in solution* are annually transported by rivers from land to sea. While all soil constituents are in some degree soluble in soil water and therefore liable to removal from soils, some are much more so than others. Among those most readily soluble are some of the most important plant-food constituents whose liability to loss in drainage water are considered on pages 118 and 119.

Losses by leaching depend upon various conditions, most important of which are: (1) Amount and time of rainfall, (2) absorbing and holding power of the material in soil and subsoil, (3) amount of organic matter in the soil, and (4) form or compounds in which the plant-food constituent is present.

(1) *Loss by leaching in relation to amount and time of rainfall.*—Large amounts of rainfall increase the amount of soluble plant-food removed in drainage water. Rainfall during summer, when the amounts of soluble plant-foods in soils are largest, causes greater loss by leaching than at other times of year, other conditions being the same.

(2) *Loss by leaching in relation to the absorbing and holding power of soil particles.*—The tenacity with which soils absorb and hold moisture, due to colloidal condition (p. 000), is influential in controlling the amount of plant-food material lost in drainage. On this account, clay soils lose plant-food less easily than do sandy soils. Some of the plant-food compounds have more or less chemical action in many soils, especially clay, which tends to decrease or prevent leaching (p. 113).

(3) *Loss by leaching in relation to organic matter in soils.*—One of the important properties decaying organic matter has in soils is its solution-absorbing and -holding power. Keeping in the soil generous amounts of organic matter is one of the most effective means of decreasing losses of plant-food materials in drainage water. The lighter the soil, the greater the need and the more pronounced the effect.

(4) *Loss by leaching in relation to form of plant-food constituents in soil.*—Some compounds are not easily held by any soil constituent; others are. All nitrates are easily soluble in water so do not form insoluble compounds with any soil constituents; therefore, of all compounds of nitrogen, nitrates are most easily

lost in drainage water. Other compounds of nitrogen, potassium and phosphorus are considered under another head (p. 118).

Remedies for loss of plant-food by leaching.—In addition to keeping soils supplied with abundant humus and making the surface of cultivated soils open to rainfall, the most effective method of diminishing the losses of plant-foods due to leaching is to keep the ground occupied with growing plants as much as possible. The roots are able to use the plant-food as it becomes soluble during the growing season. During periods when the ground would otherwise long lie bare, cover-crops should be grown when practicable. In this way the soluble plant-food will be taken up by the cover-crop and its loss by leaching largely prevented.

Loss of plant-food by mechanical agencies.—Soil particles are carried bodily from soils into streams, especially in bare, hilly regions. The amount of plant-food material thus removed is large in the aggregate. The portion thus removed by water running over the surface consists of the finest soil particles which contain the largest part of the quickly available plant-food of the soil. Their removal also affects most seriously the physical condition of the soil. Loss of soil particles by flowing water is promoted when land lies bare and uncultivated, unprotected by any covering of trees or other vegetable growth. Losses which occur thus may be largely reduced or prevented by one or more courses of treatment, according to special conditions: (*a*) By keeping the surface of the soil under cultivation well stirred so rains sink in instead of running off the surface; (*b*) keeping the soil occupied with crops as much as possible during the time when rains are most abundant; (*c*) keeping the soil well supplied with organic matter; (*d*) proper under-drainage; (*e*) in case of hilly land, handling the soil so as tc make furrows, ridges or terraces, according to special conditions, around the hill in such a way as to catch and hold the rain-water. Steep hills should be covered with trees or fine-rooted crops, like some of the grasses, as a protection against mechanical loss.

Soil particles on the surface are also lost through transportation by wind. In some regions, this is serious. It may be remedied (*a*) by covering the surface with growing plants, (*b*) maintaining a good supply of organic matter, (*c*) windbreaks, and (*d*) any form of mulch.

(4) *Loss of plant-food by conversion into gases.*—The only form of plant-food liable to appreciable loss as gas is nitrogen. When nitrate nitrogen is decomposed into free nitrogen (p. 144), the plant-food value is lost. It is possible also that ammonia nitrogen may be lost to soils by passing into the air when the soil is hot, dry and lacking in humus. Ordinarily, under agricultural conditions that commonly prevail, the amount of nitrogen lost either

from decomposition of nitrate or from escape of ammonia into the air may be regarded as insignificant. Another source of nitrogen loss is the burning of animal or vegetable matter. When stubble or any other form of vegetable material or any animal substance is burned the nitrogen is rendered useless as plant-food by being converted into free nitrogen.

How plant-food compounds are held by soils.—If a plant-food compound, such as potassium chloride, calcium acid phosphate, ammonium sulfate, etc., is dissolved in water and this solution poured upon a 6-inch layer of ordinary soil so as to enable one to catch the solution that drains through the soil, it will be found that most of the plant-food constituents are retained in the soil. This retention is due in part to chemical and, in part, to physical causes. It varies with the compounds applied and with those in the soil, as well as with the fineness of the soil particles.

Obviously these processes are of practical interest in connection with a study of the plant-food losses by leaching. The process by which soluble compounds are held in soils is called *fixation*. It may be due to chemical or physical causes or both.

(1) *Chemical fixation* applies to cases in which soluble compounds, applied to the soil, undergo chemical reaction with one or more constituents of the soil and are thereby changed into other compounds less soluble.

(*a*) Nitrogen compounds. *Nitrate nitrogen* is not chemically held in soils because all nitrates are easily soluble in water, and, therefore, cannot form insoluble compounds by reaction with any soil constituent.

Ammonia nitrogen, whether as gaseous ammonia, ammonium carbonate or a neutral ammonium salt like ammonium sulfate, is capable of reacting with certain soil constituents and entering into combinations not easily leached from the soil. The same is true of soluble organic nitrogen compounds. The fixation of ammonia and of soluble organic nitrogen compounds is associated with colloidal material in soils. While compounds of ammonia and of organic nitrogen are retained in soils so long as they maintain their insoluble condition, they are sooner or later changed into nitrate nitrogen (p. 139) and then they are easily removed by leaching.

(*b*) Phosphorus compounds. The amount of phosphorus compounds found in drainage water is small. This is easily explained by the fact that phosphoric acid can form insoluble compounds with several constituents common in soils—calcium carbonate and compounds of iron and aluminum in clay. When we put soluble calcium phosphate ($CaH_4(PO_4)_2$, p. 33) on a soil that contains carbonate of calcium, it first goes into solution and is distributed through the soil water near the surface; soon the calcium of the carbonate combines with the soluble calcium phosphate and forms throughout the soil fine particles of the di-calcium or "reverted" phosphate ($Ca_2H_2(PO_4)_2$, p. 33); or, on soils that contain large amounts of calcium carbonate, more calcium may unite with the phosphate to form tri-calcium phosphate ($Ca_3(PO_4)_2$, p. 33). Whether the di- or tri-calcium phosphate is formed, the phosphate is soon distributed throughout the upper layer of soil in a state of fine division, to form a coating on the soil particles with which the soluble phosphate comes in contact before being changed to the insoluble form. In this condition it is soluble in water that contains in solution carbon dioxide or weak organic acids. Under normal soil conditions, it goes into solution gradually to meet the demands of growing plants.

In acid soils and soils that contain only small amounts of calcium carbonate, the application of a soluble phosphate is followed by the formation of iron and aluminum phosphates (p. 83), which are only slightly and slowly soluble in soil solutions. The foregoing facts suggest the desirability of keeping soils well supplied with calcium carbonate in order to prevent the formation of the too slowly available iron and aluminum phosphates. Recent investigations have, however, shown marked losses of phosphorus in case of light, open textured soils treated with large amounts of farm manure for such crops as cabbage, asparagus, tobacco, etc.

(*c*) Potassium compounds. Most soils contain constituents which by means of chemical action have the power to change soluble potassium compounds into forms that do not dissolve too easily in soil water. Soils that contain abundance of clay and humus produce this change in greater degree than do light soils. When soluble potassium compounds, such as the chloride, sulfate and carbonate, are applied to soils, they remain largely in the surface soil, though some portions pass into the subsoil, but little goes into the drainage. In this condition the potassium remains until dissolved gradually in the soil water to supply the roots of growing plants.

(2) *Physical fixation.*—When finely powdered charcoal is shaken up in a colored liquid, the dissolved coloring-matter may be so completely removed that the solution becomes colorless. Charcoal has the power to attract and condense upon its surface the dissolved coloring-matter. This is not a chemical but a purely physical action because none of the substances that take part in the action undergoes any change in composition. This process is technically known as *adsorption.* Fine powders in general, including soil particles, have the power to withdraw from solution and hold dissolved substances, the degree of power varying with the substance dissolved and the adsorbing solid. This process is of importance in soils, since in fine-grained soils considerable dissolved

material is held, instead of being leached out. Plant-food compounds held in this way are in condition to permit of their ready use by the root-hairs that come into contact with the soil particles. While physical fixation by adsorption may play a more or less active part in holding any dissolved soil constituent, it is of especial interest in connection with nitrates which are held to some extent by adsorption. This is shown by the fact that when we add to a clay soil a given quantity of nitrate salt, such as sodium nitrate, and then treat the soil with successive portions of distilled water, we are able to leach out only a small part of the nitrate applied. It may be added that the finer the soil particles the greater is the adsorbing power; for this reason clay soils possess this power to a greater extent than do sandy soils.

Kinds and amounts of plant-foods removed from soils.—In the case of loss by mechanical agencies, such as erosion by water and transportation by wind, the soil particles, with all the kinds of plant-food contained in them, are carried away bodily; there is no selective action to remove one constituent more than another. But in case of loss by leaching, there is some selective action, which is dependent upon the solubility of the plant-food compounds in soil water. While all soil constituents are in some degree soluble and, therefore, likely to be removed from soils in soil water by leaching, some are much more so than others. The soil compounds most soluble are the chlorides, nitrates and sulfates of sodium, magnesium and calcium, and also calcium as bicarbonate.

(1) *Calcium compounds.*—Of all soil constituents, calcium is usually found in drainage water in largest amounts. In case of cultivated soils well supplied with calcium carbonate, the annual loss of this compound per acre in drainage water is often over 500 pounds (equivalent to 280 pounds of calcium oxide) and may exceed 1,000 pounds (equivalent to 560 pounds of calcium oxide). These large losses are due, in part, to the solubility of calcium carbonate in soil water that contains in solution carbon dioxide and, in part, to the important chemical reactions in the soil in which calcium carbonate takes part. Thus, when any acid material is formed in or applied to soils, calcium carbonate is usually the first compound to be acted upon, with the resulting formation of calcium salts that dissolve in water. The formation of organic acids in the decay of vegetable matter and the production of nitric acid by nitrification (p. 139) convert calcium carbonate into soluble forms. The application of materials that contain ammonium or potassium sulfate, chloride, nitrate, etc., results in the decomposition of insoluble calcium compounds with the formation of the corresponding calcium salt. All of these salts are soluble in water.

(2) *Nitrogen compounds.*—Little nitrogen is lost from soils

except in the form of nitrate. Nitrate nitrogen may be lost annually in drainage water under various conditions from a few pounds an acre of crop-covered soils to 100 pounds in case of bare soils. Organic and ammonia nitrogen are seldom found in drainage water in significant amounts, but are sooner or later converted into nitrate nitrogen, which then may be lost.

(3) *Phosphorus compounds.*—For reasons already considered (p. 116), relatively small amounts of phosphorus compounds are lost in drainage water, only a few pounds an acre at most. If considered with reference to large areas, however, the aggregates are large, amounting for the area of the United States to 400,000 tons annually.

(4) *Potassium compounds.*—The amount of potassium compounds in drainage water may exceed 75 pounds an acre annually, but is usually less. The loss is increased when large amounts of soluble potassium compounds are applied, and especially on sandy soils. As compared with phosphorus compounds, the loss of potassium compounds is considerably greater under ordinary conditions.

GAINS OF PLANT-FOODS BY SOILS

To make up the losses of plant-food which soils suffer from various causes considered, various sources of supply may be utilized, chief of which are: (1) Application of fertilizers, (2) plant-food contained in rain-water, (3) nitrogen added by micro-organisms, (4) plant-food brought from subsoil, (5) plant-food added through manure from purchased feeding-stuffs.

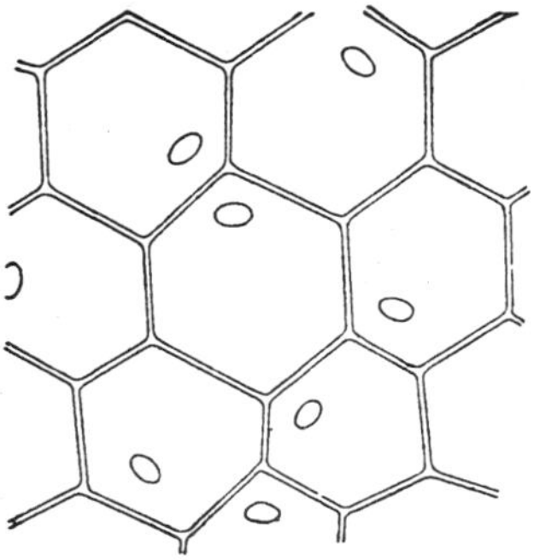

Application of fertilizers.—The application of substances that contain any form of plant-food material, whether a product of the farm or a commercial product, serves to furnish plant-food to take the place of that removed. The amount of plant-food thus supplied depends, of course, upon the composition and amount of the material added.

Diagram of plant cells showing nuclei.

Plant-food in rain-water.—Dew and rain bring to the soil annually, from 3 to 10 pounds of nitrogen an acre, usually 5 or 6 pounds, chiefly in the form of ammonia and nitrate nitrogen. The variations in the amount of nitrogen thus supplied must necessarily be wide, according to rainfall, proximity of large towns, climate, etc.

Sulfur in the form of sulfuric acid is added annually to the soil in rain-water in varying amounts, from a few pounds to 40, averaging about 25 pounds.

Nitrogen added by means of micro-organisms is treated in the next chapter (p. 121).

Plant-food from subsoil.—Considerable amounts of plant-food are brought to the surface soil where plants can use it, (*a*) as the result of growing deep-rooted crops, (*b*) by means of gradually increasing the depth of plowing, and (*c*) through the usual upward and downward movements of the soil water, dry periods causing the soil solution to ascend to the region of growing plant-roots.

Plant-food added through purchased feeding-stuffs.—When a farmer purchases feeding-stuffs for his animals and applies the resulting manure to his soil, he obtains the double feeding and fertilizer value.

Of these various methods of making up losses of plant-food constituents, the materials in most mineral fertilizers, in rain-water and in the nitrogenous products of certain micro-organisms are the only ones actually added to the soil, the others coming from the subsoil or the soil itself. What plant-food is applied as animal or vegetable materials is simply restoration of what was previously a part of the soil.

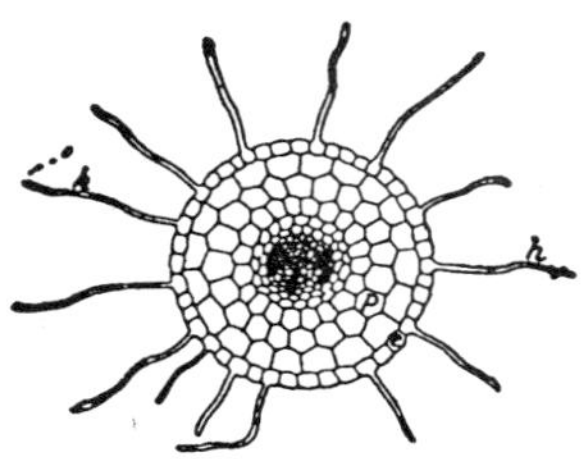

Cross-section of root magnified. *p*, layers of soft, thin-walled cells; *e*, cells forming outer covering of root or epidermis layer; *h h*, *root-hairs*.

XI

HOW PLANTS TAKE AND USE THEIR FOOD

In order to understand more clearly the direct relation of nutrient materials to plant growth, supplied either naturally or artificially, it is desirable to know something of the way in which plants take and make use of these materials.

We have already seen (p. 12) that plants obtain their food from two sources, air and soil. We will now study in more detail how the plant obtains and uses its food supplies, under the following divisions: (1) Plant-cells as manufactories, (2) how plants obtain and use carbon, (3) relation of roots to food supply, (4) absorption of soil solution, (5) turgor of plant-cells, (6) selective power of plants, (7) forms of food constituents used, (8) feeding-power of crops.

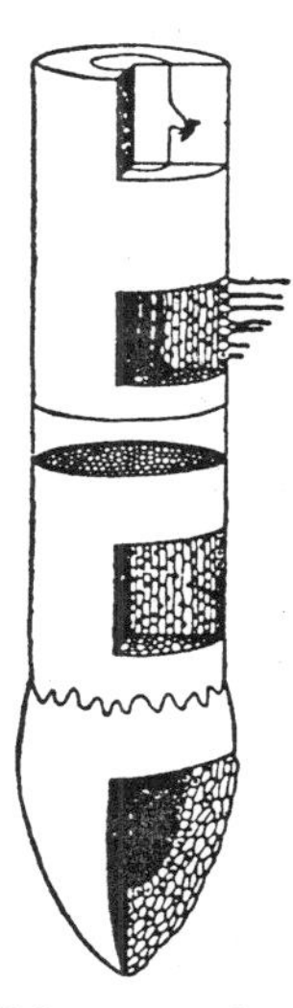

Diagram of root, showing various cells and structures with root cap at the tip.

Plant-cells as manufactories.—The ability of a living plant to make use of nutrient materials and to develop through various stages of growth from seed to maturity depends upon the work performed in the millions of the small individual cells of which plants are made up. We may regard an ordinary living plant as a chemical laboratory or factory that contains immense numbers of small rooms, which we call plant-cells, each of which contains everything essential for its work of production, and in which there is at one time or another intense industrial activity. While plant-cells vary in size, shape and other details, it will answer our purpose to describe certain general properties of fundamental importance that belong to living cells in common.

We can think of a plant-cell as a small, bladder-like sac, consisting essentially of two parts, the covering and the cell contents. The important thing to know in this connection about the cell-covering or wall is that, in the young, growing portions of plants, it is a thin, firm, strong and elastic membrane, through which water, gases and dilute solutions of various compounds, but no solid particles, however small, can pass. Each active plant-cell is filled with a semi-liquid substance, known as *protoplasm,* which is a complex mixture of many compounds, such as water, proteins, carbohydrates and other organic compounds, together with inorganic compounds in solution. Protoplasm is the living part of plant-cells—which makes life processes possible. In and through this substance chemical changes take place in cells, resulting in processes of assimilation, breaking down of one set of compounds to form new ones. In the presence, and from the substance of protoplasmic material are formed all the organic compounds found in plants—sugar, starch, cellulose, proteins, oils, acids, poisons, flavoring compounds, etc.

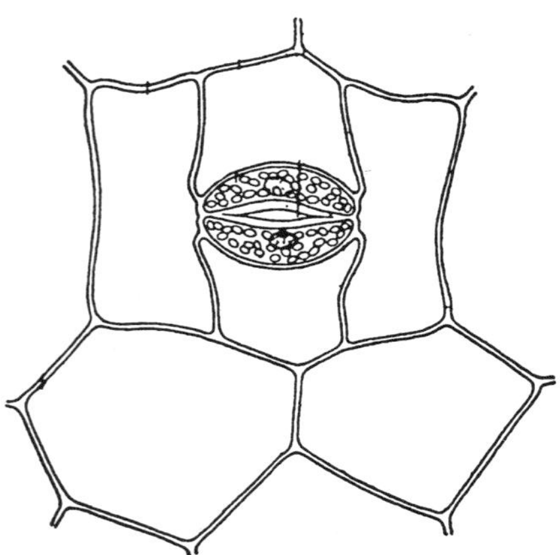
Stoma or breathing mouth through which air passes into leaf and other green tissues.

How plants obtain and use carbon.—As previously stated (p. 63), nearly one-half of the dry matter of plants consists of carbon from the carbon dioxide of the air. The air is absorbed by the plant through special small openings or breathing-pores distributed thickly on the under surface of leaves. The air with its carbon dioxide passes into these openings and then through the plant into the spaces that surround the plant-cells, which absorb the gas as needed. Inside the plant-cell the carbon is separated from the oxygen and made to combine with other elements to form new compounds, while most of the separated oxygen passes back into the air.

The ability of the cell-protoplasm to utilize the carbon of carbon dioxide depends mainly upon two agents, *chlorophyl* and *sunlight.* Chlorophyl is the name of the green coloring-matter in plants. It is contained most abundantly in the cells of leaves, mixed as small green particles with the protoplasm. It has the power to absorb sunlight, and through the energy thus obtained and with the aid of protoplasm, carbon dioxide and water are decomposed into their constituent elements, carbon, oxygen and hydrogen; the carbon then unites with other elements to form new compounds, which appear in the cell first as a carbohydrate, probably starch. This process of assimilating or fixing carbon is known as photosynthesis. Its action does not take place in the absence of sunlight, as shown by the white color of plants grown in the dark. The rapidity of fixation of carbon depends upon the intensity of sunlight, temperature, supply of water, the amount of carbon dioxide in the air, the number of breathing-pores in the leaves and the supply of other kinds of plant-food (nitrogen, phosphorus, potassium, etc.). As previously stated (p. 60), the presence of a small amount of some soluble iron compound is necessary for the formation of chlorophyl.

Relation of roots to food supply.—The plant-food constituents contained in soil (water, nitrate nitrogen, compounds of potassium, phosphorus, calcium, iron, magnesium, etc.) enter plants through the roots. Plants can use these constituents only in the form of solution because solid particles, however small, are unable to pass through the thin walls of plant-cells. The soil water dissolves some of the plant-food material with which it comes in contact and this nutrient solution passes into the root-hairs, but not through all portions of the root surfaces. If a growing plant is carefully removed from the soil so as to disturb its roots as little as possible, slender, white, hair-like branches are found near the tips of the smallest rootlets that branch out from the larger roots and penetrate the soil in all directions. These root-hairs are so numerous that over thirty thousand may in some cases be found on one square inch of root surface. They are often so numerous and fine as to have a fuzz-like appearance. It has been estimated that the area of root surface is commonly increased five to twelve, and in extreme cases over fifty, times by the root-hairs. In a soil well supplied with water and other available plant-food, the number of root-hairs is much larger than in the case of soils lacking in moisture and other nutrients. The feeding-power of a plant is therefore, greater in rich, well-watered soils than in dry, poor ones.

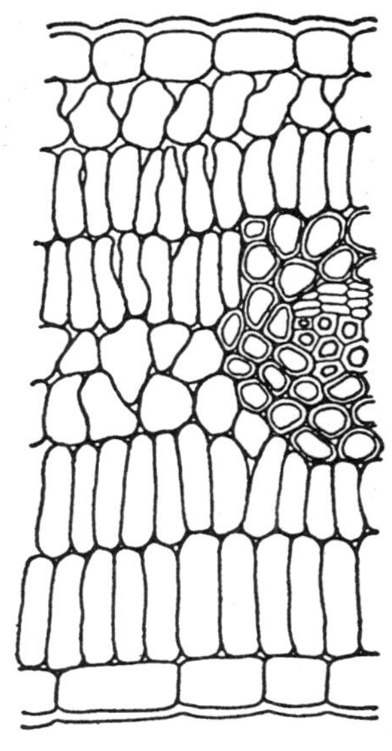

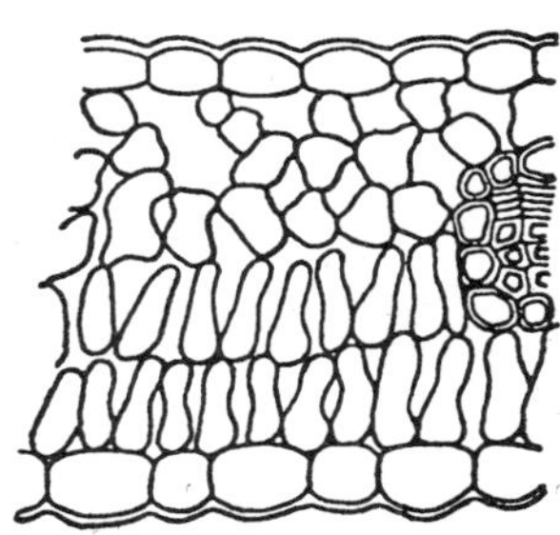

Cross sections of leaves. *Above,* grown in full light; *below,* grown in partial shade.

Each root-hair is a long, single-celled tube, the walls of which are so exceedingly thin and delicate as to allow solutions but not solid particles to pass through. The tip of each rootlet is constantly pushing through the soil in search of food, and new hairs are continually being formed just back of the growing tip, the old ones ceasing their activity and disappearing as the result of death and decay.

Another fact of interest and importance is the intimate way in which these little hairs come into the closest contact with the colloidal coating of the soil particles. They not only grow between, but actually apply themselves closely to the soil particles which they frequently almost inclose. In this way the cell-wall of the root-hair is brought into closest possible contact with the coating of colloidal material and the thin film of water that surrounds the soil particles. After being taken into the root-hairs, the nutrient soil solution is passed on to the roots and carried through other sets of cells up to the leaves, where its constituents are utilized in various manufacturing operations.

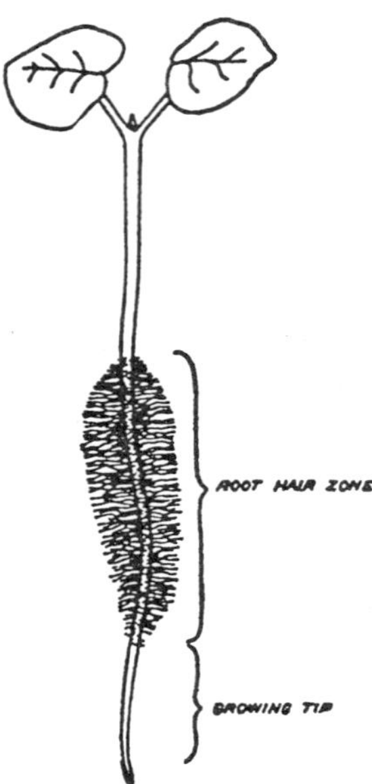

Seedling which shows root hairs and growing tip.

Absorption of .soil solution.—The movement of gases, pure water, or that of solutions through a membrane, is called diffusion. It applies to the passage of soil water into the root-hairs; also to the movement of gases and solutions between cells. The one main condition which determines whether water, or a solution will pass through the membrane into or out of a cell; namely the relative concentration of the solution in the cell and the one outside. Under ordinary conditions, the solution or cell-sap in the root-hairs is more concentrated than the soil solution; that is, it contains a smaller proportion of water to the substances dissolved in it; water then passes from the soil into the root-hairs. The solution passes by the same process, from root-hairs into other cells in the roots and from these into still other sets of cells, until finally it reaches the leaves which are using materials in the cell-sap to manufacture needed materials for plant tissues and, other products.

The water not used or needed is passed from the leaves into the air by transpiration (p. 105). This transpiration concentrates the cell-sap in the upper portion of the plant and thus fosters the movement of water upward to replace the water lost. At any time if the soil solution is more concentrated than the cell-sap, water passes from the root-hairs into the surrounding soil water.

A fact of practical importance to remember in this connection is that when the solid matter in a soil solution is greater than 1 part in 500 parts of water (for example, 1 pound in 60 gallons of water), the solution is approaching a concentration beyond which it may cause many plants to suffer. This is shown when leaves turn yellow on the edges, become spotted and drop off, or growth is checked, shortened and dwarfed, and leaves often become puckered and twisted. The roots and root-hairs are also shortened and deformed.

Turgor of plant-cells is intimately connected with the process of diffusion. The passage of water by diffusion causes cells to fill up to such a degree as to produce pressure from within on the cell-wall and to make a more or less rigid, swollen condition. This is maintained by continued absorption of water, which takes place so long as the cell-sap is more concentrated than the solution outside the cells. On the other hand when soil water is more concentrated than cell-sap water leaves the plant-cells, passes into the soil, the plant-cells lose their fullness, the tissues wilt or become limp, and, if the withdrawal of water is carried too far, the plant may die. We have practical illustrations of loss of turgor in plant-cells when a strong solution of nitrate of soda, for example, comes in contact with a green leaf; water passes by diffusion from the cells to the stronger solution outside to such an extent as to cause the leaf to wilt and present a scorched or burned appearance. A similar result takes place when a concentrated fertilizer, rich in salts such as sodium nitrate, ammonium sulfate, potassium chloride, sulfate, etc., is applied in too large amounts to a soil. Under these conditions, more water leaves the plant than enters and the plant wilts.

Root-hairs on wheat when very young and four weeks later. All the water and food from the soil enter the plant through the root-hairs. Note how closely the root-hairs adhere to the soil particles. (After Sachs)

In hot, dry weather, when there is insufficient water in the soil to supply the place of that transpired from the leaves, plant-cells lose their turgor and the leaves show more or less wilting; in such cases, the wilting acts as a protection in preventing more water from going off too rapidly, since the openings in the leaves, or breathing-pores, through which the water passes into the air, become much smaller through loss of turgor in the cells around the openings.

Selective power of plants.—It is well known that plants use some soil constituents in much larger amounts than others, so it is a matter of practical interest to consider briefly how a plant obtains its food supplies in the proportions needed. The soil solution contains various constituents, some of which nourish plants and some do not. Through the process of diffusion, plants exercise a certain kind of power of selection, distinguishing between nutritive and non-nutritive constituents and between those used in smaller and those used in larger amounts. The amount of each constituent present in the soil solution taken into a plant is determined primarily by the amount of each constituent used by the plant in its manufacturing work; the process of diffusion automatically acts to keep up the supply.

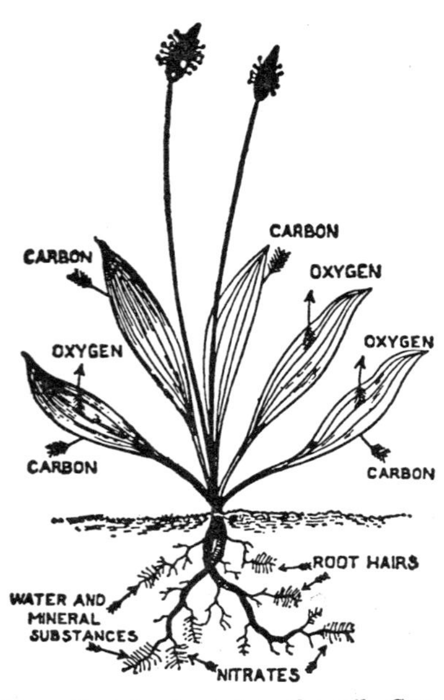

How plant-food gets into the soil. Carbon is taken in through the stomata, or mouths, on the underside of the leaf. All the mineral elements, including the nitrates, are in solution in the water and pass in this way into the plant through the root-hairs near the tip end of the growing root. Later most of this same water passes out of the leaves as vapor.

When a plant-food constituent is taken from its solution within the plant-cell and built by the protoplasm into some new form, a new supply of that particular constituent passes into the cell from the solution outside as soon as the amount of that constituent in the solution within the cell becomes less than in the solution outside the cell. For example, when a single food element like potassium present in the cell-sap is taken from the solution by the protoplasm in the cell then the solution contains less potassium. Then there will be a fresh inflow of potassium from the weaker solution outside until the proportion of potassium is the same in the solution within and without the cell. Should the potassium be used as fast as it is absorbed, then absorption will continue rapidly enough to maintain the supply.

In the case of a constituent like sodium, of which plants use only small amounts at most, new supplies are called for to a much more limited extent, the cell-sap holds all it can and no more is admitted to the plant from the soil. In the case of potassium and sodium, the soil solution usually contains more sodium than potassium salts, but in spite of this more potassium enters the plant. Therefore, the amounts of various plant-food constituents that come into roots through the root-hairs are regulated not merely by the amounts present in the soil solution but more largely by the amount of each constituent utilized in manufacturing operations. This power of taking from soil solution any plant-food constituent or constituents needed is called the *selective power* of the plant. It exists within the living cells themselves. All materials dissolved in soil water pass by diffusion through the walls of the root-hairs into the plant, but they will not continue to enter unless they are made use of by the protoplasm.

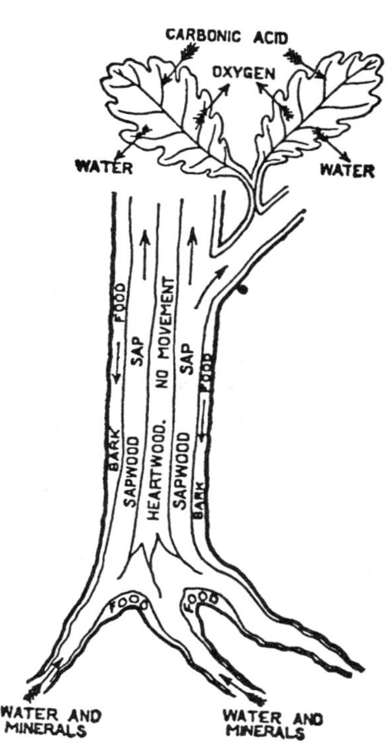

How the sap current moves.

Forms of constituents used.—We ordinarily think of compounds in soil solutions being absorbed by plants in the exact form in which they exist in the soil solution. For example, we usually suppose that the sodium nitrate in a soil solution passes as such into the plant, but it does not necessarily. When a plant is grown in a water solution and sodium nitrate ($NaNO_3$) is supplied, the compound is split into sodium (Na) and nitrate radical (NO_3); the nitrate is used by the protoplasm, while the sodium is rejected. It combines with any carbon dioxide present and appears as sodium carbonate in the solution outside the plant. As a result, the surrounding solution soon becomes alkaline. In other cases, as with potassium chloride (KCl) or sulfate (K_2SO_4), the plant uses the potassium and rejects most of the acid portion (Cl or SO_4), the presence of which results in making the surrounding solution acid. We must regard each individual element or radical in the soil solution as being taken up and used independently by plants; the supply of each by itself comes in accordance with the demand made on each for manufacturing operations. Each element or base or acid is used and supplied more or less independently of the others.

Feeding-power of crops.—We are in the habit of speaking of some crops as "weak-feeding" and others as "strong-feeding." In these expressions we recognize the fact that some crops obtain abundant supplies of plant-food from soils on which other crops suffer for lack of nutrients. For example, corn, oats and cabbage often grow well on soils where onion or wheat crops give poor yields. Stated in the briefest way, a crop depends for its feeding power on its ability to absorb plant-food from the soil. This depends upon several factors, chief among which are the extent or area of the root-hairs and the activity of the plant's chemical processes. Plants which have large root systems and a proportionately large number of root-hairs are in condition to absorb more plant-food than those which have small root systems. The rapidity with which plants use their food in making new growth and in elaborating various compounds affects their feeding power. The larger the amount of constituents used in the plant, the larger must be the amount absorbed. A crop in its most rapid stage of growth absorbs more plant-food material than at other times.

Effects of plant-food distribution on root growth.—Experiments have proved that root development varies according to the location of the plant-food. For example, when the plant-food is uniformly mixed through the soil, the roots grow evenly throughout the entire soil mass; when placed about one inch below the surface, the roots form a mat in this layer and those that extend beyond are not much branched; when placed about the middle, between top and bottom, the root system shows great expansion at this point; when at the bottom, the roots are slender and with few branches until they reach the food where they form a mat; when placed in a layer around the sides of the vessel holding the soil, the outside roots are greatly branched, but the inner ones are only slightly developed; when placed as a core down the middle of the cylinder of soil, the inner roots are greatly developed, the outer ones are less so. Root growth takes place where the food supply is.

Shallow rooted trees seriously rob the soil of plant food. Soft maple roots often spread fifty feet or more on each side of the trunk of a large tree.

XII

THE RELATIONS OF MICRO-ORGANISMS TO PLANT-FOOD

THE soil was once regarded as the abiding place of mostly non-living matter. Such changes as were known to occur, especially obvious ones like the decay of organic matter, were regarded as purely chemical without any relation to biological activities. We now know that soils, especially cultivated soils, teem with unseen life, and that many of the most important chemical processes that take place within are due to the action of living organisms found mostly upon the colloidal film that surrounds the inorganic soil particles. These living beings, though minute, are subject to the same general conditions as prevail among higher organisms in relation to the biochemical problems of nutrition, reproduction and death.

Because the use of a microscope is required to enable one to see them satisfactorily, they are spoken of under the general term of micro-organisms. Their relations to crop production form a field of investigation which has yielded results of far-reaching practical value to plant nutrition. These relations appear to be of increasing importance as our knowledge enlarges.

The numerous soil micro-organisms in which we are here interested may be considered under four general divisions: 1, Bacteria; 2, Actinomyces; 3, Fungi; 4, Protozoa. Of these bacteria are, in their beneficial effects, of the greatest nutrition importance. Bacteria, actinomyces and fungi belong to the vegetable kingdom, protozoa represent the simplest form of animal life.

BACTERIA

Bacteria are the simplest and smallest known forms of plant life. Each individual consists of a single cell. On account of their power of rapid multiplication under favorable conditions, they are capable of producing extensive chemical changes, especially in animal and vegetable materials. Before taking up their detailed study, it is desirable to obtain some knowledge of their general characteristics, such as (1) size, (2) forms, (3) reproduction, (4) effects, and (5) conditions of growth.

Size.—Individual bacteria average in diameter about one twenty-five thousandth of an inch and vary from one thousandth to less than one fifty-thousandth of an inch. This means that if twenty-five thousand bacteria of average size were placed side by side, they would extend only one inch.

Forms.—Bacteria appear in three general varieties of form, round or ball-shaped (*coccus*), rod-like (*bacillus* or *bacterium*) and cork-screw or spiral form (*spirillum*). Each of these shapes varies more or less with diverse kinds of bacteria. In the first shape they may exist as separate balls, or in pairs, chains or square and cubical masses. Those that appear in rod-like form may be much like lead-pencils in shape, long and slender, or relatively thick and short. The rods may be separate or strung together, end to end. The cork-screw shaped variety may closely resemble a cork-screw or may be much like a comma, a letter S, or a watch-spring. There are, however, numerous variations from these regular forms. Some bacteria have one or more little threadlike, external appendages, which by working rapidly back and forth enable the organisms to move about.

Reproduction.—Bacteria reproduce by simple division; that is, when a cell grows it increases more in one direction, so as to lengthen out slightly. A partition forms across the cell, thus producing two cells in place of the old one. Then each of these subdivides, and so on continuously. In some kinds the whole cell is transformed into a spore, which bears much the same relation to bacteria as a seed to a higher plant, or an egg to an animal. Spores are not so easily killed as are bacteria, therefore, spore-producing bacteria reproduce under conditions which would kill fully-developed bacteria. Under favorable conditions the rapidity of growth of bacteria is beyond comprehension. In some cases, one cell divides into two cells every thirty minutes, that is, passes through its cycle of life from birth to reproduction; if such a rate were kept up for twenty-four hours, the one cell would multiply into more than three hundred trillions (300,000,000,000,000). Owing to the exhaustion of food supply, accumulation of poisonous products and the consequent death of many cells, such rapidity of increase does not occur under ordinary conditions.

Effects produced by bacteria.—In the course of their growth, bacteria produce many kinds of change in the materials in which they live, and which they use as food. The immediate cause of each change is a specific substance known as an *enzym,* produced by the bacteria, which is a non-living, chemical substance, capable of causing marked changes in many complex organic compounds, the enzym itself undergoing little or no change.

The special function of enzyms is generally to assist bacteria in procuring food. Bacteria, like the hair-root cells, take their food materials only in soluble form through the cell-walls; substances not in solution cannot enter, so in such condition are useless as food. In order to utilize insoluble materials, such as cellulose and insoluble proteins, many bacteria produce enzyms internally. These pass out of their cells, act on insoluble materials which when made soluble pass into the bacterial cells for use as food. Bacteria convert such food materials into new forms of combination, and produce various kinds of substances, sugars, acids, alcohols, soluble nitrogen compounds (peptones, amino acids, ammonia, nitrous acid, nitric acid, free nitrogen), aromatic compounds and numerous other products. Some bacterial products are intensely poisonous, others highly useful.

Conditions of growth.—The life and growth of bacteria are affected by (1) food, (2) temperature, (3) oxygen, (4) moisture, (5) light, (6) chemical and physical conditions, (7) products of bacterial action, (8) presence of other forms of life.

(1) *Food.*—The bodies of bacteria consist of proteins, carbohydrates, oils, etc., and, therefore, like higher plants, for satisfactory growth they require as food soluble compounds that contain nitrogen, carbon, hydrogen, sulfur, oxygen, and, in addition, small amounts of inorganic or mineral compounds (phosphorus, potassium, magnesium, etc.). Compounds contained in vegetable and animal matter form the usual sources of supply of food for bacteria. Various kinds of bacteria show decided preferences for one or another kind of food compounds. For example, some thrive only on organic nitrogen, others on ammonium compounds, others on nitrate and still others on nitrite nitrogen. When bacteria live in or on some other organism and derive their nourishment therefrom and at the expense of their "host" they are called *parasites* (*para,* beside, and *sitos,* food); many plant and animal diseases are due to the parasitic action of bacteria.

(2) *Temperature.*—Various bacteria flourish best within definite narrow limits of temperature. The most favorable range of temperature for bacterial activity lies between 70° F. and 105° F. (21° and 40° C.), or, perhaps, for most soil bacteria within the narrower range of 60° to 80° F. (15.5° to 25.5° C.). In general, activity decreases as the freezing point of water (32° F. or 0° C.) is approached, when all signs of active life cease; but intense cold, unless applied for a long time, usually does nothing more than render bacteria inactive; on warming again, they become active. Above 110° F. (43° C.) bacteria gradually lose their vigor and

many are killed at 130° to 140° (54° to 60° C.) in ten minutes, or in less time at higher temperatures. At the temperature of boiling water (212° F. or 100° C.) most bacteria are killed, but many bacterial spores are not killed except at considerably higher temperatures and may even then require one to three hours. Dry heat is less effective than moist. When bacteria are not actually killed by heat, their vitality may be greatly weakened, as manifested by diminished activity.

(3) *Oxygen.*—Bacteria vary greatly in respect to their ability to live with or without oxygen. Some called *aërobes* or *aërobic* can live only when well supplied with air. They cause the elements of organic matter to unite with oxygen; that is, they oxidize organic matter, so are useful in making available the plant-food materials in organic matter. Bacteria that cause rapid decomposition and also those that produce nitrification, as well as certain kinds that fix nitrogen are aërobic.

Other bacteria called *anaërobes* or *anaërobic* soon die or cease being active if exposed to the air. These take oxygen away from (reduce) organic matter, decompose it and tend to change it into gaseous compounds. Bacteria that convert cellulose into soluble compounds are anaërobic. Both aërobic and anaërobic bacteria are found in soils, manure heaps and other places. In many cases, as in manure piles, the air is used by aërobes and conditions thus made suitable for anaërobes.

(4) *Moisture.*—Many bacteria become dormant though still retaining life, on being dried, then become active again when moisture is supplied. Some are killed at once by drying. In the decomposition of vegetable matter in soils bacterial action is said to be little when the percentage of moisture drops below 25 per cent.

(5) *Light.*—Sunlight kills most bacteria when exposed directly to the sun's rays for a few hours. Spore forms are less easily affected. In the case of bacteria in soil and water, direct sunlight affects only those bacteria in the topmost layer. Diffused sunlight and artificial light may lessen the activity of bacteria.

(6) *Chemical and physical conditions.*—The presence of an acid destroys, or prevents the growth of, those kinds of bacteria which flourish only in a neutral or slightly alkaline medium. Hence, a soil that contains free acid or acid salts is unfavorable to some bacteria; in strongly acid condition, molds flourish instead of bacteria. Under conditions that prevail in such places as brackish swamps, humus-free soils and sand beaches, certain kinds of bacteria are not found. Many chemical compounds have the effect of either killing or checking the growth of bacteria. Those compounds that simply retard the rapidity of growth of bacteria are called *antiseptics;* those that destroy bacterial life are called *dis-*

infectants. The addition of acid phosphate, kainite and similar substances to stable manure is usually for the primary purpose of checking or preventing bacterial action. Many compounds which, in dilute solutions, serve as food, act as poisons when present in concentrated form; for example, nitrifying bacteria use soluble organic matter as food, but their activity is retarded when too much is supplied.

Certain physical surroundings are unfavorable. For example, bacteria do not flourish in pure clay or sand or in soil filled with stagnant water. These are, however, connected with conditions of temperature, air supply, etc.

(7) *Products of bacterial action* tend to accumulate as results of chemical changes due to the work of bacteria in transforming chemical compounds. When they reach a certain degree of concentration, they often act as poisons in reducing or stopping continued bacterial activity. In many cases the products of one kind of bacteria are poisonous to other kinds. To illustrate, lactic acid bacteria, characteristic of sour milk, convert milk-sugar into lactic acid, but cease to act when about 1 per cent. of lactic acid has been formed, even though only a part of the milk-sugar is changed. Lactic acid stops the activity not only of these organisms but also that of many other bacteria that require a neutral or slightly alkaline reaction, such as change insoluble protein, milk-casein into soluble nitrogen compounds. If we add to sour milk some calcium carbonate and thus neutralize the lactic acid, the bacteria renew their activity and produce acid until the mixture again contains about 1 per cent., when they stop; but, if enough calcium carbonate is added, the action will be continued until all the milk-sugar is changed. Many soil bacteria act in the same way; they produce acids until the accumulation is sufficient to check or stop their activity, but in soils well supplied with calcium carbonate the action can go on as long as food supply and other conditions are favorable.

(8) *Presence of other forms of life.*—The soil may contain protozoa (p. 154) which are believed to feed upon living bacteria to such an extent as to interfere seriously with the proper performance of certain bacterial functions in the soil.

Soil bacteria.—One of the important functions of soils is to furnish a favorable living place for bacteria, especially for forms that do useful and necessary work in connection with nutrition of crops. Many kinds of bacteria are present in soils; some useful, others injurious, many apparently neither.

(1) *Distribution.*—In fertile soils bacteria are exceedingly numerous; in dry, sandy soils and those that contain no humus, their numbers are few. At the surface of soils, are comparatively

few bacteria; in the upper layer, just beneath the surface, are many, three feet down only a few, below five or six feet practically none. In open soils, such as sandy soils and sandy loams, that permit free circulation of air, those organisms abound that require air (aërobic); in heavy clay soils, in which circulation of air is limited, aërobic bacteria are not abundant but rather those that thrive in the absence of air (anaërobic). Other conditions that influence the kind and number of bacteria in soils are food supply, altitude, exposure, plant growth, methods and extent of cultivation, kinds of fertilizers applied, drainage, liming, crop-rotation, supply of organic matter, etc. Crops with heavy foliage, such as cabbage, beets, etc., by shading the soil, affect conditions of temperature, moisture and light differently from such crops as wheat. Generally speaking, bacteria are found most abundantly in soil where there is vegetable matter to furnish food, provided other conditions of chemical and physical environment are favorable to their existence.

(2) *Changes produced.*—Soil bacteria effect important changes in the mineral and organic plant-food constituents of soil. They are also instrumental in making atmospheric nitrogen available as plant-food. Some bacteria also convert available forms of plant-food into less available condition. The effects of various kinds of soil bacteria include: (1) Changes in the organic soil constituents, (2) fixation of atmospheric nitrogen, (3) changes in the mineral soil constituents.

CHANGES IN ORGANIC SOIL CONSTITUENTS

Of the so-called organic constituents, carbon, hydrogen and oxygen are obtained directly or indirectly from the air (p. 12) without the intervening action of bacteria. But the nitrogen of the air, as such, is not directly available to plants, although it constitutes the sole and original source of supply. The means of changing portions of atmospheric nitrogen into plant-food material have depended largely upon bacteria. The nitrogen thus originally obtained from the air has been transformed into plant and animal tissues and these have gone into the soil, there to be acted upon by micro-organisms, largely bacteria, and worked over into forms which plants can use again as food.

The changes by which plant and animal substances are made useful in the soil as sources of available plant-food are not only highly interesting but extremely important in their applications to practical crop production. Thus, the conversion of the vegetable matter of green crops, roots, and all vegetable waste material into soil humus depends chiefly upon the action of bacteria (pp. 130-

153). Likewise, the fermenting or "rotting" process by which animal and vegetable materials in farm manures are profoundly changed in composition is due chiefly to the work of bacteria. The bacterial changes produced in organic matter are largely in the direction of converting insoluble into soluble plant-food material in one or another of the following ways: (1) Decomposition, (2) ammonification, (3) nitrification, (4) denitrification, (5) sulfofication.

Decomposition.—The term decomposition is used here to include those chemical and physical changes which vegetable and animal matter undergo when their complex compounds (proteins, carbohydrates, etc.) are changed into simple compounds by means of the action of various living organisms, especially bacteria. The subject has already been briefly discussed in relation to the organic matter in soils (Chapter VIII, p. 87). The processes of bacterial decomposition vary greatly, depending on the kind of organic substance and various other conditions, but especially on the presence or absence of oxygen (air). As previously stated, (p. 136), they may be divided into two general kinds: (1) Decomposition in air (aërobic), and (2) decomposition away from air (anaërobic). The character of the chemical changes produced by each process is different.

(1) *Decay or decomposition in presence of air (aërobic).*—Going more into detail than on page 136, the following changes occur:

(*a*) Non-nitrogenous organic compounds (the carbohydrates, oils and organic acids). (1) The carbohydrates, *sugar* and *starch*, are decomposed rapidly by various kinds of bacteria and fungi, being converted into various intermediate simpler compounds, among which are organic acids, such as acetic, the characteristic acid of vinegar. But whatever intermediate changes occur, the final products of decay are carbon dioxide and water. (2) *Lignocellulose,* the carbohydrate that constitutes the bulk of vegetable cell-walls, is less quickly attacked by decay bacteria than are sugar and starch. While several species of aërobic bacteria can dissolve the celluloses some fungi (p. 154) are also capable of doing the same, and, under ordinary conditions, the decay of the celluloses is due to the combined action of fungi and aërobic bacteria, mainly through enzyms. The intermediate and final products of cellulose decay are much the same as those of sugar and starch. In connection with the decay of vegetable matter in both soils (p. 93) and manure piles (p. 232), the dissolving of the tough cellulose that covers the cells is a matter of practical value, because it sets free the various carbohydrate and nitrogenous food constituents contained in the cells and makes them accessible to other bacteria. (3) *Oils* are changed into acids and these finally into carbon dioxide and water. (4) *Organic acids* ultimately form carbon dioxide and water. Some of the intermediate decomposition products of the non-nitrogenous organic compounds are used as food by other bacteria.

(*b*) *Proteins.* When protein compounds decay they form a succession or series of simpler nitrogen-containing compounds (1) albumoses or proteoses, (2) peptones or peptids and (3) amino acids. Sooner or later the nitrogen becomes ammonia (ammonification, p. 138). The carbon, oxygen and hydrogen of proteins are finally converted into carbon dioxide and water, the sulfur into sulfuric acid and the phosphorus into phosphoric acid.

(*c*) Mineral constituents originally present in vegetable matter are left as a residue, comparable to ashes, mainly as salts of calcium, magnesium, potassium, sodium, etc., such as phosphate, sulfate, carbonate and chloride.

(2) *Decomposition in absence of air (oxygen, anaërobic).*—Bacteria, different in kind from the aërobic and different also in the results of chemical change produced, are responsible for the process of anaërobic decomposition of organic matter in soils, known as *humification.* The general differences in the results are as follows:

(*a*) The anaërobic process is shown by the difference in the rate of decomposition in organic matter in loose, as compared with close, soils, or at the surface of a soil as compared with that at considerable depths, or on the outside of a compacted manure pile as compared with the inside, or in a well-drained, as compared with a water-logged soil.

(*b*) Chemical changes less complete. In anaërobic decomposition the chemical changes do not go so far as in aërobic; the complex organic compounds, whether nitrogenous or non-nitrogenous, are not so completely decomposed into simple compounds. While some carbon dioxide, water and ammonia are formed, the carbon, hydrogen, oxygen and nitrogen are more largely found in the form of intermediate compounds of considerable complexity of composition.

(*c*) Products resistant to further change. The dark-colored, intermediate organic compounds formed by anaërobic decomposition are not easily decomposed by other bacteria, even when exposed to air, as already explained in connection with organic matter in soils (p. 92).

(*d*) Offensive and poisonous compounds. Many compounds formed by anaërobic decomposition of organic matter, and particularly of nitrogenous substances that contain sulfur and phosphorus, are peculiarly offensive in odor (giving rise to the application of the term *putrefaction* to the process). Many of them are poisonous, for instance: Hydrogen sulfide or gaseous sulfureted hydrogen (H_2S), phosphine or gaseous phosphureted hydrogen (PH_3), skatol and indol (which furnishes characteristic odors in human excrements), etc.

(*e*) Difference in gases formed. While in anaërobic decomposition some of the same gases are formed as in anaërobic—carbon dioxide, water, ammonia and nitrogen—they are formed in much smaller amount, while other gases, such as marsh gas (CH_4), phosphine, hydrogen sulfide, etc., which do not occur appreciably, if at all in decay are present.

Owing to the variety of conditions which influence one or another kind of bacterial activity, the general process of decomposition of organic matter in soils, is one of extreme complexity. In each kind of decomposition many kinds are at work at one time or another. While the chemical changes during the processes of decomposition go through an orderly progression, many of the changes in the various classes of compounds go on at the same time under the action of different kinds of bacteria, working approximately side by side, the products of some retarding or hastening the activity of others.

Ammonification is a term applied to the general process by which organic or protein nitrogen is converted into ammonia. In the decomposition of organic matter the nitrogen compounds are sooner or later changed into ammonia partly or completely; therefore ammonification is essentially the process of decomposition, limited to protein compounds, under conditions favorable to the formation of ammonia, this product being the result of the last stage of the decomposition of nitrogen-containing organic compounds or proteins. The production is accomplished by many kinds of *ammonifying* bacteria. Some are aërobic, others anaërobic. The various kinds differ in their chemical processes and rapidity of action. Numerous fungi found in soils also produce ammonia.

The formation of ammonia from organic nitrogen compounds in soils depends upon various conditions, such as the kind of protein, the kind of soil bacteria or fungi, and all the physical and chemical conditions that modify or influence micro-organic activity. The bacteria that cause ammonification are usually abundant in air, rain-water, stables, manure piles and surface soils. The process is one of extreme importance in crop growing because upon it depends the formation of nitrate nitrogen in soils. While some agricultural plants appear to be able to use compounds of ammonia as food directly, it is generally believed the most effective form of nitrogen as a direct source of plant-food is the nitrate. The ammonia produced by the action of micro-organisms combines, under conditions present in normal soils, with carbon dioxide and water to form ammonium carbonate,

Ammonia		Carbon dioxide		Water		Ammonium carbonate
$2NH_3$	+	CO_2	+	H_2O	=	$(NH_4)_2CO_3$

Nitrification is the name of the bacterial process by which nitric acid (HNO_3) or nitrate nitrogen, is formed. The term is ordinarily used in its broadest sense to indicate the conversion of organic nitrogen into nitrate; therefore, when so used, it includes the process of ammonification. It is also used in a more limited sense to indicate the conversion of ammonia into nitrate. In any case, ammonia is the form of nitrogen directly used in the process of nitrification.

(1) *Chemical changes of nitrification.*—The conversion of ammonia into nitric acid takes place as the result of two distinct bacterial chemical operations, one kind of bacteria being responsible for the first operation and another for the second. The operations consist of (1) the change of ammonia into nitrous acid (HNO_2) or nitrite nitrogen and (2) the change of nitrous acid into nitric acid (HNO_3) or nitrate nitrogen, represented as follows:

	17		48		47		18
(1)	NH_3	+	$3O$	=	HNO_2	+	H_2O
	Ammonia		Oxygen		Nitrous acid		Water

	47		16		63
(2)	HNO_2	+	O	=	HNO_3
	Nitrous acid		Oxygen		Nitric acid

These reactions show that the process of nitrification is, chemically, an *oxidation process, adding oxygen to nitrogen.* In the production of nitrous acid from ammonia, one pound of nitrogen in ammonia combines with about 2.3 pounds of oxygen. In the final stage about 1.15 pounds more oxygen is added to produce nitric acid. Stated in another way, 1 pound of nitrogen in ammonia combines with about 3.5 pounds of oxygen to form nitric acid. On the basis of ammonia, instead of nitrogen alone, 1 pound of ammonia is capable, in nitrification, of producing 2.75 pounds of nitrous acid and 3.7 pounds of nitric acid. It will be observed that the addition of oxygen has the effect of diluting the nitrogen, or reducing its percentage, as shown by comparing the percentage of nitrogen in ammonia (82.35), in nitrous acid (29.8) and in nitric acid (22.2).

As shown above, the final product of nitrification is free nitric acid (HNO_3). In a soil well supplied with calcium carbonate, the nitric acid quickly combines with the calcium to form calcium nitrate, as represented by the following reaction:

126		100		164		44		18
$2HNO_3$	+	$CaCO_3$	=	$Ca(NO_3)_2$	+	CO_2	+	H_2O
Nitric acid		Calcium carbonate		Calcium nitrate		Carbon dioxide		Water

Therefore, in good agricultural soils, the nitric acid is present as a salt of calcium, potassium (KNO_3), magnesium ($Mg(NO_3)_2$), or sodium ($NaNO_3$), according to special chemical conditions present in a soil.

It is to be kept in mind that if the best conditions for further formation of nitrate nitrogen and for the growth of plants are to be maintained, the nitrification process results in producing an acid, which calls for some constituent to combine with and neutralize it. For example, each pound of nitric acid formed calls for nearly a pound of calcium carbonate.

(2) *Character of bacteria of nitrification.*—The nitrous or nitrite bacteria capable of changing ammonia into nitrous acid, are known as *Nitrosomonas* and *Nitrosococcus.*

The nitric or nitrate bacteria that change nitrous acid (HNO_2) into nitric acid (HNO_3), are called *Nitrobacter.* These two groups are together often called *nitrifying bacteria* or *nitrifiers.*

The nitrous and nitric organisms appear to be influenced largely by the same conditions, favorable and unfavorable, and, therefore, are usually found working together, because nitrous acid, almost as rapidly as it is formed from ammonia, is changed into nitric acid. Only under abnormal soil conditions are appreciable amounts of nitrous acid found.

Nitrifying bacteria are found in all cultivated soils, water, manure, and sewage. They are not abundant in soils either excessively poor or excessively rich in organic matter as, for example, in forest soils, muck deposits, arid deserts, etc. In soils of high productive power the nitrifying ability of the bacteria is often greater than in soils of less productive power.

While much remains to be learned in regard to the conditions under which nitrifying bacteria work with greatest effectiveness, many soil conditions favor or retard the process; for instance, (*a*) location, (*b*) moisture, (*c*) food, (*d*) oxygen, (*e*) temperature, (*f*) light, (*g*) non-acid condition or soil reaction.

(*a*) Location. Nitrifying bacteria are most numerous in the upper layers of soils, because the conditions are there most favorable for their growth. Probably two-thirds of the nitrate nitrogen produced in soils is in the first foot of the upper surface and the remainder is in the next foot below. The depth at which these bacteria occur depends upon the looseness and warmth of the soil.

(*b*) Moisture. In a dry soil nitrification does not take place. Therefore, in drouth, this process is likely to cease in the upper layers of soil. In a wet soil, on the other hand, where water is stagnant, the process does not occur actively.

(*c*) Food. The nitrogen of organic matter in soils forms the primary source of nitrogenous food for the nitrifying bacteria; in the more limited sense of nitrification, ammonia constitutes the nitrogen food supply. The complete absence of organic nitrogen does not stop the process of nitrification so long as there is a supply of ammonia. It is possible, however, to have so much soluble organic nitrogen present that the nitrifying organisms will stop growing, as in the case of a manure pile. An excess of soluble carbohydrate material also interferes with activity. In cultivated soils organic matter is rarely present in such amounts as to interfere with the process of nitrification, even when applications of 20 tons of farm manure per acre are applied two or three years in succession; interference may happen in greenhouse and market-garden soils, where immense applications of stable manure (50 to 100 tons an acre) are often made. The use of sewage-irrigation also furnishes a striking illustration of the effect of soluble organic materials on nitrification. Phosphoric acid compounds are essential for the satisfactory growth of these organisms and probably also potassium and other mineral compounds.

(*d*) Oxygen. The formation of nitrous and nitric acids from ammonia consists essentially in combining oxygen with nitrogen chemically. Hence, nitrifying organisms cannot work vigorously without a generous supply of air. Therefore they are most abundant in those layers of the soil where there is most air. Water-saturated soils prevent nitrification by shutting off the air. Limited air supply is one of the important factors which explains slow nitrification on heavy sod land and in clay soils. All operations connected with the cultivation of soils greatly promote the change of ammonia into nitrate nitrogen, because the oxygen of the air is brought into closer contact with the nitrifying bacteria.

(*e*) Temperature. Nitrifying bacteria are most active between the temperatures of 54° and 99° F. (12° and 37° C.), the most favorable temperature being 80° F. to 90° F. The process may go on slowly as low as 37° F. (3° C.) and as high as 120° F. (49° C.). Activity rapidly increases with rise of temperature above 54° F. (12° C.), so that at 99° F. (37° C.) nitrification takes place about ten times as rapidly as at 54° F. (12° C.). In northern latitudes, nitrification slows down during the fall and stops during the winter, starting up again in spring and becoming most active during mid-summer. In warm climates, the process continues at all times of year. One reason why crops start earlier in spring on light soils because the greater warmth of such soils favors the more rapid formation of nitrate.

Response to applications of nitrogen. This Baldwin tree has received ten pounds of nitrate of soda annually since 1921. The yield in 1929 was three and five-eighths bushels. Compare with facing picture.

(*f*) Light. Darkness is essential to the activity of nitrifying bacteria, since the direct rays of sunshine destroy bacterial life. These organisms are, therefore, not found on the exposed surface of bare fields. They are more active in soil shaded by growing crops than in unprotected soil. Nitrification is promoted on warm nights, and in warm dark places, as under boards, stones, etc.

(*g*) Non-acid condition or soil reaction. Nitrifying bacteria are extremely sensitive to acids, which act injuriously like poisons on them. The final product of nitrification, free nitric acid, if allowed to accumulate without being neutralized, stops further nitrification.

Similarly, in the presence of any other free acid, nitrifying bacteria fail to work. For this reason, nitrification cannot take place in soils that have an acid reaction. In order that there may be no accumulation of free nitric or other acids in soils, it is necessary to have an abundance of some basic compounds, which can unite with

Response to applications of nitrogen. This Baldwin tree has received no nitrogen since 1921. The yield in 1929 was one-eighth bushel. Compare with facing picture.

the free acids and form salts that do not act injuriously on nitrifying organisms. The compound most available for neutralizing acids is calcium carbonate, though the carbonates of magnesium, potassium, sodium, etc., are also efficient. Calcium carbonate and nitric acid, combine to form calcium nitrate, carbon dioxide and water (p. 140). Such compounds as nitrate of calcium, magnesium, sodium, etc., even when present in considerable quantities, are not injurious to nitrifying bacteria. It is, therefore, highly important to keep agricultural soils well provided with acid-neutralizing or basic materials in order to favor the formation of nitrate nitrogen. When slaked lime (calcium hydroxide) is applied to soils in amounts so large as to make the reaction of the soil solution distinctly alkaline, the action of nitrifying bacteria is suspended, but will begin again when the calcium hydroxide has been changed into carbonate and the alkaline condition thereby changed to neutral.

Denitrification.—The bacterial process of denitrification, chemically considered, produces results the reverse of those produced by nitrification. Denitrification, a *deoxidation* or *reduction* process, consists in taking oxygen away from nitric acid (HNO_3), first to form nitrous acid (HNO_2) and then either ammonia (NH_3) or nitrogen gas (N). There is also another form of denitrification by which oxygen is removed and the nitrate nitrogen is changed into insoluble organic nitrogen compounds or proteins. These three general forms may be represented as follows:

(1) Complete denitrification: (*a*) Nitric acid (HNO_3), (*b*) nitrous acid (HNO_2), (*c*) nitrogen gas (N).

(2) Partial denitrification: (*a*) Nitric acid (HNO_3), (*b*) nitrous acid (HNO_2), (*c*) ammonia (NH_3).

(3) Assimilation or conversion into organic nitrogen compounds: (*a*) Nitric acid (HNO_3), (*b*) protein compounds.

The term denitrification strictly applies only to the first form, the process by which nitrate nitrogen is changed into uncombined or free nitrogen gas, which in this form is lost as plant-food to soil and crops. When nitrate nitrogen is converted into nitrous acid, ammonia or organic nitrogen, these compounds remain in the soil, and, under favorable conditions, are again changed back into nitrate nitrogen.

Generally in soils and in animal manures and on straw are found several kinds of bacteria which have the power to decompose nitrate nitrogen in one of the ways mentioned. It is desirable to know something of the conditions under which complete denitrification occurs and thus learn how to avoid large losses of nitrate nitrogen. Two general conditions favor the process of denitrification: (1) Lack of air and (2) superabundance of soluble organic matter.

(1) *Lack of air in relation to denitrification.*—Denitrifying bacteria require some oxygen for their activity. When this need is not supplied by air, they take the oxygen from the nitrates in the soil. This fact affords an explanation of the decomposition of nitrate nitrogen in the lower layers of soils, as well as in surface soils saturated with stagnant water. It explains also why heavy, compact soils tend to suffer loss from denitrification to a greater extent than open, sandy loams. Of course, no denitrification can occur, even in the absence of air, when the soil contains no nitrate and no readily decomposable organic matter. Any condition that promotes circulation of air in soils, such as thorough tillage, good drainage and application of lime, diminishes the danger of denitrification. In the use

of fertilizers it is obvious that the application of a nitrate to a soil in which, for any reason, the supply of air is deficient is in danger of being followed by conversion of the nitrate nitrogen into free nitrogen.

(2) *Superabundance of organic matter in relation to denitrification.*—An important condition for active denitrification is the presence of an abundant supply of soluble organic matter, especially carbohydrates, in soils. When large amounts of nitrate are applied to soils along with an excessive quantity of fresh farm manure, there is likely to be large loss of nitrate by denitrification. Similar loss may come from plowing under heavy crops of green-manures. Generous applications of partly rotted farm manure along with nitrate may cause no loss when the soil is kept well supplied with air. Denitrification becomes a source of serious loss only under certain conditions, as when a large amount of nitrate is applied to undrained or sour land, when used with excessive amounts of fresh farm manure, or whenever large amounts of fresh organic matter, rich in soluble carbohydrates, accumulate in the soil.

Denitrification by assimilation.—One form of denitrification has been given above as assimilation by conversion of nitrate nitrogen into proteins, though this process may not be regarded as of denitrification proper. All micro-organisms, like other plants, require some nitrogen as an essential constituent of their bodily substance. In supplying this need they take soluble, nitrogen-containing compounds (nitrate, nitrite, ammonia, proteins with their decomposition products) and convert them into the insoluble organic compounds of which their bodies are composed. It is inevitable, therefore, when such processes as bacterial decomposition, ammonification, nitrification, denitrification and many others go on actively in soils, that the bacteria greatly increase and that all these organisms appropriate for the growth of their own bodies soluble nitrogen compounds, which are changed into insoluble organic nitrogen compounds. Under favorable conditions of growth, the enormous increase of bacteria present in a soil may in this way render insoluble fairly large amounts of soluble nitrogen. In addition, fungi and actinomyces similarly appropriate nitrate nitrogen. Nitrogen thus taken from soluble forms is held mostly until the micro-organisms die and their bodies decompose, yielding again ammonia and nitrate. This process may be desirable under conditions where considerable nitrate accumulates in a soil and be in danger of leaching before being used by crops. The change then serves to hold insoluble organic compounds for future use and decreases loss by leaching.

FIXATION OF ATMOSPHERIC NITROGEN

Nitrification does not increase the total amount of nitrogen present in a soil. The amount of nitrate produced is strictly limited to the amount of nitrogen compounds already in the soil; no more nitrogen is present in the soil after than before the process occurs. The form of combination is changed; that is all. But other kinds of bacteria have the ability to *add nitrogen* to the soil's supply, transforming the atmospheric free nitrogen into combined forms of nitrogen that can be used as plant-food. The process is called *nitrogen fixation.* The free nitrogen of the atmosphere can undergo fixation through the action of two classes of nitrogen-fixing or nitrogen-gathering bacteria: (1) Bacteria that live in the soil itself, independent of the presence of growing plants, and (2) bacteria dependent upon the presence of certain higher plants upon whose roots they live and work. Those in the first class are known as *non-symbiotic;* those in the second class as *symbiotic.* The word symbiotic means *living together* and is applied to cases where two different organisms live together in mutual relations that are of advantage to both.

Nitrogen fixation by independent or non-symbiotic bacteria.—Soils that lie unused, either without any vegetation or covered with grass, appear to accumulate appreciably increased amounts of nitrogen in one or two years. This fact has been explained by the discovery that certain bacteria are able to make use of atmospheric nitrogen contained in the soil for their growth, and to combine it with other elements in their bodies. This process is known as *azofication.* The bacteria most active in thus fixing atmospheric nitrogen in normal soils belong to the *Azotobacter* group. As the result of several investigations the amount of nitrogen fixed by azofication has been estimated to vary between 15 and 40 pounds an acre annually.

Among the conditions that influence azotobacter are: (1) Plentiful supply of carbohydrates favor azofication; (2) so does shortage of nitrate; large amounts appear to lessen it. (3) Soluble compounds of calcium, phosphorus, potassium, sulfur, iron and manganese in the soil furnish needed inorganic constituents. (4) The organisms are unfavorably affected by acid soil, working best near the neutral points in acid soils, another group of nitrogen-fixing bacteria (known as *Clostridia*), may function; (5) when vegetation like grass is grown continuously for years and permitted to decay on the soil surface, azofication takes place to a marked degree.

Nitrogen fixation by dependent or symbiotic bacteria.—The practical experience of hundreds of years led farmers to believe that leguminous crops (clovers, alfalfa, beans, peas, etc.) possess some peculiar power to make succeeding crops give better yields. This belief has been proved true. We can now explain why leguminous crops are of such value as has been claimed. The benefit noticed was first traced to the fact that the amount of nitrogen in a soil can be definitely increased by the growth of leguminous plants, and it was learned also that the free nitrogen of the air can be used by such plants. Leguminous plants contain in both tops and roots large proportions of nitrogen compounds, as compared with most other plants. It has been shown that the ability of leguminous plants to utilize the free nitrogen of the air depends upon certain kinds of bacteria, as follows: (1) Root-nodules and bacteria, (2) relation of bacteria and legumes, (3) varieties of legume bacteria, (4) conditions of action of legume bacteria, (5) amount of atmospheric nitrogen used, (6) distribution of nitrogen gathered by legume bacteria, (7) relation of legumes to the amount of acid nitrogen, (8) effect of legumes on succeeding crops, (9) independent action of legume bacteria, and (10) nitrogen-fixing root-nodules in relation to non-leguminous plants.

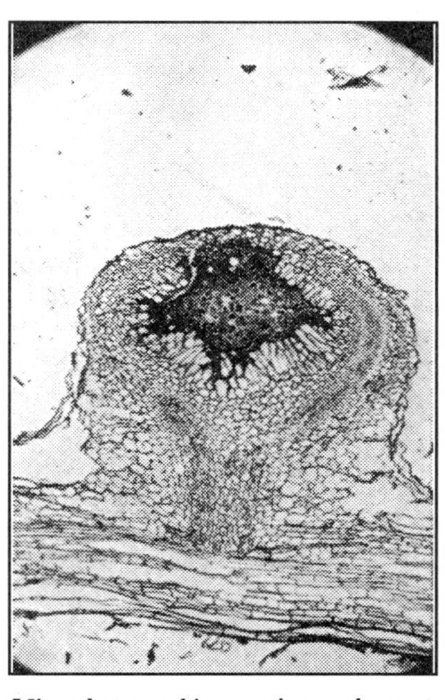

Microphotographic section of root nodule. Magnified about one hundred times.

(1) *Root-nodules and bacteria*.—If we carefully remove from the soil the roots of a vigorous leguminous plant, we usually find distributed on the younger roots little ball-like bunches, called *nodules* or *tubercules*. These vary in size and shape on diverse kinds of plants. For example, on red clover, they sometimes are as large as a pea and more or less ball-shaped; on cowpea and soy-bean they are much larger, and on velvet-bean may even reach the size of a baseball; on vetches they vary irregularly in both size and shape.

These root growths were noticed and described over two hundred years ago, but their true nature was not understood until 1886. The nodules are the dwelling-places of nitrogen-fixing bacteria characteristic of leguminous plants, many millions occupying each tubercle. The tubercles are also the direct result of bacterial action; when there are no legume bacteria, there are no root nodules, and there is no utilization of free nitrogen by the plant.

Legume bacteria are comparatively common in soils, especially where leguminous crops have grown well; they are found also in the water of streams and lakes. They gain entrance into plant-roots by first working their way through the thin cell-walls of root-hairs. Then they make their way into and through the tissues of the nearest root-branch, where they multiply rapidly and accumulate in enormous numbers, causing the growth of the nodule for their accommodation. In reality root-nodules are rootlets developing in abnormal forms on account of the presence of nitrogen-fixing bacteria within their cells. Nodules are found only on the younger parts of roots. When legumes mature, the root-tubercles decay and any bacteria that remain in them go back to the soil, where they remain, if conditions are favorable, until an opportunity comes to enter again the roots of a new leguminous crop. The percentage of nitrogen in root nodules is generally high as compared with other portions of the plant.

(2) *The relation of nitrogen-fixing bacteria to leguminous plants* is one of mutual helpfulness or of real symbiosis (living together). For their nourishment the bacteria require considerable amounts of sugar or other soluble carbohydrate and also mineral salts, which are generously supplied in the plant juices of legumes. The bacteria, in some way not yet understood, supplement this food supply from the soil air with nitrogen which they build into nitrogenous compounds within their cells. It is supposed that these compounds are diffused through the cell-walls and absorbed into the general circulation of the leguminous plant where they are used for building protein compounds characteristic of legumes. It is thus seen that leguminous plants furnish the nitrogen-fixing bacteria a needed food and home, and the bacteria reciprocate by changing free nitrogen, not usable directly by legumes, into usable nitrogen compounds, which they pass over to the plant in exchange for the bacterial food and lodging.

(3) *Kinds of nitrogen-fixing legume bacteria.*—The root-nodules on leguminous plants are produced by organisms, known as *Bacillus radicicola,* which exist in several varieties or strains. The organisms that promote the growth of nodules on clover roots do not form nodules on roots of alfalfa, cowpeas and most other leguminous plants. It appears, in general, that the bacteria that grow on the rootlets of one kind of legume are not well suited to the growth of other legumes, and that, therefore, best results are obtained when, for example, organisms growing on clover are used for that crop, those growing on alfalfa for alfalfa, etc. In some cases the same

organism appears to work equally well for various, closely related legumes. Thus, alfalfa, bur clover, the sweet clovers, and yellow trefoil use the same strain; the various clovers (mammoth, red, alsike, crimson, hop, white) use one strain in common, etc. Pieters, A. J., "Green Manuring," p. 140. N. Y. 1927.

(4) *Conditions of action of legume bacteria.*—The soil conditions that in general favor nitrification also favor the presence of nitrogen-fixing legume bacteria—abundance of calcium carbonate, soluble carbohydrates, air and moisture. In soils ill-drained, strongly acid, lacking in organic matter, the bacteria tend to be weakened or killed. The nodule-forming bacteria require abundance of sugar or other carbohydrate food, soluble potassium and phosphorus compounds, but they develop to best advantage when provided with a minimum supply of available nitrogen-containing compounds (nitrates, ammonia, proteins, etc.). The bacteria use available forms of nitrogen compounds, when within reach, in preference to using free nitrogen. Therefore, when supplied with available nitrogenous compounds, they fail to use atmospheric nitrogen; so in order to make them use free nitrogen, it is necessary to deprive them of available nitrogen compounds. Moreover, in a soil that contains an abundance of available nitrogen, or in one to which a nitrate has been generously applied, the formation of root-nodules is prevented in part or entirely. As a matter of practical application other conditions being favorable, it follows that the nitrogen-gathering powers of legume bacteria are fully realized only when the soil contains a minimum amount of available nitrogenous plant-food.

(5) *Amount of atmospheric nitrogen gathered by bacteria.*—Numerous investigations have been made to ascertain the amount of nitrogen obtained from the air by legume bacteria in the growing of leguminous crops. Under most favorable conditions, the results indicate that it is possible for a leguminous crop, including roots, to secure from the atmospheric supply as a maximum 400 pounds of nitrogen an acre in one season. The amount varies greatly according to several conditions, such as kind of legume, number of legume bacteria in the soil, amount of available nitrogenous plant-food in the soil, seasonal conditions, etc.

(6) *Distribution of nitrogen gathered by legume bacteria.*—The nitrogenous compounds which the legume bacteria produce with the use of atmospheric nitrogen may be utilized in three different ways. First, the legume absorbs them largely in promoting growth. Second, to some extent, they may go into the soil and be utilized by some other crop. Third, when the legume crop dies or is plowed into the soil, the nodules readily undergo decomposition, their nitrogen being made available for the following crop.

(7) *Relation of legumes to the amount of soil nitrogen.*—The effect of legumes on the amount of nitrogen in the soil is not definite. There may be a gain of soil nitrogen or a loss or no change, depending on various conditions. When leguminous crops, grown under conditions that favor the assimilation of atmospheric nitrogen, are incorporated in the soil, there is usually a gain of nitrogen, which may reach 50 to 100 pounds an acre in one year. When a legume crop is removed, especially if grown on soil rich in nitrate, there may be a loss of soil nitrogen, since the nitrogen in the legume is then taken from the soil rather than from the air. In some cases, the amount of nitrogen that remains in the roots and stubble, after the removal of the crop-producing portion, has been found equal to the amount of nitrogen furnished by the soil. This means that the amount of nitrogen removed by the harvested crop is equal to the nitrogen obtained from the air. Under such circumstances, the soil is no poorer in nitrogen, for the growth and removal of the crop because the nitrogen in the crop residues again becomes available through decomposition and nitrification. In other words, it is possible to make legume bacteria furnish from the air practically all of the nitrogen in the harvested portion of the crop. The relative amounts of nitrogen taken by legume crops from the soil and from the air vary greatly according to many conditions, such as the kind of legume, the number of nitrogen-gathering bacteria in the soil, the amount of nitrogenous plant-food in the soil, seasonal conditions, etc. As compared with non-leguminous crops, legumes generally deplete soil nitrogen less.

(8) *Effect of leguminous crops on succeeding crops.*—It has been found that non-leguminous crops usually grow better on a soil previously occupied by a legume. This is probably because the organic nitrogen in the parts left in the soil is more readily changed to nitrate nitrogen than in the case of non-leguminous crops, a property that emphasizes the value of legumes in crop rotation.

(9) *Independent action of legume bacteria.*—Bacteria from the root-tubercles of legumes are able to fix free nitrogen under favorable conditions with abundant supply of available carbohydrate and inorganic nutrients, even when they grow independently in the soil and are not connected with leguminous plants, but the amount thus fixed is small compared with that when they work in the root-nodules of leguminous plants.

(10) *Nitrogen-fixing root-nodules in relation to non-leguminous plants.*—Root-nodules that contain nitrogen-gathering bacteria are not confined to the Leguminosæ but occur on some plants outside this family, such, for example, as the alder (*Alnus mitchelliana*).

Soil inoculation for legumes.—The use of nitrogen-fixing legume bacteria has been made a practicable means of furnishing nitrogen to crops in place of nitrogen compounds obtained from commercial sources. Methods have been successfully used in connection with green-crop manures (p. 260). Here we confine our attention chiefly to a single one of the requisite conditions, viz., the method of supplying to leguminous crops the nodule-forming bacteria.

In many cases, the special kind of bacteria needed is present in soils, especially where a given legume has been grown successfully for years. For example, in the central and eastern states, clover

Austrian winter peas following continuous corn. At this stage the growth amounted to 12,000 pounds of green material to the acre.

bacteria are common in most cultivated soils, because these crops have been generally grown for many years. Similarly, alfalfa in the western, and cowpeas in the southern, states can be grown often without making any special effort to supply the soil with the required bacteria. On the other hand, alfalfa in the eastern states commonly requires an artificial supply of its bacteria on soils where it is grown for the first time. The same is true of soy-beans.

It has been demonstrated that, when failure to grow a leguminous crop is due to the absence of the nitrogen-fixing bacteria characteristic of the particular crop, good crops can be produced as the result of furnishing the soil with the required organisms. Supplying a soil with bacteria, known as *inoculation,* can be accomplished in various ways, among which are the following: (1) By use of soil that contains desired bacteria and (2) by use of special preparations or pure cultures.

(1) *Inoculation by use of soil.*—The most effective method of soil inoculation is to use soil taken from a field on which has been growing a clean, thrifty crop of the kind desired, one which is well

provided with root-nodules. In obtaining the inoculating soil, one removes the upper layer of earth for two or three inches and takes the soil below this for several inches, or that portion of the soil in which the root-nodules are most plentiful. The amount of soil used for inoculation is usually 300 to 500 pounds an acre. This is applied in any convenient manner to the surface of the field to be inoculated and should be harrowed into the upper layer of the soil promptly, preferably in cloudy weather, because, as previously stated (p. 133), bacteria are weakened or killed by exposure to direct sunlight. The disadvantage of this method of inoculation is the possibility of carrying undesirable weeds and germs of plant diseases from one soil to another. The precaution should always be observed, therefore, to use only soil from a field free from such defects. It is obvious that only soil which is in good tilth and easily permits reduction to fine powder can be used advantageously for inoculation.

Another plan, useful especially in case of large seeds such as peas and beans is to moisten the seeds with a 10 per cent. solution of glue and while still wet to sift over them a coating of dry, powdered, inoculating soil. The seeds are shoveled over a few times, screened to keep them from sticking together, and planted within a day; or they may be spread out to dry and kept away from direct sunlight if it is desired to prepare them some time before planting.

(2) *Inoculation by pure cultures.*—A pure culture is a preparation of one kind of micro-organism, free from the presence of every other kind. Many efforts have been made to prepare cultures for each of the leguminous crops and commercial preparations are offered for sale. The difficulty to be overcome is to prepare the pure cultures so the bacteria will remain alive and retain their full vigor for a reasonable length of time. Pure cultures are applied directly to seed, by sprinkling after dilution with the prescribed amount of water. Specific directions are furnished with each preparation as to the method of use.

Inoculation is not a means of providing crops with a complete fertilizer, as some people believe; it simply helps, under certain conditions, to add to a crop's available nitrogen supply. Inoculation is only one of several factors necessary to the successful growth of leguminous crops; practically all other conditions essential for the growth of any crop are required.

Inoculation may be expected to give good results on soils well aërated and abundantly supplied with available phosphorus, potassium, calcium and other mineral compounds: (*a*) Where leguminous crops do not grow well, (*b*) where the roots of such crops do not have nodules, or (*c*) where a leguminous crop is to be grown for the first time.

It usually fails to produce results: (*a*) Where the leguminous crop inoculated flourishes already, (*b*) where the soil is strongly acid, (*c*) where the supply of available potassium and phosphorus compounds is deficient, (*d*) where the soil is supplied with too much available nitrogen, (*e*) where the physical conditions of the soil are unsatisfactory in respect to texture, aëration, drainage, etc., or (*f*) where seasonal conditions are not favorable.

Sulfofication.—Vegetable and animal matter contains compounds of sulfur. When such material undergoes bacterial decomposition, the sulfur is converted from its organic combination, first, into hydrogen sulfide (H_2S) and free sulfur (S); second, these products are further changed by other bacteria (in some cases by fungi) into sulfurous acid (H_2SO_2) and sulphuric acid (H_2SO_4). This second change is a process of oxidation (p. 153) known as *sulfofication*. Chemically, the change is of the same type as takes place in nitrification, when nitrogen combines with oxygen to form nitrous acid (HNO_2) and nitric acid (HNO_3). Sulfuric acid, whenever formed, may combine with some of the soil constituents to form sulfates of calcium, magnesium, potassium, etc. In some cases the sulfur in calcium sulfate (gypsum) is changed into hydrogen sulfide gas by anaërobic bacteria.

CHANGES IN INORGANIC CONSTITUENTS BY SOIL BACTERIA

Through direct action of bacteria or through the action of their products, extensive chemical changes occur in the inorganic constituents of soils, as well as in fertilizers applied to soils. Attention has already been called to the fact that acids of various kinds are prominent among the products of the bacteria that cause decomposition, nitrification, sulfofication, etc. The most common acid product is carbon dioxide gas, others are nitrous, nitric, sulfuric acids, and numerous organic acids. Carbon dioxide when dissolved in water, and all free acids formed by bacterial action bring about the following changes: (1) Calcium and other insoluble carbonates are made soluble. (2) Insoluble phosphates are made soluble, whether native or applied as ground rock (floats), ground bone, slag phosphate, farm manure, ground fish, tankage, tobacco stems, green-crop manure, etc. (3) In some cases, bacteria use soluble phosphorus compounds to build their bodies and thus convert insoluble organic forms into soluble ones by their death. (4) Insoluble potassium compounds are made soluble.

These statements serve to show that extensive and far-reaching chemical changes are wrought by bacterial action. It is obvious that the more favorable the soil conditions are for the growth the more intense will be the activity of bacteria and the greater the changes brought about.

ACTINOMYCES

Actinomyces are thread-like micro-organisms which resemble bacteria in some respects and fungi in others. They are abundant in the soil, constituting on an average, about one-fifth of the micro-organisms present (Conn). They are usually more abundant in meadows and pastures than in tilled soils. Actinomyces are prominent in the work of decomposing both nitrogenous and non-nitrogenous organic compounds in the soil. They are more effective than many other organisms in decomposing resistant materials such as lignin.

RELATIONS OF FUNGI TO SOILS

The chlorophyl-free microscopic plants known as fungi, commonly called molds, are plants considerably larger and more complex in structure than bacteria. Fungi are of interest in connection with soils on account of their relations to processes of decomposition.

Many soil fungi assist in decomposing organic matter, especially in the early stages of decomposition and are particularly active in acid soils in which many bacteria cannot work. They can convert insoluble into soluble nitrogen compounds and make these into the insoluble protein of their own bodies and so temporarily reduce the amount of available nitrogen in soils, but they give it back again when they die and undergo bacterial decomposition. The decomposition of carbohydrates, including cellulose, is carried on vigorously by fungi. When farm manure is allowed to become loose and dry, the action of bacteria is replaced by that of fungi, an undesirable condition commonly known as "fire-fanged" (p. 232).

RELATIONS OF PROTOZOA TO SOILS

Protozoa, the simplest forms of animal life, consist of single cells, but much larger than and differing from bacteria in certain fundamental ways. What part protozoa have in activities that take place in the soil is not definitely known. It has been suggested that some kinds feed upon bacteria that produce ammonia, diminishing the amount of this nutrient to such an extent that crop growth is appreciably lessened.

PART III

SOURCES AND COMPOSITION OF MATERIALS USED AS FERTILIZERS

XIII

CLASSIFICATION AND DEFINITION OF PLANT-FOOD MATERIALS

THE terms employed to describe and classify fertilizers and fertilizing materials are often used loosely and without discrimination. For example, we find the words "fertilizer" and "phosphate," used with the same meaning. Many use the terms "fertilizer" and "manure" with different meanings, restricting the former to commercial plant-food materials and applying the latter exclusively to farm manure. By many the expressions, "plant-food" and "fertilizer" are always used as synonymous. It is desirable that the use of these terms should be reduced to some kind of system and definiteness. We shall attempt to define them as clearly and sharply as possible in accordance with what we regard as the best usage.

A fertilizer *is any substance, which under favorable conditions when added to a soil, will produce a better growth of crops, whether through direct or indirect action on the crop or on the properties of the soil.*

By reference to the definition of a plant-food (p. 5), it will be seen that the terms, plant-food and fertilizer, do not necessarily mean the same. Thus, all fertilizers are not plant-foods; indirect fertilizers which are added to soils to modify physical or chemical conditions are not necessarily plant-foods; for example, a large part of the material we add to soils in farm manures and green-crop manure is not used as plant-food at all, but it is useful in supplying the soil with organic matter, which performs important functions. (pp. 97-99).

The materials which come under the head of fertilizers are numerous in kind, different in source, forms and the manner in which they act.

CLASSIFICATION OF FERTILIZING MATERIALS

It is impossible to arrange materials used as fertilizers in sharply divided classes. In any system that can be devised there are unavoidably cases where some materials belong to more than one

TABLE 21—CLASSIFICATION OF FERTILIZING MATERIALS

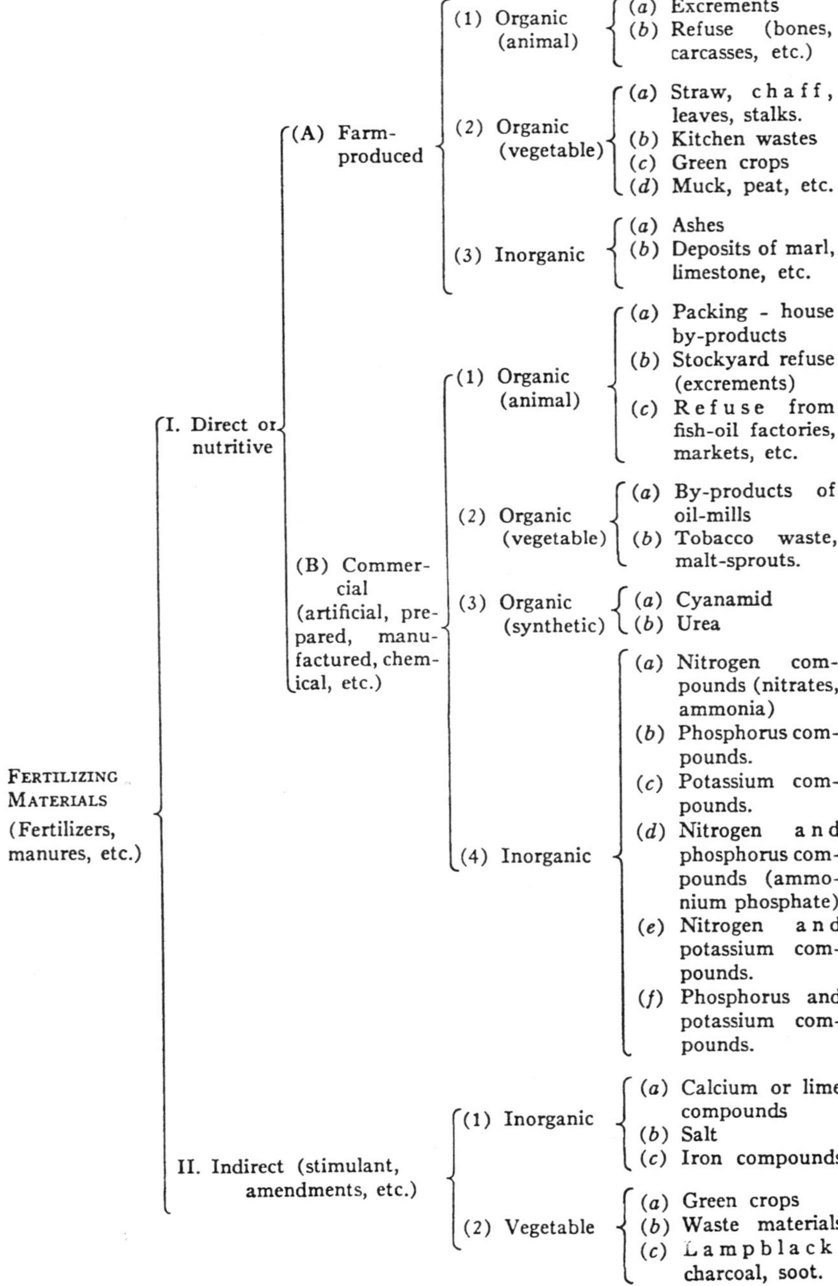

FERTILIZING MATERIALS (Fertilizers, manures, etc.)

- I. Direct or nutritive
  - (A) Farm-produced
    - (1) Organic (animal)
      - (a) Excrements
      - (b) Refuse (bones, carcasses, etc.)
    - (2) Organic (vegetable)
      - (a) Straw, chaff, leaves, stalks.
      - (b) Kitchen wastes
      - (c) Green crops
      - (d) Muck, peat, etc.
    - (3) Inorganic
      - (a) Ashes
      - (b) Deposits of marl, limestone, etc.
  - (B) Commercial (artificial, prepared, manufactured, chemical, etc.)
    - (1) Organic (animal)
      - (a) Packing - house by-products
      - (b) Stockyard refuse (excrements)
      - (c) Refuse from fish-oil factories, markets, etc.
    - (2) Organic (vegetable)
      - (a) By-products of oil-mills
      - (b) Tobacco waste, malt-sprouts.
    - (3) Organic (synthetic)
      - (a) Cyanamid
      - (b) Urea
    - (4) Inorganic
      - (a) Nitrogen compounds (nitrates, ammonia)
      - (b) Phosphorus compounds.
      - (c) Potassium compounds.
      - (d) Nitrogen and phosphorus compounds (ammonium phosphate)
      - (e) Nitrogen and potassium compounds.
      - (f) Phosphorus and potassium compounds.
- II. Indirect (stimulant, amendments, etc.)
  - (1) Inorganic
    - (a) Calcium or lime compounds
    - (b) Salt
    - (c) Iron compounds
  - (2) Vegetable
    - (a) Green crops
    - (b) Waste materials
    - (c) Lampblack, charcoal, soot.

division. The tabulated classification (Table 21), while not free from this objection, and also otherwise imperfect, will be found helpful in giving a fairly clear, general idea of the number and relations of the various terms used in speaking of fertilizers. At the same time, it furnishes in outline a suggestive though incomplete statement of the general sources of fertilizing materials.

A direct or nutritive fertilizer *is one that contains essential plant-food or nutrient elements available (p. 6) at once or that are easily made so under favorable soil conditions.* The plant-food materials are in such forms or compounds that they can be immediately absorbed and used by plants, or else change into such forms in time

Partial view of experimental plots at the Nebraska Experiment Station.

to be used by crops during the growing season. Direct fertilizers dissolve immediately in soil water, or readily change in the soil under ordinary conditions into such soluble form, and are therefore in condition to be taken up by plants either immediately or within a short time. For example, meat tankage contains organic nitrogen largely in an insoluble form, which plants cannot use, but, under normal soil conditions, this insoluble nitrogen undergoes bacterial decomposition, ammonification and nitrification, sooner or later appearing in the quickly available form of nitrate nitrogen (p. 139).

An indirect or stimulant fertilizer (known also as a *soil amendment) is one which favorably influences plant growth, not by furnishing any needed element of plant-food, but chiefly by producing in the soil some effect beneficial to plant growth.* Among such beneficial effects are: (1) The conversion of unavailable into available plant-food in the soil; (2) changing the physical conditions of the soil so that favorable factors of structure (p. 76) are substituted for unfavorable ones, as illustrated by the flocculation (p. 78) of puddled clay by calcium compounds; (3) neutralizing acids and other injurious substances in soils (p. 200); (4) making neutral or alkaline soils acid and therefore favorable to certain plants such as rhododendron and blueberry.

Indirect fertilizers may not contain any essential element used in plant growth, or they may contain essential plant-food elements already in the soil in sufficiently large amounts to supply the needs of crops indefinitely. For example, common salt (sodium chloride), when used, is scarcely ever applied because crops are in actual need of sodium or chlorine. The various forms of lime are usually applied, not because soils contain an insufficient amount to furnish crops the needed calcium but to make acid soils neutral or prevent soil acidity, and perform other functions beneficial to crop growth. In a similar way, green-crop manures give beneficial effects often much more largely from their influence upon the mechanical and chemical factors of the soil (making conditions more favorable for plant growth) than from the amounts of nitrogen, potassium and phosphorus compounds they supply.

Farm-produced fertilizers include all materials commonly obtained on farms. They include (1) solid and liquid excrements of farm animals, (2) materials used for bedding, (3) unused portions of crops, including crop residues left in the soil, (4) all kinds of animal and vegetable waste, such as kitchen garbage, carcasses of animals, bones, human excrements, etc., (5) green-crop manures, (6) ashes, (7) muck and marl (found on many farms).

Commercial fertilizers are preparations or mixtures of plant-food materials, commonly sold under special trade-names. They are also known as *artificial, prepared* or *manufactured*. When they consist only of commercial chemical compounds they are known as *chemical* fertilizers or manures. The term "chemicals" is often applied to such materials as sodium nitrate, ammonium sulfate, and salts of potassium.

The materials used in making commercial fertilizers include substances found in, or prepared from, natural deposits, synthetic products and, in addition, by-products of numerous industries, obtainable only through the channels of trade. While commercial fertilizers usually contain numerous constituent elements, their chief value depends on the presence of only three forms of plant-food, *nitrogen, phosphorus* and *potassium* compounds. Their agricultural value varies in accordance with the forms and amounts of these plant-food compounds. Commercial fertilizers are known as *complete* and *incomplete*.

Complete fertilizers, also called *general* fertilizers, are those that contain the three plant-food constituents, nitrogen, phosphorus and potassium compounds.

Incomplete fertilizers, known also as *special* fertilizers, contain only one or two of the three important plant-food constituents.

Plan of treatment.—In the remaining chapters of Part III we shall discuss the sources and general character of the materials used as fertilizers, considering them in the following order:

(1) Commercial fertilizing materials that contain compounds of
- (*a*) Nitrogen,
- (*b*) Phosphorus,
- (*c*) Potassium.

(2) Farm-produced fertilizing materials
- (*a*) Farm manure,
- (*b*) Green-crop manure.

(3) Indirect fertilizers:
- (*a*) Calcium or lime compounds,
- (*b*) Miscellaneous materials.

In treating the subjects of farm manure, green-crop manure and indirect fertilizers, it has seemed better to consider the practical applications along with the more general features of interest instead of postponing it to later chapters.

XIV

MATERIALS THAT CONTAIN NITROGEN

THE general character and relations of the nitrogen-containing materials used as fertilizers will be discussed in the order indicated by the following outline:

1. **Processes of manufacture:** (1) The nitric oxide or arc process, (2) the cyanamid process, (3) the synthetic ammonia process, (4) relation of synthetic nitrogen compounds to plant-food, (5) efficiency of production by different processes.

2. **Inorganic compounds:** (1) Sodium nitrate, (2) sodium-potassium nitrate, (3) potassium nitrate, (4) calcium nitrate, (5) ammonium nitrate, (6) ammonium sulfate, (7) ammonium sulfate-nitrate ("Leunasaltpeter"), (8) ammonium chloride, (9) ammonium nitrate and calcium carbonate ("Calnitro"), (10) ammonium phosphate, (11) ammonium nitrate, ammonium sulfate and ammonium phosphate ("Leunaphos").

3. **Organic compounds:** (1) Calcium cyanamid, (2) urea, (3) "Calurea."

4. **Vegetable materials:** (1) Cottonseed-meal, (2) castor-bean pomace, (3) linseed-meal, (4) vegetable pomaces.

5. **Animal materials:** (1) Dried blood, (2) dried meat, (3) tankage, (4) dried fish, (5) hoof-meal and horn-dust, (6) garbage-tankage, (7) sewage-sludge, (8) commercial dried manures, (9) nitrogenous guanos, (10) hair and leather, (11) wool and silk waste.

The nitrogen-containing materials used as fertilizers are known commercially as "ammoniates," an inaccurate term based originally on the erroneous supposition that the nitrogen in all such materials is present as ammonia. Some of the more important facts of general interest in relation to these materials follow.

MANUFACTURED COMPOUNDS

At first the materials in this class came from natural deposits and from the by-products of some manufacturing operations. At present they are, and will continue to be in increasing amounts, the products of direct synthetic processes of manufacture from atmospheric nitrogen.

Processes used in producing synthetic compounds from air nitrogen.—Among the several processes developed for utilizing air nitrogen in the production of nitrogenous compounds are (1) the nitric acid arc process, (2) the cyanamid process, and (3) the direct synthetic ammonia process.

(1) *The Nitric Acid Arc Process* was first developed commercially in Norway by Birkeland and Eyde about 1903. Briefly stated, the general process consists of the following steps: (1) Ordinary air is passed through a powerful electric arc or flame, which combines more or less of the oxygen and nitrogen in the form of nitric oxide ($N_2 + O_2 = 2NO$), (2) this nitric oxide, on exposure to air, combines with more oxygen to form nitrogen dioxide ($2NO + O_2 = 2NO_2$). (3) Though there are some intermediate steps, which need not be considered here, before all of the NO_2 is converted into HNO_3, treatment with water converts this last oxide into dilute nitric acid ($3NO_2 + H_2O = 2HNO_3 + NO$). (4) The last step is to concentrate the nitric acid.

The nitric acid is used in the preparation of the nitrate compounds, calcium nitrate, sodium nitrate, potassium nitrate and ammonium nitrate.

(2) *The Cyanamid Process,* devised by two German chemists, Frank and Caro, was put into operation commercially in 1906.

The two principal materials used in the preparation of cyanamid are calcium carbide (the substance familiar in the production of acetylene gas) and pure nitrogen obtained by liquefaction of air and distillation. The carbide, in the condition of a fine powder, is placed in a heated retort, into which the nitrogen is passed. The cyanamid produced is removed from the retort in the form of a hard cake, which is cooled out of contact with the air, then crushed to a powder and lightly sprayed with water and oil before packing. The chemical change that takes place in the manufacturing operation may be represented as follows:

$$\underset{\text{Calcium carbide}}{CaC_2} + N_2 = \underset{\text{Calcium cyanamid}}{CaCN_2} + \underset{\text{Carbon}}{C}$$

(3) *The Synthetic Ammonia Process,* known in its first form as the Haber-Bosch or German process, was first operated commercially in 1913. Several modifications of the process have been developed in various countries, but the fundamental reaction in all is essentially the same, as expressed in the equation,

$$\underset{\text{Nitrogen}}{N_2} + \underset{\text{Hydrogen}}{3H_2} = \underset{\text{Ammonia}}{2NH_3}$$

A mixture of hydrogen and nitrogen is passed under pressure (1500 to 15,000 pounds, per square inch) at a temperature of 400° to 700° C. over some special substance like reduced iron oxide, which promotes the chemical combination of the two gases. The ammonia formed is condensed as liquid ammonia or dissolved in water to form commercial aqua ammonia (ammonium hydroxide, NH_4OH).

Relation of Synthetic Nitrogen Compounds to Plant-Food.—The three compounds produced by fixation of air nitrogen, cyanamid, nitric acid and ammonia, may be regarded as fundamental compounds in relation to the commercial supply of nitrogen—containing plant-food. Calcium cyanamid, when treated with steam under pressure, forms ammonia, as shown by the equation,

$$\underset{\text{Calcium cyanamid}}{CaCN_2} + \underset{\text{Water}}{3H_2O} = \underset{\text{Calcium carbonate}}{CaCO_3} + \underset{\text{Ammonia}}{2NH_3}$$

Ammonia, whether formed from cyanamid or by direct synthesis, can be converted into nitric acid by passing it over some special substance promoting the reaction in the presence of oxygen, as expressed by the equation,

$$\underset{\text{Ammonia}}{NH_3} + \underset{\text{Oxygen}}{2O_2} = \underset{\text{Nitric acid}}{HNO_3} + \underset{\text{Water}}{H_2O}$$

Ammonia is used to form all the ammonium compounds (sulfate, nitrate, phosphate and chloride) used in fertilizers. Nitric acid is employed in the production of nitrates, such as calcium, sodium, ammonium and potassium.

Either cyanamid or ammonia may be used in the production of urea.

Efficiency of production by different processes.—In the processes by which air nitrogen is converted into different nitrogen-containing compounds, electric power is required. The synthetic-ammonia process requires least power, the cyanamid process uses about three times as much and the arc (nitric acid) process about fifteen times as much. At present the largest production comes from the synthetic-ammonia process and the least from the arc process. Production by the arc process is mostly confined to Norway. Plants for the production of cyanamid and synthetic ammonia are found in America, Japan and several countries of Europe. At present there are over forty cyanamid and over sixty synthetic ammonia manufactories. The production capacity of world increased from 1,500,000 tons in 1927-8 to 3,500,000 in 1930. The world consumption for 1931 has been estimated at 2,000,000 tons.

Sodium nitrate (commercial *nitrate of soda*) is obtained from two sources of supply, deposits in Chile and manufacture from air nitrogen. Chilean nitrate, commonly known as *Chile saltpeter;* is found in immense beds of a rock-like deposit, called *caliche,* mostly 2 to 10 feet below the surface. In this the sodium nitrate is present in varying proportions, mixed with numerous other compounds. The nitrate is separated from the other materials by dissolving in water and crystallizing, after which the product is drained, dried and put in bags, ready for market. Improved processes of preparation have been recently introduced, lessening the cost.

Sodium nitrate is being manufactured from air nitrogen in increasing amounts in the United States. The Chilean nitrate contains nitrogen varying from 15 to 16 per cent., as compared with 16.47 per cent. in pure sodium nitrate ($NaNO_3$, p. 31). The manufactured nitrate contains 16 per cent. or more of nitrogen. While sodium nitrate tends to absorb moisture from the air and become sticky, it can be treated so that, under ordinary conditions, it remains dry and granular, resembling granulated sugar in appearance. In this physical condition the grains remain separate and run freely when applied as a fertilizer by hand or machine.

Sodium-Potassium nitrate, sold under the trade name *Nitra-Po,* is prepared from some of the crude Chilean caliche containing potassium nitrate in considerable amount. The refined product is a mixture of sodium nitrate and potassium nitrate which contains about 14.8 per cent. of nitrogen and 10.8 per cent. of potassium (equal to 13 per cent. of potash).

Potassium nitrate (KNO_3, p. 31) is known under such names as nitrate of potash, niter and saltpeter. Its use in fertilizers has been limited on account of the demands for various manufacturing purposes, especially gunpowder. With the decreased cost of nitric acid, as a result of the synthetic process of manufacture from air nitrogen, potassium nitrate is more extensively manufactured and used in fertilizers. It can be made by treating potassium chloride with nitric acid. It is a valuable plant-food, furnishing both nitrogen and potassium. The product sold as fertilizer contains about 13 per cent. of nitrogen and 44 per cent. of potash (equal to 36.5 potassium), while pure potassium nitrate contains 13.85 per cent. of nitrogen and 46.6 per cent of potash (equal to 38.7 per cent. of potassium).

Calcium nitrate ($Ca(NO_3)_2$, p. 31), in the commercial form, known also as Norwegian saltpeter, lime nitrate, nitrate of lime, basic calcium or lime nitrate, etc., is prepared by combining nitric acid, prepared by the use of air nitrogen, with calcium oxide (lime) or carbonate. Chemically pure calcium nitrate contains about 17 per cent. of nitrogen and 24.4 per cent of calcium; the commercial calcium nitrate used as a fertilizer in the United States contains over 15 per cent. of nitrogen and about 20 per cent. of calcium. Calcium nitrate attracts moisture rapidly and dissolves. The commercial calcium nitrate used as a fertilizer is a white salt, granular in texture, somewhat coarser than granulated sugar, running freely when handled.

Ammonium nitrate (NH_4NO_3, p. 31), when pure, contains 35 per cent. of nitrogen, one-half in the form of nitrate and one-half in the form of ammonia. It is made by combining ammonia and nitric acid (and produced from air nitrogen), the ammonia being absorbed in 30 to 50 per cent. nitric acid and the resulting solution concentrated ($NH_3 + HNO_3 = NH_4NO_3$). Ammonium nitrate has the advantage over other inorganic nitrogen-containing compounds of furnishing the most concentrated form of plant-food nitrogen. It has the disadvantage of absorbing moisture from the air rapidly, but this can be largely prevented by mixing a saturated solution of ammonium nitrate with powdered limestone or precipitated calcium carbonate and spraying the mixture into a granulating chamber. This mixture contains 20.5 per cent. of nitrogen and about 35 per cent. of calcium carbonate. In America it is known as *Ammonium Nitrate Limestone;* in Germany, as *Lead-Nitro,* and in England, as *Nitro-Chalk.* The German product is colored green for ready identification.

Ammonium sulfate or sulfate of ammonia ($(NH_4)_2SO_4$, p. 31) contains, when pure, 21.2 per cent. of nitrogen (equal to 25.75 per cent. of ammonia, NH_3). The commercial product used in fertilizers contains about 20.5 per cent. of nitrogen (equal to 25 per cent. of ammonia).

The ammonium sulfate of commerce was originally produced as a by-product in the manufacture of illuminating gas and coke from coal, one ton of coal yielding 8 to 25 pounds of ammonium sulfate, containing 1.7 to 5.3 pounds of nitrogen.

Ammonium sulfate is being made in increasing amounts by absorbing ammonia in sulfuric acid ($2NH_3 + H_2SO_4 = (NH_4)_2SO_4$).

Ammonium sulfate used in fertilizers is a grayish-white substance, consisting of small needle-shaped crystals. It does not absorb moisture readily. In physical condition, it is free-running and well adapted to use in sowing by hand or machinery.

Ammonium sulfate-nitrate, commercially known as *Leuna-saltpeter,* consists of ammonium sulfate and ammonium nitrate, containing about 26 per cent. of nitrogen. About three-fourths of the nitrogen is present as ammonia and one-fourth as nitrate. The material absorbs moisture from the air, but much less readily than untreated ammonium nitrate does.

Potassium and ammonium nitrate, made in Germany and sold under the trade-name, *Kali-ammon-saltpeter*, contains about 16 per cent. of nitrogen, half as nitrate and half as ammonia nitrogen, with about 28 per cent. of potash. This is equal to 23 per cent. of K.

Ammonium Chloride (NH_4Cl, p. 26) is formed by the reaction of hydrochloric acid and ammonia ($HCl + NH_3 = NH_4Cl$). Pure ammonium chloride contains 26.2 per cent. of nitrogen (equal to 31.8 ammonia). It absorbs moisture easily. The commercial product contains about 24 per cent of nitrogen.

Leunaphos is the trade name of a mixture of ammonium nitrate, ammonium sulfate and ammonium phosphate, containing about 25 per cent. of ammonia (equal to 20.5 nitrogen) and 15 per cent. of phosphoric acid, P_2O_5, (equal to 6.6 phosphorus).

Leunaphoska, Nitrophoska, Ammo-Phos-Ka (see p. 180).

Calcium Cyanamid ($CaCN_2$, p. 32), sometimes called nitro-lime and lime-nitrogen, is a synthetic, non-protein organic compound. When chemically pure it contains 35 per cent. of nitrogen, but the commercial product is not pure; its percentage of nitrogen averages about 21. It usually contains as impurities over 25 per cent. of calcium hydroxide (p. 294), 12 per cent. of carbon, some calcium carbonate and other compounds and it is alkaline in reaction. It is a fine grayish-black powder treated with oil in order to prevent being dusty and disagreeable to handle during application. It does not absorb moisture, for which reason it is well adapted to distribution by machine.

When calcium cyanamid is applied to soils, its nitrogen is not directly used as cyanamid nitrogen, but the compound undergoes gradual change in the presence of soil moisture, forming ammonia and calcium carbonate, as follows:

$$\underset{\text{Calcium cyanamid}}{CaCN_2} + \underset{\text{Water}}{3H_2O} = \underset{\text{Calcium carbonate}}{CaCO_3} + \underset{\text{Ammonia}}{2NH_3}$$

Urea (CON_2H_4) is a non-protein organic compound, one of the normal constituents of the urine of mammals. It is one of the commercial products of synthetic ammonia manufacturing plants for use in fertilizers. It can be made directly from cyanamid or from ammonia and carbon dioxide. Urea has the highest percentage of nitrogen in any of the nitrogen-containing compounds used in fertilizers, about 46 per cent. (equal to nearly 56 per cent. of ammonia). It absorbs moisture quickly. Urea is a white, crystalline material, resembling granulated sugar in appearance, and is readily soluble in water.

Calurea is a mixture of urea and calcium nitrate, containing 34 per cent. of nitrogen, about four-fifths of the nitrogen coming from urea and one-fifth from nitrate. It contains about 9 per cent. of calcium. It is a white, crystalline material in a granular condition that enables it to be distributed easily.

VEGETABLE MATERIALS THAT CONTAIN NITROGEN

The materials in this class are the seed residues obtained in oil factories and are limited in amount by the yield of the crops from which the seed comes.

Cottonseed-meal is the ground residue from cottonseed, after the oil is extracted. It has been extensively used as a fertilizer, especially in the South. When of good quality, it contains 6.5 to 7 per cent. of nitrogen and 3 per cent. of phosphoric acid (equal to 1.3 phosphorus) and 2 per cent. of potash (equal to 1.65 potassium). When mixed with cottonseed-hulls, these percentages are lower.

In addition to its plant-food constituents, it is valuable in fertilizers, because when used in sufficient quantity it produces a desirable mechanical effect in a mixed fertilizer, absorbs moisture, prevents caking and insures a loose condition of particles. When used as a food for cattle, for which it is highly valued, a large proportion of the plant-food constituents is recovered in the manure, thus making this the most economical way of utilizing it as a fertilizer. When applied directly to the soil, it is a valuable fertilizer, but cannot be generally used in this way profitably when it costs over $25 a ton.

Castor-bean pomace, a by-product of castor-oil factories, is chiefly used for fertilizing purposes. It is generally less rich in plant-food constituents than cottonseed-meal, containing nitrogen in percentages usually varying between 4.75 and 5.25 and averaging near 5 per cent. along with 2 per cent. of phosphoric acid (equal to 0.9 per cent. phosphorus) and 1 per cent. of potash.

Linseed-meal, the residue left after extracting oil from flaxseed, is most extensively used as cattle-food. It is too high-priced for common use as a source of nitrogen in commercial fertilizers. It contains about 5.5 per cent. of nitrogen, together with nearly 2 per cent. of phosphoric acid (equal to 0.9 per cent. of phosphorus) and 1.5 per cent. of potash (equal to 1.35 per cent. of potassium).

Tobacco-Waste.—See p. 190.

Other vegetable nitrogen-containing meals obtained in limited amounts are soy bean meal, the ground cake of the soy bean;

copra meal from cocoanut; rape meal from rape seed; cocoa cake, the material remaining after pressing out the oil of dried and powdered seed kernels of the cocoa or chocolate tree in the manufacture of cocoa-butter; various vegetable or fruit pomaces, such as apple, tomato, pumpkin, etc.

ANIMAL NITROGENOUS MATERIALS

The materials in this class come mostly from the meat-packing industries; they are animal by-products more or less rich in nitrogen and are known in general terms as blood, tankage and offal. In the United States much of the nitrogen used in commercial fertilizers formerly came from this source.

Dried blood used in fertilizers is prepared for market by evaporating, drying and grinding. The color varies from red to black, according to the temperature and method employed in drying. Red dried blood is of higher grade than black, contains more nitrogen, is more uniform in composition and has only a small amount of phosphorus compounds. It usually contains 13 to 15 per cent. of nitrogen. Black dried blood is often mixed with hair, dirt and various kinds of animal refuse and therefore varies greatly in composition, the percentage of nitrogen ranging from 6 to 12. It also contains more or less phosphoric acid, sometimes as much as 4 per cent. (equal to 1.75 per cent. of phosphorus). Black dried blood is somewhat tough or leather-like in texture and decays in the soil less rapidly than the red. This grade is most extensively used in fertilizers.

Dried meat is known under various trade-names, such as animal matter, ammonite, azotin, meat-meal, etc. It consists of the dried, ground, inedible soft portions of animal carcasses, such as tendons, membranes and other portions of meat, from which fat has been extracted, and also of by-products from beef-extract manufacture. It comes mainly from slaughter-houses and rendering-plants and varies considerably in composition, usually containing 10 per cent. of nitrogen and often as much as 13 or 14 per cent. It also usually contains 3 to 4 per cent. of phosphoric acid (equal to 1.30 to 1.75 per cent. of phosphorus).

Tankage consists of a mixture of miscellaneous slaughter-house refuse unavailable for other commercial purposes—bone, flesh, tendons, intestines, lungs, blood, hair, hides, horns, hoofs, etc. The mixed materials are treated with gasoline or cooked in tanks under pressure to remove fat, after which they are dried and ground. Tankage varies greatly in composition because the proportions of the various materials used vary greatly. When bone materials are in excess, it is usually called bone-tankage; and when the softer, more highly nitrogenous materials predominate, it is often called meat-tankage. In the case of small slaughter-houses, the tankage varies much more in every way than that prepared by large packing-houses and rendering plants.

Two general varieties of tankage are recognized, nitrogenous or concentrated tankage and crushed or mixed tankage. In the first kind, the waste water or liquid that results from the cooking of bones and meat refuse is evaporated. It makes a product high in nitrogen, often over 12 per cent., but with little phosphorus. This product, which by itself is usually sticky and difficult to handle, may be ground and mixed with dry, inert materials to give it a good mechanical texture, or it may be used in mixing with ordinary tankage to increase the percentage of nitrogen. In mixed or crushed tankage, the composition varies greatly, the nitrogen ranging from 4 to 10 per cent. and the phosphoric acid from 7 to 12 per cent. (equal to 3 to 5 per cent. of phosphorus). Owing to the extreme variations in composition, tankage should be bought only on a guaranteed percentage of nitrogen and phosphoric acid.

Processed tankage is the name applied to a product made from inert nitrogenous materials by treatment under steam pressure, with or without the use of sulfuric acid, to increase the availability of nitrogen as plant-food. The process is used with such materials as horns, hoofs, leather, etc.

Base-Goods Ammonia is the name applied to the nitrogen which comes from such waste products as hair, wool waste, feathers, all kinds of leather waste, garbage tankage, peat and any other nitrogen-containing materials, which have been mixed with phosphate rock and sulfuric acid, in the process of making superphosphate, and allowed to react in a closed place (den). The materials lose their identity completely and the nitrogen is converted into readily-available form.

Dried ground fish and acidulated fish (sometimes called fish-guano) consist of by-products coming mainly from fish-oil works, fish-canning factories, fish-salting plants, etc. The material obtained from the manufacture of fish-oil is the residue left after extracting the oil, a process which removes from the fish little of value as fertilizer. The fresh fish are cooked by steam, then pressed to remove the oil, and dried either in the air or by steam, after which the mass is ground. The scrap is sometimes sprinkled with dilute sulfuric acid (oil of vitriol) by which decomposition is checked, the bones softened and to some extent dissolved. Dried fish varies in composition, containing 8 to 9.5 per cent. of nitrogen with an average of about 8.80 and 5 to 9 per cent. of phosphoric acid (equal to 2.2 to 4 phosphorus). The product coming from canning-works consists more largely of skin, bones, etc., and is therefore more variable in composition, containing usually more phosphorus and less nitrogen. Acidulated dry fish usually contains slightly smaller percentages of nitrogen and phosphorus than the untreated fish.

Hoof-meal and horn-dust, by-products of slaughter-houses. They contain 10 to 15 per cent. of nitrogen and about 2 per cent. of phosphoric acid (equal to 0.9 per cent. of phosphorus). These materials are usually processed to make the nitrogen more available.

Garbage-tankage is material obtained by drying city garbage, which consists of mixed animal and vegetable refuse. Sometimes it is prepared for use as fertilizer by charring, but usually the grease is extracted and the residue dried and ground. Its composition varies greatly. In some cases the tankage is treated with dilute sulfuric acid, then mixed with calcium phosphate and potash salts to produce a complete commercial fertilizer. Garbage-tankage may contain 2.5 to 3 per cent. of nitrogen, 1.50 to 3 per cent. of phosphoric acid (equal to 0.65 to 1.3 per cent. of phosphorus) and 0.75 to 1.50 per cent. of potash (equal to 0.6 to 1.25 per cent. of potassium). This material is regarded as low-grade plant-food. In both composition and availability, its value is variable, so it should be purchased only on a guarantee of composition.

Sewage Sludge is a product of sewage-disposal plants, resulting from the treatment of sewage by bacterial and chemical processes. The dried product from some municipal plants is sold as a fertilizer. It may contain as much as 5 per cent. of nitrogen and 3 per cent. of phosphoric acid.

TABLE 22—COMPOSITION OF COMMERCIAL PULVERIZED ANIMAL MANURES

	Sheep *Per cent.*	*Goat* *Per cent.*	*Pig* *Per cent.*	*Cattle* *Per cent.*	*Poultry* *Per cent.*
Nitrogen	2.50	1.35	1.75	1.34	4.50
Phosphoric acid ...	1.50	1.40	1.75	0.90	3.20
Potash	1.50	3.60	1.00	0.85	1.35

Commercial manures from animal excrements, dried and ground, consist largely of manure obtained at stock-yards, with some admixture of dirt and refuse. They comprise cattle, pig, sheep, goat and poultry manures. In some cases, the potash is increased by addition of potassium salts. They should be purchased only on a guaranteed analysis. Table 22 shows about how they run in composition, the average and variations being given in case of sheep and pig manures.

Nitrogenous guanos are formed in warm, dry regions from the excrements and bodies of dead birds, fragments of fish, feathers, seaweed, etc. The birds live largely on fish so the resulting excrements are rich in nitrogen. Peruvian guano, deposited on groups of islands off the coast of Peru, was once used extensively and still comes into market in limited amounts. It contains about 10.5 per cent. of nitrogen and 10 per cent. of phosphoric acid (equal to 4.4 per cent. of phosphorus).

Hair and leather.—Hair is often found mixed with dried blood and tankage. It is rich in nitrogen—about 15 per cent.—decomposes with extreme slowness so may be regarded as practically useless for feeding crops. The same is true of fur-waste and feathers.

Leather in the form of scraps and ground leather is a waste product of various industries. When treated with superheated steam, dried or partially roasted, ground fine, it is then called *leather-meal.* Roasted leather, finely ground, has been used to adulterate dried blood, which it resembles in color. Leather contains 7 to 8 per cent. of nitrogen. It undergoes decomposition in soils with extreme slowness.

Wool and silk wastes include by-products of industries which are wool and silk, such as manufacture of woolen cloth, carpet-making, felt materials, silk goods, etc. Pure silk and wool contain about 15 per cent. of nitrogen, but the waste materials are usually mixed with so much cotton or other materials that the percentage of nitrogen may be reduced as low as 5.

Such materials as horns, hoofs, hair, leather, wool, silk, etc., contain nitrogen in a form which obstinately resists the action of micro-organisms and which therefore is practically unavailable as plant-food during the growing season of a crop. However, such materials are utilized and their nitrogen made available by mixing them with rock phosphate and sulfuric acid when superphosphate is made.

XV

MATERIALS THAT CONTAIN PHOSPHORUS

THE materials that furnish the phosphorus compounds used in fertilizers are, for the most part, phosphates of calcium (lime). (pp. 32-37.) Most of the phosphates used in America come from three sources, (1) mineral or rock phosphates, (2) basic slag and (3) bones of slaughtered animals, the two latter furnishing, however, only a small proportion compared with the first. The manufactured product called *superphosphate* is the form which, whatever the source of the original material, is most extensively used in commercial fertilizers. Rock phosphate when ground so fine as to be dust-like is called *"floats"*; it is used in comparatively small amounts for direct application to soil.

We will, for convenience, study the phosphorus-containing materials in accordance with the following outline:

1. *Inorganic materials:* (1) Rock-phosphates, (2) apatite, (3) guano phosphates.

2. *Manufactured products:* (1) Superphosphate (acid phosphate) (*a*) plain, (*b*) concentrated, (*c*) nitrate-superphosphate (2) potassium phosphate, (3) "Nitrophoska," (4) "Leunaphoska," (5) Ammo-Phos-Ko.

INORGANIC PHOSPHORUS-CONTAINING MATERIALS

By far the largest part of the phosphorus compounds used as fertilizers comes from inorganic materials.

Rock-phosphates are commonly known as Florida, Tennessee, South Carolina rock or phosphate because these states have been the main sources of supply. Under the term of rock phosphates we include those deposits which consist more or less largely of tri-calcium phosphate combined with calcium fluoride ($3Ca_3(PO_4)_2.CaF_2$). These are found in many different places, varying in form, condition and purity. They contain tri-calcium phosphate varying in amount from below 45 to 90 per cent. (equal to 20 to 40 per cent. phosphoric acid (P_2O_5) or 9 to 18 phosphorus), the usual proportions are between 55 and 77 per cent. tri-calcium phosphate (equal to 25 and 35 phosphoric acid or 11 and 15.5 phosphorus). The materials mixed as impurities with rock-phosphate are usually and mostly sand, clay and limestone (calcium carbonate).

According to estimates made by the United States Geological Survey in 1924, the reserve deposits of phosphate rock known to exist in the United States aggregate over *seven billion tons,* distributed in nine states. About 80 per cent. of this amount is located in Idaho, 16 per cent. in the combined states of Montana, Utah and Florida, and 4 per cent. in the remaining states of Wyoming, Tennessee, Arkansas, South Carolina and Kentucky. The present reserve supply of rock-phosphate in Florida is estimated to be over 300,000,000 tons and that of Tennessee, over 90,000,000.

How various forms of phosphorus affect millet.

Outside of the United States, the most important deposits of rock-phosphate are in Tunis, Morocco, Algeria and Egypt. The combined output of these countries is about 5,500,000 tons, that of the United States about 3,500,000. The world's estimated consumption of rock phosphate, chiefly as superphosphate, is about 10,000,000 tons annually.

A few of the more important facts in regard to the composition, character, formation and extent of the various American phosphate fields follow:

The mining of American phosphate rock began in South Carolina in 1867, in Florida in 1888, and in Tennessee in 1894.

South Carolina was for some years the main source of mineral phosphates. The present production is relatively small. The phosphate rock is fairly uniform in composition, usually 26 to 28 per cent. of phosphoric acid (equal to 57 to 62 tri-calcium phosphate or 11.5 to 12.5 phosphorus), with a small amount (2 to 4 per cent.) of iron and aluminum compounds.

Florida phosphate beds at present furnish over 80 per cent. of the phosphate materials produced in the United States. The phosphate is found in four forms, "Land Pebble," "River Pebble," "Hard Rock" and "Soft Rock."

The "Land Pebble" deposits furnish most of the phosphate now mined. Pebble phosphate is made up of white, gray, brown, amber and black, hard, irregular-shaped and rounded granules embedded to the extent of 10 to 50 per cent. in a matrix of clay and sand. They vary from small grains to one and one-half inches or more in diameter. A small part is colloidal. The deposits vary in thickness from 5 to 50 feet and are overlaid with a covering of sand and clay of varying thickness. The process of mining is the open pit. The covering of sand and clay is removed by steam-shovels or by streams of water under high pressure. The layers of pebble phosphates are broken up, usually by hydraulic streams, the pebbles are washed free from sand and clay and finally dried at a high temperature in rotating ovens. The phosphate thus prepared contains from 68 to 77 per cent. of tri-calcium phosphate (bone phosphate of lime, B.P.L.) equal to 31.3 to 35.4 phosphoric acid (P_2O_5) or 13.6 to 15.4 phosphorus. This material is shipped to factories where it is converted into superphosphate (p. 34).

"Hard Rock" phosphate occurs in layers having a maximum thickness of 50 feet, being made up of fragmentary rock, boulders of varying size and pebbles. It is mined by the open-pit process. When washed free from impurities and dried, it averages somewhat richer in tri-calcium phosphate (B.P.L.) than the pebble phosphate, varying from 76 to 81 per cent. equal to 35 to 37 phosphoric acid or 15 to 16 phosphorus.

The Tennessee deposits, two kinds, brown and blue, occur in veins and pockets of phosphate rock. The brown phosphate contains 70 to 79 per cent of tri-calcium phosphate (B.P.L.), equal to 32 to 36 phosphoric acid or 14 to 16 phosphorus. The blue variety contains from 65 to 70 per cent. of tri-calcium phosphate.

Colloidal Phosphate is the name applied to finely-divided material obtained in the preparation of Florida hard-rock phosphate.

Apatite is a crystalline mineral containing calcium phosphate and, in addition, one of the elements, fluorine or chlorine ($3Ca_3(PO_4)_2.CaF_2$). When pure it contains about 40 per cent. of phosphoric acid (equal to 88 per cent. of tri-calcium phosphate or 17.5 per cent. of phosphorus). Apatite is a definite chemical compound and is probably the original source of phosphorus in all other phosphorus compounds. It is more widely distributed than any other form of phosphate. Its use in making fertilizers is limited because it is extremely hard and does not dissolve easily in acids. It is mined in a small way in Canada and Norway. Apatite, as taken from the ground, is mixed more or less with other substances present in the earth; it therefore varies in composition according to the amount of impurities, usually ranging from 32 to 41 per cent. of phosphoric acid (equal to 70 to 90 tri-calcium phosphate or 14 to 18 phosphorus). Canadian apatite is usually high in phosphorus.

Guano phosphates.—Guanos containing considerable percentages of phosphorus were for some years extensively used until largely displaced by superphosphate. Small amounts still come into the market, mainly from the West Indies and from islands in various parts of the Pacific Ocean.

MANUFACTURED PHOSPHATE PRODUCTS

Superphosphate (acid phosphate).—Under this head we consider the products made by converting an insoluble phosphate into a soluble one by means of an acid. The process was first applied to bones, but at present rock-phosphate is most extensively employed. While formerly the term superphosphate applied only to mixtures consisting chiefly of soluble or mono-calcium phosphate ($CaH_4(PO_4)_2$) and hydrate calcium sulfate (gypsum), the use of the term now includes more concentrated products containing no admixture of gypsum. Specific names are used to designate such compounds as ammonium phosphate and potassium phosphate.

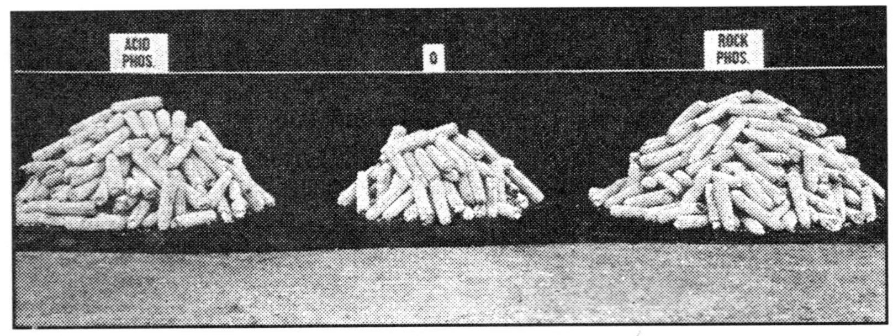

Lasting effect of phosphorus in clay soils. Fertilizer applied three years previously (right and left), none in center. *Left,* thirty-three bushels to the acre; *right,* thirty-eight; *center,* eighteen.

Superphosphate is made by treating some form of tri-calcium phosphate (usually rock-phosphate) (1) with sulfuric acid, (2) with sulfuric acid and phosphoric acid or (3) with phosphoric acid, producing two grades or concentrations of superphosphate, (1) *plain* superphosphate, containing 16 to 20 per cent. of phosphoric acid and (2) *concentrated* superphosphate, containing 40 to 46 per cent. of phosphoric acid.

Plain Superphosphate is made by mixing the finely ground phosphate material with about an equal weight of sulfuric acid of a definite dilution. The first chemical action to take place is the formation of free phosphoric acid and calcium sulfate (gypsum), as represented by the following reaction:

(1)	$2Ca_3(PO_4)_2$	+	$6H_2SO_4$	=	$4H_3PO_4$	+	$6CaSO_4$
	Tri-calcium phosphate		Sulfuric acid		Phosphoric acid		Calcium sulfate

In the second chemical change, the phosphoric acid acts upon an additional portion of tri-calcium phosphate to form the soluble or mono-calcium phosphate, the calcium sulfate remaining unchanged, as indicated in the following equation:

(2)							Superphosphate		
	$Ca_3(PO_4)_2$	+	$4H_3PO_4$	+	$6CaSO_4$	=	$3CaH_4(PO_4)_2$	+	$6CaSO$
	Tri-calcium phosphate		Phosphoric acid		Calcium sulfate		Mono-calcium phosphate		Calcium sulfate (gypsum)

The mixed mass, more or less wet and hot, is allowed to stand until the chemical changes are completed; the calcium sulfate takes up water and the whole mass solidifies, after which it is crushed fine and stored. In those phosphate materials in which the impurity is largely calcium carbonate, more acid is used, the product contains more gypsum, is therefore drier and more easily powdered. In order to avoid the presence of free sulfuric acid in acid phosphate, somewhat less sulfuric acid is used than is actually required to dissolve the phosphate completely. Superphosphate generally, therefore, contains some undissolved tri-calcium phosphate, which, on standing, reacts with the soluble phosphate and forms some di-calcium (reverted) phosphate as shown by the following equation:

(3)	$CaH_4(PO_4)_2$	+	$Ca_3(PO_4)_2$	=	$2Ca_2H_2(PO_4)_2$
	Mono-calcium phosphate		Calcium tri-phosphate		Calcium di-phosphate

The impurities in rock-phosphates that interfere most seriously with their conversation into satisfactory superphosphate are compounds containing iron and aluminum. These interfere with satisfactory results in two ways: (1) The resulting mixture is of jelly-like consistency, and is hard to handle or convert into a dry, fine powder, (2) iron and aluminum compounds tend to cause the formation of reverted phosphate (p. 34), in which the very insoluble iron and aluminum phosphates form an undesirably large proportion. This chemical reaction is illustrated thus:

(4)	$2CaH_4(PO_4)_2$	+	Fa_2O_3	=	$2FePO_4$	+	$2CaHPO_4$	+	$3H_2O$
	Mono-calcium phosphate		Ferric (iron) oxide		Ferric phosphate		Calcium di-phosphate		Water

Superphosphate is, therefore, a mixture of soluble, reverted and insoluble phosphate of calcium, containing also about one-half its weight of gypsum, together with some free phosphoric acid and small amounts of other compounds.

With pure tri-calcium phosphate ($Ca_3(PO_4)_2$), 100 pounds would require about 63 pounds of pure sulfuric acid, and would produce about 75 pounds of soluble calcium phosphate and 88 pounds of gypsum (calcium sulfate and water). Therefore, in 100 pounds of the completed superphosphate made from pure materials, there would be about 54 pounds of gypsum and 46 pounds of soluble calcium phosphate (equal to 28 per cent. of phosphoric acid or 12 of phosphorus). But we usually have a considerably larger proportion of calcium sulfate and a correspondingly smaller proportion of soluble calcium phosphate, because the phosphate material usually contains some calcium carbonate and other constituents that reduce the proportion of soluble phosphate and increase that of gypsum and other compounds in the completed product. In the actual manufacture of superphosphate containing 16 per cent. of phosphoric acid, 1,000 pounds of insoluble phosphate material make 1,500 pounds of superphosphates.

In making insoluble into soluble phosphate, it is seen, therefore, that the process is one of *dilution* so far as the percentage of phosphorus is concerned. Whatever the percentage of phosphorus in an insoluble phosphate, whether rock, bone, or other form, the percentage of phosphorus in the completed product is less than two-thirds what it was before dilution by treatment with sulfuric acid.

There are two well-recognized grades of the ordinary or plain superphosphates in common use, one containing about 16 per cent. of phosphoric acid (equal to 7 of phosphorous) and the other, 20 per cent. of phosphoric acid (equal to 8.8 of phosphorus). Formerly the usual percentage was 14. Both grades may contain 1 to 3 per cent. of phosphoric acid in the citrate-soluble (reverted) form and about the same amount of insoluble.

In appearance superphosphate is generally a grayish powder, the shade varying with the color of the original phosphate material used in the manufacture.

Concentrated Superphosphate is known as *double* or *treble* or *triple* superphosphate. There are two methods of preparation. In the first method, the phosphate, generally high-grade rock, is treated with dilute sulfuric acid, as in the preparation of plain superphosphate, producing dilute phosphoric acid and calcium sulfate (gypsum), as represented in equation (1), page 177. The gypsum, being in the form of a finely-divided solid or precipitate, is easily separated by setting up filtration from the solution of phosphoric acid. The phosphoric acid is then concentrated and mixed with more phosphate rock, when the following reaction takes place:

$$\underset{\text{Tri-calcium phosphate}}{Ca_3(PO_4)_2} + \underset{\text{Phosphoric acid}}{4H_3PO_4} = \underset{\text{Mono-calcium phosphate}}{3CaH_4(PO_4)_2}$$

In the second method of preparation, no sulfuric acid is used but only commercially prepared phosphoric acid which is added directly to the phosphate material. This method has the advantage that there is no calcium sulfate to get rid of. In either case, the final product is the same, mono-calcium phosphate, free from calcium sulfate. The percentage of phosphoric acid in concentrated superphosphate, thus prepared, depends upon the percentage of phosphoric acid in the phosphate material used. In most of these products now on the market, the percentage of phosphoric acid varies from 40 to 47 (equal to 17.6 to 20.7 phosphorus), as compared with 60.7 per cent. of phosphoric acid (equal to 26.5 phosphorus) in chemically pure mono-calcium phosphate. Liquid phosphoric acid is commercially prepared in any degree of concentration by treating finely-ground phosphate rock with sulfuric acid and concentrating the products, or by the electric-furnace process, or by smelting phosphate rock in a blast-furnace, volatilizing the phosphorus and then converting it into pure phosphoric acid.

The commercial manufacture of pure phosphoric acid has made possible the preparation of ammonium phosphate (p. 37) and potassium phosphate (p. 37), as well as the concentrated superphosphate.

Ammoniated Superphosphate.—Superphosphate plain or concentrated, alone or in a mixed fertilizer can be treated with ammonia, either anhydrous liquid ammonia or ammonium hydroxide containing 25 to 30 per cent. of ammonia, the ammonia being sprayed into the material. The ammonia reacts quickly with the superphosphate and with any free sulfuric acid present, forming ammonium sulfate and mono-ammonium phosphate. This treatment is advantageous in various ways. It adds several per cent. of nitrogen at a minimum cost without appreciable dilution of superphosphate or increased formation of tri-calcium (citrate-insoluble) phosphate unless excessive amounts of ammonia are used. It improves in a marked degree the mechanical condition of superphosphates and of mixed fertilizers containing superphosphates, giving a dry product that drills readily and that can be stored without caking. The rotting of bags by superphosphate is prevented or greatly decreased. In the manufacture of superphosphate, time of curing may be lessened by using an excess of sulfuric acid in dissolving the phosphate rock and then treating with ammonia.

When ammonia solution (containing 25 to 30 per cent. of ammonia) is used in excess of 2 or 3 per cent. with superphosphate, the resulting product contains water to such an extent that its mechanical condition is unsatisfactory. This is avoided by using anhydrous liquid ammonia. The use of anhydrous ammonia in excess of 3 per cent. in treating plain superphosphates or in excess of 7 per cent. in treating concentrated superphosphates, causes the formation of tri-calcium phosphate.

Nitrate-superphosphate.—Nitric acid is used to dissolve insoluble phosphate and form a superphosphate, as indicated by the following reaction:

$Ca_3(PO_4)_2$	+	$4HNO_3$	=	$CaH_4(PO_4)_2$	+	$2Ca(NO_3)_2$
Tri-calcium phosphate		Nitric acid		Mono-calcium phosphate		Calcium nitrate

The resulting product is a mixture of the soluble compounds of mono-calcium phosphate and calcium nitrate.

Potassium phosphates can be prepared by the combination of phosphoric rock and a potash silicate acid and potassium hydroxide. Experimental work has been done to produce potassium phosphate by smelting phosphate.

Ammo-Phos-Ka is the name of a series of complete fertilizers, containing from 40 to 50 per cent. of plant-food. In the preparations now offered, the percentage of nitrogen varies from 9 to 12; the available phosphoric acid from 16 to 30; the potash, from 10 to 18.

Nitrophoska is a trade-name applied to a series of complete German fertilizers of which eight grades are now on the American market. They contain nitrogen in the form of nitrate and ammonia, phosphoric acid as ammonium phosphate, and potassium as chloride or sulfate or a mixture of the two. All are readily soluble in water. They are all highly concentrated, containing from 45 to 60 pounds of plant-food in 100 pounds of material. The percentage of nitrogen varies from 10 to 16.5 (equal to 12 to 20 of ammonia); the phosphoric acid, from 11 to 30 (equal to 4.8 to 13.2 of phosphorus); the potash, from 12 to 26.5 (equal to 10 to 22 of potassium).

Leunaphoska is a trade-name used for a mixture of compounds containing leunasaltpeter (ammonium nitrate and sulfate), together with ammonium phosphate and a potassium salt.

Ammonium Phosphate is made by combining ammonia and liquid phosphoric acid thus:—($3NH_3 + H_3PO_4 = (NH_4)_3PO_4$). The product sold under the trade name of *Ammophos,* is a mixture of ammonium phosphate and ammonium sulfate and is made in two grades known as "20-20" and "13-48." It is a grayish-white, granular material. It does not absorb moisture. The first grade contains more ammonium sulfate and less ammonium phosphate, the percentage of ammonia being 20 per cent. (equal to 16.5 nitrogen) and that of phosphoric acid (P_2O_5) being 20 per cent. (equal to 8.8 phosphorus). The second grade contains more ammonium phosphate and less ammonium sulfate, the percentage of ammonia being 13 (equal to 10.7 per cent. of nitrogen) and that of phosphoric acid 48 (equal to 21 phosphorus).

Another commercial product is known as *Diammonphos* ($(NH_4)_2HPO_4$), containing about 25.5 per cent. of ammonia (equal to 21 nitrogen) and 53 per cent. of phosphoric acid (equal to 23.3 phosphorus).

Basic-slag phosphate, known also as Thomas phosphate powder, basic-iron slag, phosphate slag, etc., is a by-product of steel manufactured from iron ores containing a considerable amount of phosphorus compounds, when an excess of calcium oxide (lime) is used in the process. It is a heavy, black powder, of a fineness that enables 70 per cent. or more to pass through a one-hundredth inch mesh sieve and 90 per cent. or more through a 50-mesh sieve. It usually contains about 16 to 20 per cent. of phosphoric acid (equal to 7 to 8.8 phosphorus).

The exact composition of basic-slag phosphate is not satisfactorily settled as yet. Among the chief constituents, other than phosphorus, are compounds containing calcium (35 to 45 per cent.), magnesium (3.5 to 4 per cent.), iron (13 to 14 per cent.), manganese (6 to 7 per cent.), silicon (3 per cent.). The percentages given are only approximate. The exact form of combination in which the phosphorus occurs is a matter of some dispute. So far as present evidence goes, it is probable that most of the phosphorus is a double silicate and phosphate of calcium $(CaO)_5P_2O_5.SiO_2$, while smaller amounts exist as tetra-calcium phosphate $(CaO)_4,P_2O_5$ (p. 33). In regard to the form in which calcium is combined in basic-slag phosphate, generally less than 2, and often less than 1 per cent., is present as calcium oxide or so-called free lime, while the amount of carbonate is insignificant. The calcium therefore exists largely as phosphate and silicate. The use of basic slag-phosphate as a fertilizer is considered on page 290.

ANIMAL PHOSPHORUS-CONTAINING MATERIALS

In large packing-houses, all bones of appreciable size are utilized for various commercial purposes, and only the discarded fragments and small bones that cannot be otherwise utilized go into fertilizing material. Bones furnished to agriculture practically all of the phosphorus applied as fertilizer until the advent of rock-phosphate.

Bones and bone products used for fertilizing purposes are known under such general names as *bone phosphate, bone phosphate of lime, animal phosphate, bone-tankage, bone-black, bone-ash,* etc.

The materials, aside from water, of which bones consist can be divided into two kinds, inorganic and organic. The inorganic portion constitutes the hard framework, making up 50 to 60 per cent. of the whole weight of raw bone (about 70 per cent. of dry bone), and consisting of tri-calcium phosphate-carbonate ($3Ca_3(PO_4)_2 \cdot CaCO_3$). The organic part of bones consists mainly of two substances: (1) A soft, flesh-like substance called *ossein* or *collagen,* distributed through the entire mass of the bone (forming 25 to 30 per cent. of the weight); this material, commonly known as *bone-cartilage,* is rich in nitrogen. (2) The other organic substance is animal fat.

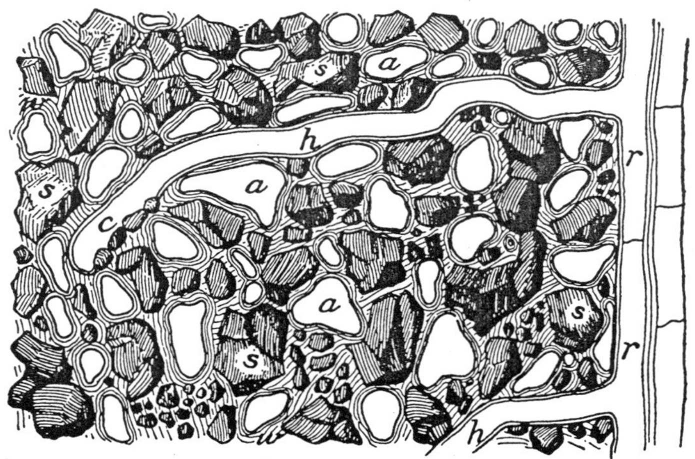

Root-hair in soil magnified, showing absorption of soil solution; *h,* is a root-hair forcing its way between soil particles (*s*) which are the dark portions; the large, rounded, white spaces (*a a*) represent air bubbles; the waved lines (*w*) surrounding the particles of soil and inclosing air bubbles represent water held to the grains by surface attraction. At points marked *c,* near the end of the root-hair, the soil particles are in close contact with the root-hair.

When a bone is allowed to lie in a dilute acid for some time, the inorganic framework (calcium phosphate) is completely dissolved, leaving the soft, cartilaginous organic matter still holding the original shape of the bone. The familiar substance, *gelatin,* is a direct product of the bone-cartilage, formed by boiling bones a long time. Gelatin by further treatment can be converted into *glue.* When bones are burned, the fat and nitrogenous compounds are destroyed, and only the mineral portion remains.

The amount of phosphorus and nitrogen in bones varies with different conditions, such as kind of animal, age, location of bone, etc. For example, fish bones are much richer in nitrogen than beef bones. Bones of young animals contain more nitrogen and less phosphorus, than those of older animals. The large, hard bones, the ones bearing greatest strain, are richer in phosphorus and less rich in nitrogen than the smaller, more compact bones. However, the differences found in the percentages of nitrogen and phosphorus in the bones used for fertilizing purposes are due more largely to differences in methods of treatment than to differences of composition in the original bones.

Raw bone includes any bone that has been ground in its natural state without previous treatment by steaming, boiling, etc. It is usually marketed under the names bone-meal, raw bone-meal, coarse ground bone, etc. Raw bone contains the fat naturally present in bones, objectionable for the following reasons: (1) The presence of fat makes it practically impossible to grind bone fine enough for use as a fertilizer. (2) The fat protects the particles of bones from bacterial action and makes decomposition in the soil slow. (3) It is believed that some of the decomposition products of fat, when present in soils, are poisonous to plants. (4) The fat itself has no value as plant-food, but by its presence reduces the percentage of nitrogen and phosphorus in bone.

Raw bone of good commercial quality should contain 3.5 to 4.5 per cent. of nitrogen, and 20 to 25 per cent. of phosphoric acid (about equal to 45 to 55 per cent. of tri-calcium phosphate or to 9 to 11 of phosphorus).

Steamed bone products include materials sold under various commercial names, such as bone-meal, fine bone-meal, pure ground bone, bone-flour, bone-dust, precipitated bone phosphate, etc. The various terms refer usually to fineness of division rather than to differences of composition. Fineness affects value as fertilizing material; the finer the particles, the more quickly do the plant-food constituents become available. Most bone products used for agricultural purposes have been treated by some process which removes the fat, to be used in soap-making, and usually more or less nitrogen for making gelatin or glue, according to the process employed. On account of the large demand for bones for various purposes and their consequent increase of value, there has been some temptation to adulterate bone products, both raw and steamed, with such substances as calcium carbonate, gypsum, coal-ashes, ground oyster-shells, ground rock-phosphate, etc.

Fat is removed from bones by extraction either by steam or gasoline or both. Extraction by treatment with gasoline has the advantage of removing little or no nitrogenous material, but treatment with boiling water or with steam takes out some of the nitrogen compounds. By treating the fat-free bones at a much higher temperature and pressure (50 to 60 pounds), the cartilaginous part of the bone is converted into gelatin, which dissolves in water and separates from the mineral portion of the bone. The removal of fat makes it much easier to grind the bone fine, so does removal of part of the cartilage from the fat-free bone. Steamed bone-meal or bone-flour generally has a somewhat gritty feeling and a characteristic odor.

The composition, as well as the fineness, of bone products is affected by the character of treatment. Fat-free bone-meal contains, according to samples on the market, 1.8 to 3.5 per cent. of nitrogen and from 22 to 28 per cent. of phosphoric acid (about equal to 48 to 62 per cent. tri-calcium phosphate or 9.7 to 12.3 phosphorus). In case of bone treated for the manufacture of gelatin or of glue, the nitrogen is often reduced below 2 per cent. and sometimes below 1 per cent., while the phosphoric acid may go as high as 28 to 30 per cent. (about equal to 62 to 66 tri-calcium phosphate or 12 to 13 phosphorus). Bone-flour prepared from bone after the extraction of fat with naphtha may contain as much as 6 per cent. of nitrogen. Some products appear on the market under the name of bone-meal, containing less than 2 per cent. of nitrogen and only 16 to 17 per cent. of phosphoric acid. In view of the possibility of so great variations, it can be readily understood how important it is to purchase these materials only under definite guaranteed percentages of nitrogen and phosphorus. The degree of fineness should also be guaranteed.

Precipitated Bone is the product formed by dissolving bone in hydrochloric acid (one of the processes in the manufacture of gelatin) and precipitating the dissolved phosphate with enough calcium carbonate or hydroxide to form the reverted calcium phosphate (p. 36).

Bone-tankage, while more or less rich in nitrogen, contains varying amounts of phosphorus, according to the relative proportions of bone and meat present. Tankage consisting almost entirely of bone is, of course, about like bone in composition. It also follows that in a genuine bone-tankage the higher the percentage of phosphorus, the lower is the percentage of nitrogen. The bone in tankage may be usually regarded as having the same agricultural value as boiled or steamed bone of the same degree of fineness. The materials of tankage are often found to be rather too coarse.

The phosphoric acid in tankage may vary from 6 to 20 per cent. (about equal to 13 to 44 per cent. tri-calcium phosphate or 2.6 to 8.8 per cent. phosphorus). Bone-tankage proper may be regarded as containing 15 to 20 per cent. of phosphoric acid, (about equal to 44 per cent. of tri-calcium phosphate). Tankage should be purchased only on a guaranteed percentage of nitrogen and phosphoric acid.

Bone-black, known also as *animal charcoal* and *bone-charcoal,* manufactured from bone, finds its chief use in sugar-refineries. It is made by heating carefully selected bones in closed vessels, with exclusion of air. By this process, the fat, nitrogen and water are removed leaving mostly tri-calcium phosphate and carbon. When ground it is coarse, granular and thus becomes commercial bone-black. The presence of the carbon, coating the particles of phosphate, makes it slow to change to available plant-food, offering resistance to bacterial or other dissolving agents. When bone-black has been used repeatedly it becomes useless in sugar-refining and is sold as a fertilizer either for direct use or to be made into superphosphate, known as *dissolved bone-black.* Good bone-black may contain 30 per cent. of phosphoric acid (about equal to 66 per cent. of tri-calcium phosphate or 13 per cent. phosphorus). At present it forms an insignificant part of the phosphorus used in fertilizers.

XVI

MATERIALS CONTAINING POTASSIUM

THE largest amounts of materials containing potassium compounds come from Germany and France; less extensive supplies are furnished by Poland, Spain, Italy and the United States.

PRODUCTION IN GERMANY AND FRANCE

After having been used for the purpose of mining rock-salt, the mines near Stassfurt in northern Germany were abandoned about 1839. Later a further exploration of the beds by means of borings revealed large quantities of potassium compounds, especially chloride (muriate), and, about 1861, operations were begun to work the deposits for the commercial production of potassium compounds. As they exist in the deposits, these are more or less impure, being mixed with chlorides and sulfates of calcium, sodium, and magnesium. The average grade of the crude material contains about 10 per cent. of potassium (equal to 12 per cent. of potash). The materials after removal from the mines are ground, dissolved in water and allowed to crystallize for purification and concentration. By this process the impurities are more or less removed, depending on the degree of concentration and purity desired in the finished products, of which there are several grades. These are, therefore, more or less manufactured rather than natural products and include chiefly chloride and sulfate, though some of the materials used for fertilizers are the crude products as taken from the deposit, ground fine, or mixtures of the refined and crude salts.

In general, the German potassium products, as they appear in the market, are gritty powders, varying in fineness and color (white, dirty-white, gray or pink). They dissolve readily in water. Most of them have a tendency to absorb water and so become moist when exposed in damp places. They should therefore be stored only in dry places if kept long. While these potassium materials are used to a considerable extent in the United States in chemical industries for the manufacture of other potassium compounds, much the larger proportion is used for fertilizers.

Deposits containing potassium salts were discovered in Alsace in 1904 and commercial production began in 1910. The material consists largely of a mixture of potassium chloride and common salt, the potassium varying from about 12 to 18 per cent. (equal to about 14 to 22 per cent. of potash). The French purified product averages higher in potassium than the German and, owing to the smaller admixture of other salts, cost of production is less. The chief product of the Alsatian mines is potassium chloride (muriate).

Statistics of 1930 show that the United States imported 980,000 tons of potash salts, equal to 342,000 tons of potassium oxide (K_2O), valued at $24,500,000. Most of this was used for fertilizers. During this same year the United States produced 106,000 tons of potash salts, equal to 61,270 tons of potassium oxide (K_2O).

PRODUCTION IN THE UNITED STATES

During the World War, when this country was forced to depend wholly upon domestic supplies of potassium, it succeeded in obtaining from various sources a maximum of 60,000 tons of potash, enough to furnish about one-fifth of the normal agricultural needs. At present about 82 per cent. of the potassium-containing materials used in the United States comes from Germany and France; the 18 per cent. produced by domestic manufacture comes mainly from two sources, (1) Searles Lake, Calif., where potassium and borax compounds are prepared from the natural salt brine of the lake, and (2) as a by-product from the manufacture of industrial alcohol from sugar-cane molasses.

Other sources of domestic supply utilized under war conditions, when the price of muriate of potash (commercial potassium chloride), increased to $500 a ton from a pre-war price of $40, were (1) the kelps of the Pacific coast, (2) the flue-dust of blast-furnaces and of cement factories, (3) alunite (potassium aluminum sulfate) at Marysville, Utah and (4) the natural brine of Jesse Lake, Nebraska. The return of normal prices made production unprofitable in most cases.

The future of successful potassium supply in large amount from domestic sources depends primarily upon profitable production in competition with German-French products. At present, the most promising prospect appears to be the commercial development of extensive deposits of potassium compounds in western Texas and south-eastern New Mexico in association with rock-salt deposits. These deposits lie at depths of 350 to 1800 feet below the surface. The potassium is present, mostly as *polyhalite,* a sulfate of potassium, calcium and magnesium, containing, when pure, 13 per cent. of potassium (equal to 15.6 per cent. of potash). This can readily be converted into refined compounds containing high percentages of potassium, such as sulfate of potassium, and sulfate of potassium and magnesium. Mixtures of crude and refined products can be prepared to furnish any desired percentage of potassium between 12 and 52. Experiments indicate that crude polyhalite, finely ground, is effective as a fertilizer in direct application to plants. The United States government has been carrying on extensive research work in connection with the Texas-New Mexico deposits as well as with other possible sources of production of potassium compounds.

Surface deposits of potassium-containing minerals, widely distributed in the United States in inexhaustible amounts, await successful methods of treatment for the economic production of potassium compounds for agricultural use. Among these minerals are the green-sand marls of New Jersey, Delaware, Maryland and Virginia; the potash feldspars of New England and of the Appalachian and Rocky Mountain States; the potash shales of Georgia; the leucite of Wyoming. These minerals contain potash varying 5 to 20 per cent. For example, green sands contain 6 to 8 per cent. of potash; potash shales, 8 to 9 per cent.; orthoclase feldspar, 17 per cent.; leucite, 22 per cent.

The more common sources of agricultural potassium are: (1) Potassium chloride, (2) potassium sulfate, (3) potassium and magnesium sulfate, (4) potassium-magnesium carbonate, (5) manure salts, (6) kainite, (7) potassium nitrate, (8) potassium phosphate, (9) wood-ashes, (10) tobacco waste, (11) by-products of sugar-mills.

Potassium chloride KCl, (p. 38), commercially known as *muriate of potash,* is one of the prepared products derived from the imported crude materials of the German and French potash mines, purified by special treatment. It is found in the market in two grades of purity, containing 76.8 and 80 per cent. of potassium chloride or 48 to 50 per cent. of potash (equal to 39.8 and 41.5 potassium). The chief impurity in the commercial muriate is common salt, which may vary from 7 to 20 per cent. There are also small amounts of calcium and magnesium compounds.

A high-grade commercial potassium chloride, known under the trade-name of *Trona Potash,* is prepared from the waters of Searles Lake in California. It contains an average of 96 to 98 per cent. of potassium chloride or 60.5 to 61.7 per cent. of potash (equal to 51 to 52 potassium).

Potassium sulfate (K_2SO_4, p. 40), commercially known as *sulfate of potash,* is one of the German manufactured products. It comes into the market usually in two grades, containing 90 and 96 per cent. of pure potassium sulfate, which is about equal to 48.5 to 52 per cent. of potash or 40.5 to 43 per cent. of potassium. The usual form in commerce contains about 48 per cent. of potash (K_2O). The sulfate is used in much smaller amounts than the muriate. The impurities in commercial potassium sulfate are small amounts of sulfates and chlorides of calcium, magnesium and sodium. It should not contain more than 2.5 per cent. of chlorine.

Potassium and magnesium sulfate, sulfate of potash and magnesia, known also as *double potash-manure salts,* is a product prepared in the manufacture of the high-grade sulfate, usually containing about 50 per cent. of potassium sulfate, which is equal to about 27 per cent. of potash or 22.5 potassium. It contains not less than 25 per cent. of magnesium sulfate and not more than 2.5 per cent. of chlorine. It is used for tobacco and fruits.

Potash manure salts or fertilizer salts is a term applied to low-grade potassium chloride (muriate), containing 20 to 30 per cent. of potash (equal to 16.5 to 25 potassium). In addition to potassium chloride, these materials contain large amounts of common salt, considerable amounts of sulfate and chloride of magnesium, with small amounts of other compounds.

Potassium-magnesium carbonate is a manufactured product used in fertilizing tobacco and various fruit crops. The material, being in dry, finely-powdered condition, is ready for application at any time. It contains 35 to 40 per cent. of potassium carbonate (K_2CO_3, p. 38), which is equal to 24 to 27 per cent. of potash or 20 to 22.5 per cent. of potassium. It contains about the same amount of magnesium carbonate.

Potassium carbonate in high-grade forms, 93 to 98 per cent. pure, contains 63.2 to 66.6 per cent. of potash (equal to 52.5 to 55.2 potassium). It is used in the growing of tobacco.

Kainite is a name applied to crude or untreated natural products coming from potash mines. The material is a variable mixture of several different compounds. The composition of the products included under kainite necessarily varies widely, depending upon the composition and proportions of the individual constituents. German kainite is largely a mixture of potassium chloride, magnesium chloride, magnesium sulfate and sodium chloride. French kainite contains potassium chloride mixed with a large amount of sodium chloride. The mixture that comes into market under the name of kainite contains 12 to 14 per cent. of potash, equal to 10.3 to 11.6 potassium.

Potassium nitrate.—(See p. 40.)

Potassium phosphate.—(See p. 37.)

Wood-ashes at one time constituted the chief supply of potassium in commerce and agriculture. Not only is the available supply now small but the percentage of potassium present averages less than formerly. So-called Canada, hard-wood, unleached ashes, supposed to represent the best quality, formerly contained over 5 per cent. of potash (equal to 4 per cent. of potassium); the percentage is now more often under 4 per cent. (or 3.5 per cent. of potassium) and may drop to one-half this amount. It is obvious that wood-ashes should be purchased only under a definite guaranteed percentage of potassium. In wood-ashes the potassium is mostly in the form of carbonate (K_2CO_3, p. 38). When of good quality, wood-ashes contain 5 to 7 per cent. of potash (or 4 to 6 per cent. of potassium) and 1.5 to 2 per cent. of phosphoric acid. Owing to their high price there has been considerable temptation to adulterate wood-ashes, making a mixture of potassium chloride and some form of lime (calcium hydroxide or carbonate) and leached ashes, and sometimes other products of a worthless character. Leached wood-ashes contain about 1 per cent. of potash (0.8 per cent. of potassium) and 1 to 1.5 per cent. of phosphoric acid, with 25 to 30 per cent. of calcium (lime) compounds. In genuine wood-ashes, the calcium compounds may vary from 20 to 50 per cent, but usually run between 30 and 35. In freshly burned ashes, calcium oxide (quicklime) is the chief calcium compound, but, in the case of ordinary wood-ashes, this is likely to be changed largely into carbonate by the time it is applied to soils.

Wood-ashes, even when pure and unleached, vary greatly in composition, according to the kind of wood, part of tree, age, method of burning, and care with which they are stored. Hard wood usually produces ashes containing more potassium than those from soft wood. When ashes have been exposed to the weather or leached, they contain much less potassium than when kept dry. Wood-ashes from lime-kilns and brick-kilns are often mixed with valueless refuse to such an extent as to possess little value as plant-food, though lime-kiln ashes usually contain larger amounts of calcium compounds than ordinary wood-ashes. Ashes from spent tan-bark of tanneries are low in plant-food value. The amount of potassium or phosphorus in coal-ashes is insignificant. For discussion of ashes in relation to their use as a fertilizer, see page 189.

Cotton-hull ashes are used as a fertilizer in tobacco growing. The percentage of potash varies from below 20 to over 35, averaging about 25. Phosphoric acid is present also, the average being about 3 per cent.

Tobacco-waste from the manufacture of commercial tobacco products, consists of tobacco stems, stalks and waste materials which were at one time thrown away. It contains considerable potassium as well as nitrogen. Sometimes it was burned and the ashes used for fertilizer; such ashes are rich in potassium, but the nitrogen is lost by burning. Tobacco-waste ground fine makes a desirable source of potassium and nitrogen, where immediate availability is not required. It is also useful in adding to mixed fertilizers on account of the beneficial effect on the mechanical condition of the mixture.

Tobacco-waste appears in three forms, stalks, stems and dust. Stalks include the main stalk and the branches. Stems include the leaf-stalk and the ribs that make up the skeleton of the leaf, together with such other portions of the leaves as are rendered useless by handling during manufacture. Tobacco dust includes fine particles that result from handling the leaves, mixed with more or less dirt. These materials give best results as plant-food when ground fine.

Tobacco-waste products vary much in composition. The nitrogen in stems varies from 2 to 3 per cent.; in stalks, from 3 to 4 per cent.; and in dust, from about 2 to 2.5 per cent. From one-third to one-half of the nitrogen in tobacco is present as nitrate.

Phosphoric acid is present in small amounts, usually running between 0.5 and 1 per cent. (0.22 and 0.44 phosphorus).

Potash is contained in tobacco stems in amounts ranging from 5 to 10 per cent. (about 4 to 8 potassium), the usual variations lying between 6 and 9 per cent.; while, in stalks, the percentage is smaller, usually between 4 and 5 per cent. (about 3 to 4 potassium).

Tobacco products possess, in addition, insecticidal value, which makes them especially useful in the growing of small fruits and greenhouse crops.

Sugar-factory wastes.—The unmarketable molasses produced in the process of sugar-making contains potassium compounds in considerable amount. For example, ashes from beet molasses may contain as much as 38 per cent. of potassium (45 per cent. potash), most of which is in the form of carbonate, with small amounts of chloride and sulfate. Low-grade molasses is sometimes mixed with other materials and applied directly to soils, but this practice should be accompanied with the application of generous amounts of calcium carbonate on account of the production of acids resulting from the fermentation of the sugar.

Reference has already been made to the production of potassium-containing material as a by-product in connection with the manufacture of industrial alcohol from sugar-cane molasses. The liquid remaining after removal of the alcohol from the fermented molasses by distillation is evaporated and the potassium recovered, partly as chloride, partly as sulfate and some as carbonate. This material, known under the trade-name of "Vegetable Potash" contains about 33 per cent. of water-soluble potash (equal to 27.4 of potassium).

XVII

CALCIUM (LIME) COMPOUNDS AND OTHER INDIRECT FERTILIZERS

An indirect fertilizer (p. 159), is a material which favorably influences crop growth through its effect on the soil rather than through direct addition of plant-food. Such materials are known also as soil *stimulants, tonics,* and *amendments.* The action varies for different materials, each producing effects more or less specific.

Indirect fertilizers are of increasing interest and importance. Their intelligent use is attended with satisfactory results in crop growing, but blind, indiscriminating use may work harm to soil and crops and loss to farmers. It is, therefore, desirable that the subject be treated fully.

CALCIUM (LIME) COMPOUNDS

The various calcium compounds to be considered are used as indirect fertilizers to bring about in the soil chemical, physical and biological changes that are beneficial to plant growth through the action of the calcium they contain. Calcium compounds are not often used solely to supply calcium directly as plant-food to crops. The operation of applying them to soils is commonly called *liming.*

COMPOSITION OF CALCIUM (LIME) COMPOUNDS

The chemical element *calcium* (Ca) is the one common, characteristic constituent of all so-called lime compounds, in which we find calcium combined with one or more other chemical elements. We are to keep in mind that it is this element, *calcium,* which accomplishes the desired purpose, whatever compound of calcium we apply as an indirect fertilizer. The various calcium compounds available for agricultural use as indirect fertilizers are: (1) Calcium oxide, (2) calcium hydroxide, (3) calcium carbonate, (4) gypsum or hydrated calcium sulfate. Other materials that contain calcium are chiefly used to supply plant-food other than calcium. They include ground phosphate rock or "floats" (p. 235), wood-ashes (p. 189), and basic-slag phosphate (p. 36), already discussed.

Calcium oxide or lime.—The word *lime* is ordinarily used to designate a compound consisting, when pure, of calcium (71.4 per cent.) and oxygen (28.6 per cent.). It is known in chemistry as *calcium oxide* (CaO). At one time it was erroneously supposed that it (lime) was present in all compounds of calcium, so all calcium compounds were called lime compounds. This inaccurate use has continued, especially in commercial practice.

Calcium oxide, in impure forms, is known under several commercial names, such as *lime, burned lime, quicklime, caustic lime, stone-lime, lump-lime, unslaked lime,* etc. The value of commercial lime depends, for many purposes, upon its freedom from impurities, that is, upon the amount of calcium oxide it contains.

Commercial lime is prepared by heating at a sufficiently high temperature any form of carbonate of lime (calcium carbonate), the carbon dioxide (carbonic acid) being driven off as a gas and calcium oxide remaining as a solid residue (p. 41).

The materials most commonly used as a source of commercial lime are limestone, oyster shells and shell-marl. The percentage of calcium oxide present (the purity of a lime) depends upon (1) the purity of the original carbonate material used, (2) the completeness of the burning, and (3) subsequent care in keeping dry. The impurities commonly found in commercial lime are compounds containing magnesium, silicon, iron and aluminum. The three most common forms of commercial lime are: (1) Burned-lime, (2) magnesium burned-lime, and (3) oyster-shell lime.

(1) *Burned-lime,* the ordinary lump-lime commonly used by masons, is prepared by burning limestone (calcium carbonate) of a high degree of purity. When freshly prepared it should contain 90 to 98 per cent. of calcium oxide. That sold for agricultural use contains 60 to 90 per cent. of calcium. Commercial quicklime can be obtained in coarsely pulverized or granulated form. This is much used in building and is a convenient form to use on soils.

(2) *Magnesium burned-lime* is prepared from limestone that contains more or less magnesium carbonate, the proportions varying greatly in different varieties. Commercial lime prepared from such material contains 55 to 85 per cent. of calcium oxide and 10 to 40 per cent. of magnesium oxide. The presence of considerable proportions of magnesium oxide affects unfavorably the slaking properties of lime.

(3) *Oyster-shell lime.*—Oyster shells contain 90 to 95 per cent. of calcium carbonate and, when properly burned, produce a high-grade lime, containing from 85 to 95 per cent. of calcium oxide.

In addition to the preceding forms of commercial lime, two others of less importance are *gas-house lime* and *acetylene-lime.* Quicklime is used at gas-works to remove impurities from the gas and after being so used it is often sold to farmers. Gas-house lime is not quicklime, since the calcium oxide is largely changed into other forms, such as hydroxide (slaked lime) and carbonate. In addition, there are the impurities absorbed from the gas, some of which, sulfides and sulfites, are injurious to seeds. These are changed into the harmless form of sulfate (land-plaster) on exposure to the air. Gas-house lime should, therefore, be allowed to lie exposed to the air before use or should be put into the soil some weeks or months before putting in a crop. Its composition varies greatly.

Acetylene-lime is the solid residue formed in the manufacture of acetylene gas from calcium carbide; it consists largely of a mixture of calcium carbonate and hydrate (slaked lime), the proportions depending upon the length of exposure to air. The material has the disadvantage of being wet and of holding some acetylene gas. It can be used in liming soils where its cost is slight, but in order to allow acetylene gas to escape it is well to expose it to the air for some weeks before applying to the soil or to apply to the soil a few weeks before putting in seeds, because the gas is injurious to seeds.

Calcium hydroxide or slaked lime.—When calcium oxide or quicklime comes in contact with water, it undergoes the change known as *slaking* (or, erroneously, slacking); its calcium oxide (CaO) combines with water (H_2O) and is changed into a compound known chemically as *calcium hydroxide* ($Ca(OH)_2$) or commercially as *slaked lime, water-slaked lime, caustic lime,* or *hydrated lime;* 100 pounds of calcium oxide combine with about 32 pounds of water and produce 132 pounds of calcium hydroxide (slaked lime, p. 41).

Slaked lime may be conveniently regarded, therefore, as quicklime diluted by combination with about one-third of its weight of water. Calcium hydroxide or pure slaked lime contains 24.3 per cent. of water and 75.7 per cent. of calcium oxide, which is equivalent to 54 per cent. of calcium. The chemical change that occurs when lime slakes or combines with water is accompanied by marked physical changes; the lime swells in bulk, crumbles to a very fine powder and generates much heat. The rapidity with which commercial burned lime slakes depends upon its composition and the amount of water used. The use of too much water or the sudden chilling of a partially slaked lime by cold water results in the production of a coarse, granular lime in place of the desired form of fine powder. Commercial limes that contain less than 10 per cent. of impurities slake more quickly than those that contain larger amounts of impurities. Magnesium burned limes slake imperfectly.

One bushel of burned lime of good quality, on slaking, increases in bulk to two or three bushels of slaked lime. The weight per bushel decreases roughly about one-half, for example, from 80 to 40 pounds. The bulk and weight of slaked lime depend upon the composition of the burned lime used, upon the amount of water added, and upon some special conditions of the slaking process.

Slaked lime suspended in or mixed with water to form a milky liquid, is known as *milk of lime.* Slaked lime dissolved in water forms a clear solution known as *lime-water.*

The calcium hydroxide or slaked lime easily absorbs carbon dioxide gas from the air and changes gradually into calcium carbonate. Just as commercial burned lime is impure calcium oxide, so commercial slaked lime is impure calcium hydroxide. Ordinary water-slaked lime is, therefore, a mixture of calcium hydroxide and calcium carbonate with such impurities as may have been present in the lime before slaking.

The slaking of lime also takes place when lime is exposed to the air, this process being known as *air-slaking.* When lumps of burned lime stand exposed for some time, the outer layer gradually absorbs moisture from the air and goes through the process of slaking, forming a fine powder. In the air-slaking process, more or less calcium carbonate is also formed. Hence, air-slaked lime usually contains larger amounts of carbonate and less of hydroxide than water-slaked lime. The rapidity with which limes air-slake depends upon various conditions. Dry, fine, air-slaked lime may at best contain as much as 75 per cent. of calcium oxide; but it varies much and is practically all carbonate if it has been long exposed to air.

Calcium carbonate is commonly known as *carbonate of lime.* In this form calcium oxide (CaO) is combined with carbon dioxide (CO_2), thus forming calcium carbonate ($CaCO_3$), which contains 56 per cent. of calcium oxide (equivalent to 40 per cent. of calcium) and 44 per cent. of carbon dioxide. (p. 47.)

As we may regard slaked lime as calcium oxide diluted by combination with water, so we may regard calcium carbonate as calcium oxide (lime) diluted by combination with carbon dioxide; 100 pounds of calcium oxide combine with about 80 pounds of carbon dioxide and produce 180 pounds of calcium carbonate. The carbonate is, therefore, a more dilute form of calcium than the oxide or hydroxide.

Limestone of high grade of purity, chalk and also many kinds of marble consist largely (95 to 98 per cent.) of calcium carbonate, (equal to 30 to 40 calcium). Limestone of low grade may contain as little as 60 per cent. of calcium carbonate (equal to 24 calcium). Limestone rocks contain varying proportions of magnesium carbonate, ranging from a fraction of 1 per cent. to over 40 per cent. High-calcium lime products contain not more than 2.5 per cent. of magnesium. When the percentage of magnesium carbonate exceeds 40, the limestone is called magnesian or dolomitic. Shells, coral rock, shell-marl, air-slaked lime, etc., contain calcium carbonate in varying proportions.

Ground or pulverized limestone is the main form of calcium carbonate applied to soils. Ground shell-marl is also used and, in its purest condition, may contain as much as 98 per cent. of calcium carbonate.

Calcium sulfate, hydrated, (p. 42), known also as *gypsum, hydrated sulfate of lime, land-plaster,* etc., is a compound of calcium, with sulfuric acid and crystallization water ($CaSO_4.2H_2O$); 100 pounds of pure hydrated calcium sulfate contain about 32.5 pounds of calcium oxide (equivalent to 23 pounds of calcium), 46.5 pounds sulfuric acid, and 21 pounds of water. Hydrated calcium sulfate or gypsum may be regarded as a more diluted form of calcium than either the hydroxide or the carbonate. To contain as much calcium as is present in 100 pounds of calcium oxide or pure lime requires about 310 pounds of gypsum.

Hydrated calcium sulfate in its commercial form of gypsum or land-plaster contains varying amounts of impurities, the presence of which decreases the percentage of calcium. Some land-plaster consists of only 50 per cent. pure sulfate of calcium, while the best contains over 95 per cent. It should therefore be purchased only on the basis of a guaranteed analysis. It should be used only in finely-ground condition. It is well to remember that most commercial super-phosphates contain about half their weight of gypsum.

Comparative composition of calcium compounds.—It is of practical importance to have a clear idea of the relative amounts of calcium contained in the principal compounds used to supply this constituent in agriculture, since it is the element calcium that these compounds are used to furnish. It is desirable to know the equivalent amounts of calcium in the various compounds. Knowing that calcium oxide (lime) contains 71.4 per cent. of calcium, calcium hydroxide (slaked lime) contains 54 per cent of calcium, and calcium carbonate contains 40 per cent. of calcium, we can prepare the following table, which enables one to find the equivalent amount of any of these three calcium compounds in terms of any other.

To illustrate the use of Table 23, one wishes to know how much calcium carbonate is equivalent to a ton of burned lime containing 95 per cent. of calcium oxide. First find the number of pounds of calcium oxide in the ton of burned lime by multiplying 2,000 by 0.95, which gives 1,900 pounds; then multiply 1,900 by 1.78 (the amount of calcium carbonate equivalent to 1 pound of calcium oxide), which gives 3,382 pounds as the desired amount of calcium carbonate. To take another example, how much calcium is contained in 1,000 pounds of limestone containing by analysis 90 per cent. of calcium carbonate? The stated amount of limestone contains 900 pounds (1,000 × .9) of calcium carbonate, which multiplied by 0.4 (the amount of calcium equivalent to 1 pound of calcium carbonate) gives 360 pounds as the desired amount.

TABLE 23—FACTORS SHOWING EQUIVALENT AMOUNTS OF VARIOUS CALCIUM COMPOUNDS

	To find from a given amount of		*Multiply by*
(1)	Calcium carbonate the equivalent amount of	calcium,	0.40
(2)	" " " " " "	calcium oxide,	0.56
(3)	" " " " " "	calcium hydroxide,	0.74
(4)	Calcium hydroxide " " " "	calcium,	0.54
(5)	" " " " " "	calcium oxide,	0.76
(6)	" " " " " "	calcium carbonate,	1.35
(7)	Calcium oxide " " " "	calcium,	0.71
(8)	" " " " " "	calcium hydroxide,	1.32
(9)	" " " " " "	calcium carbonate,	1.78
(10)	Calcium " " " "	calcium oxide,	1.40
(11)	" " " " "	calcium hydroxide,	1.85
(12)	" " " " "	calcium carbonate,	2.50

MAGNESIUM COMPOUNDS

Compounds containing magnesium are magnesium carbonate ($MgCO_3$), magnesium hydroxide ($Mg(OH)_2$) and magnesium oxide (MgO). In the so-called lime materials used in agriculture, they are present in smaller amounts than calcium, varying from a fraction of one per cent. up. Most of the ground limestone used on soils does not contain more than 10 per cent. of magnesium carbonate. Limestone containing 20 to 25 per cent. of magnesium carbonate is not objectionable and may even be desirable directly as plant-food on some soils deficient in magnesium. Carbonate and sulfate of magnesium have been used to a limited extent as indirect fertilizers. Their action is like that of common salt and calcium compounds in changing unavailable into available potassium. Kainite is about one-sixth magnesium sulfate. The sulfate of potash and magnesium is about one-third magnesium sulfate. The use of magnesium sulfate or carbonate cannot be recommended except in those rare cases of soils that are deficient in magnesium.

The magnesium compounds in limestone, burned lime and slaked lime neutralize acids and bring about other effects the same as calcium compounds. There are, however, quantitative differences in the chemical reaction of the compounds of magnesium as compared with those of calcium, due to difference in combining weights (p. 18). For example, in neutralizing acids, 1 pound of magnesium carbonate is approximately equal to 1.2 pounds of calcium carbonate; 1 pound of magnesium hydroxide, to 1.3 pounds of calcium carbonate; and 1 pound of magnesium oxide, to 1.4 pounds of calcium oxide. In view of these quantitative differences, it may be desirable, having a given amount of magnesium compound, to know how much calcium it is equal to in neutralizing power. This can be ascertained by use of Table 24.

TABLE 24—FACTORS SHOWING EQUIVALENT AMOUNTS OF CALCIUM AND MAGNESIUM COMPOUNDS

	To find from a given amount of		*Multiply by*
(1)	Magnesium carbonate	the equivalent amount of calcium oxide,	0.66
(2)	" "	" " " " calcium hydroxide,	0.88
(3)	" "	" " " " calcium carbonate,	1.19
(4)	Magnesium hydroxide	" " " " calcium oxide,	0.97
(5)	" "	" " " " calcium hydroxide,	1.28
(6)	" "	" " " " calcium carbonate,	1.72
(7)	Magnesium oxide	" " " " calcium oxide,	1.40
(8)	" "	" " " " calcium hydroxide,	1.85
(9)	" "	" " " " calcium carbonate,	2.50

To illustrate the use of the table, a limestone shows an analysis of 45 per cent. of calcium and 20 per cent. of magnesium carbonate. To what percentage of calcium carbonate is the sum of the two equivalent? Multiply 20, the amount of magnesium carbonate, by 1.19, the factor given in the table; the result is 23.80, which, added to 45, gives 68.80 as the answer.

In another case, a slaked magnesian lime contains 50 per cent. of calcium oxide and 25 per cent. of magnesium oxide. What is the equivalent of these two compounds, expressed as total calcium carbonate? First, convert the calcium oxide into the equivalent amount of calcium carbonate by multiplying 50 by the conversion factor 1.78 (Table 23), which gives 89. Next multiply 25 by the conversion factor 2.50 (Table 24), which gives 62.50. These two results (89 and 62.50) added together give 151.50 as the total amount of calcium carbonate equivalent to the given amounts of oxides of magnesium and calcium.

Weight of a bushel of lime.—It is the custom in some places to sell lime by the bushel. *It should be purchased only by weight,* because the weight of a bushel varies greatly according to a variety of conditions, chief among which are porosity of the limestone burned and the conditions of burning. The average weight of a bushel (heaped measure) of stone-lime is given as 75 to 93 pounds, but it may vary from 60 to 120 pounds. Magnesium stone-lime usually runs somewhat lower. Oyster-shell lime varies from 40 to 75 pounds a bushel and averages about 60.

Chemical changes and solubility of calcium compounds.—Some of the facts already stated about the changes that calcium compounds undergo need reviewing for the sake of emphasizing their importance and relations. Some additional facts regarding the solubility of calcium compounds deserve special mention at this point.

(1) *Change of quicklime to slaked lime.*—Calcium oxide (CaO) or quicklime, as previously stated, combines with water (H_2O) to form calcium hydroxide ($Ca(OH)_2$) or water-slaked lime; the attraction between quicklime and water is intense. Quicklime put on or in soils is changed into slaked lime before chemical action on the soil constituents takes place. When, therefore, we speak of the action of quicklime in soils, we usually mean the direct action of slaked lime formed from quicklime.

(2) *Change of slaked lime to carbonate.*—Slaked lime or calcium hydroxide and carbon dioxide have strong mutual attraction and combine to form the carbonate ($CaCO_3$). Perfectly dry quicklime is not acted upon by carbon dioxide; it must first be changed into hydroxide by combination with water. When quicklime *air-slakes,* the first action is combination with water to form hydroxide, after which carbon dioxide is taken from the air by the hydroxide and carbonate is formed. Whether applied to soils as quicklime or slaked lime, it is sooner or later changed into carbonate and does its principal work in the form of carbonate. In general, when we speak of the action of calcium compounds in soils, we usually mean the carbonate if we do not specify some other form.

(3) *Solubility of calcium compounds in water.*—The extent to which calcium compounds dissolve in water at ordinary temperatures is a matter of practical importance in connection with soils. Quicklime does not dissolve as such in water but changes into the hydroxide, which dissolves at the rate of one part in about 600 to 800 parts of water, the solution being known as *lime-water.* It is the calcium in this solution which combines so readily with carbon dioxide to form carbonate. Calcium carbonate dissolves only slightly in pure water, but, when water is saturated with carbon dioxide, one part of carbonate dissolves in about 1,000 parts of carbonated water, less carbonate dissolving in water containing less carbon dioxide. One part of hydrated calcium sulfate or gypsum is soluble in 400 to 500 parts of water; it is more soluble in water containing in solution common salt or nitrate of soda.

WHAT CALCIUM COMPOUNDS DO IN SOILS

When applied judiciously, calcium compounds produce several specific effects, resulting in improvement of soils and increase of crops; however, injurious effects may attend their use under some conditions, especially when certain forms are used in too large amounts. For convenience, the effects of calcium compounds may be grouped under three heads: (1) Chemical, (2) biological and (3) physical.

Chemical effects of calcium compounds.—The direct beneficial chemical effects of calcium compounds in soils may be divided into two general classes: (1) They make available some insoluble forms of plant-food, especially compounds containing potassium or phosphorus; and (2) they change into harmless forms, or neutralize the injurious effects of many compounds that interfere with the growth of plants.

(1) *Action on insoluble potassium compounds.*—One of the most important effects of calcium compounds is the conversion of insoluble into soluble forms of potassium; the soluble calcium takes the place of the potassium in the insoluble compounds and thus enables the potassium to form some new soluble form or combination. All of the conditions under which this change occurs are not understood and cases occur in which insoluble potassium compounds are not affected. The extent of this action is primarily dependent upon the amount and kind of potassium compounds in a soil. As would be expected, good results are more noticeable in clay soils, while sandy soils, which are often deficient in potassium compounds, are little affected. The effect of calcium in making potassium available is shown most prominently in the case of crops that require relatively large amounts of potassium, such as clover and other leguminous corps. It is probable that gypsum affects a soil more in this respect than in any other way.

(2) *Action on insoluble phosphates.*—Calcium compounds, especially hydroxide and carbonate, change insoluble phosphates into forms that can be more readily utilized as plant-food. This action is based on the following chemical facts: The phosphates of iron and aluminum are more or less prevalent in soils; these compounds become soluble only with extreme slowness under ordinary conditions. Calcium compounds, especially hydroxide and carbonate, change these insoluble phosphates into tri-calcium phosphate, which is much more readily soluble in water containing carbon dioxide, than in water that does not contain it. (p. 35.) The application of the calcium compounds mentioned is, therefore, of marked value in this way in case of soils rich in iron and aluminum compounds but poor in calcium carbonate.

(3) *Neutralization of acid soils.*—Soils which show acidity (p. 94), whether in the form of free acids ("sour" soils) or whether simply deficient in basic constituents (p. 20), are beneficially changed in character by the application of calcium compounds in the

form of quicklime, slaked lime, or carbonate, but *not by sulfate* (gypsum or land-plaster).

(4) *Action on soil toxins.*—In addition to specific acids, soils may contain compounds that act as poisons to plants. For example, magnesium compounds, when present in large proportions, may act in this way; so on some soils the application of magnesium com-

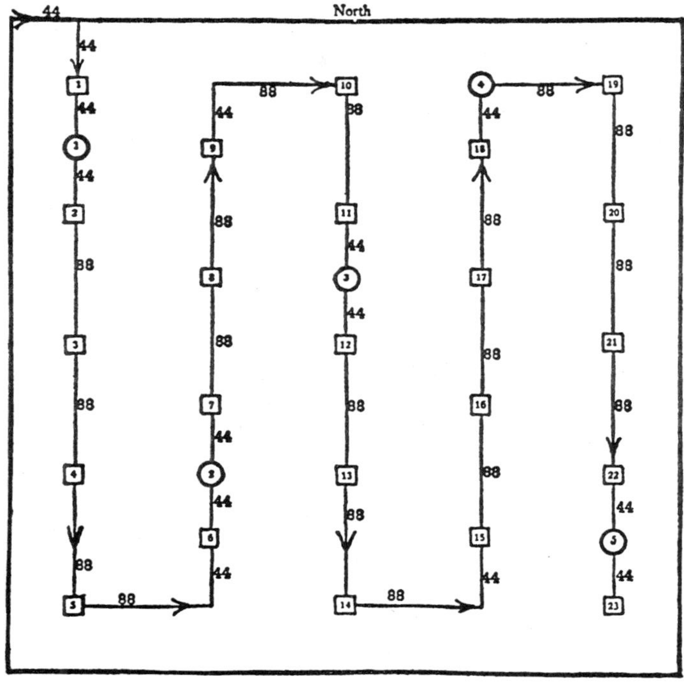

Soil sampling diagram for forty acre field. In making samples follow the direction of the arrows. The larger numbers along the lines indicate the number of three foot paces to take in locating the points where the samples are to be collected. Numbers in the squares indicate where twenty-three surface samples are taken; those in the circles where five subsoil samples are gathered.

pounds in magnesium limestone, kainite, etc., may work injury. This condition can be overcome by the use of sufficient amounts of calcium compounds. Other poisonous substances, in the form of acid organic compounds in soils, are derived from plants as products of the decay of organic matter in soils (p. 95). Their injurious effects are overcome or prevented by the presence of enough calcium hydroxide or carbonate in the soil. In some cases, injurious soluble salts in alkali soils are rendered less harmful by calcium compounds.

The "black alkali" in soils of arid regions, due largely to sodium carbonate, is changed by sulfate into (1) harmless calcium carbonate and (2) sulfate of soda, which is less poisonous to plants than is sodium carbonate.

(5) *Favorable to decomposition of organic matter.*—The decomposition of organic matter is favored by the presence of calcium compounds.

(6) *Economy of plant-food.*—It has been demonstrated that in the presence of abundance of calcium carbonate, smaller percentages of nitrogen, phosphorus and potassium compounds are generally required for crop production than when calcium is deficient.

(7) *Effect of calcium sulfate on ammonium carbonate.*—Ammonium carbonate easily escapes into the air, but in the presence of abundance of calcium sulfate (gypsum) it is changed into the stable form of ammonium sulfate, and the calcium becomes useful as carbonate. Quicklime and slaked lime do not possess this property, but, on the contrary, change all ammonia compounds into gaseous, free ammonia. Calcium carbonate does not have this unfavorable effect under agricultural conditions.

(8) *Unfavorable chemical effects of calcium compounds.*—The chemical effects thus far mentioned are for the most part beneficial, but unfavorable results, which cause loss of plant-food or result in special undesirable conditions, may attend the injudicious use of certain calcium compounds under some circumstances.

(*a*) *Loss of nitrates:* The formation of nitrates may be so hastened by calcium compounds, especially the caustic forms when used in large amounts, as to furnish nitrates more rapidly or more abundantly than crops can use. Under such conditions nitrates are likely to be lost by passing into the drainage water.

(*b*) *Acidity of soils and sulfate:* The excessive use of calcium sulfate (land-plaster)—1,000-2,000 pounds per acre—may result in adding acid to a soil. The calcium of this compound is removed from the soil, the sulfuric acid being left behind to combine with some basic material. If it continues to accumulate, in the course of time, it will obviously be present in sufficient quantities to contribute appreciably to the acidity of the soil. It must be emphasized, therefore, that *land-plaster never neutralizes soil acidity, but may help to make a neutral soil acid.*

(*c*) *Loss of sulfur:* Excessive use of calcium and magnesium compounds may cause loss of sulfur which in surface soils, may become serious when no sulfur compounds are supplied to the soil by rainfall or direct application.

Biological effects of calcium compounds.—In general, calcium compounds tend (1) to make soil conditions favorable for the growth of many crops and for certain important soil organisms, and (2) to prevent some kinds of plant disease.

(1) *Relation of calcium compounds in soils to kind of plant growth.*—The influence of calcium compounds in soils is shown by the character of growth of native trees and other plants. For example, chestnut trees, rhododrendron and arbutus show preference for soils not rich in calcium compounds; leguminous crops, like clover, alfalfa, etc., and many other crops thrive best only on soils well supplied with calcium carbonate. Many weeds, which, like sorrel, flourish in acid soils, are discouraged in the presence of calcium carbonate.

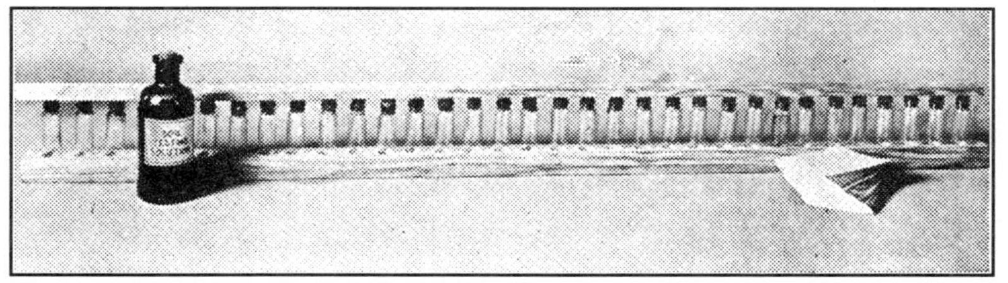

Rack, bottles, solution and paper bags ready for soil sample collecting.

(2) *Relation of calcium compounds to soil organisms.*—Calcium compounds, especially hydroxide and carbonate, are closely associated with the favorable action of certain important soil organisms, which cannot be active in the presence of acids. The effects may be grouped under four general divisions: (*a*) Promotion of the decomposition of organic matter; (*b*) making conditions favorable for nitrification; (*c*) providing essential conditions for the growth of nitrogen-gathering organisms associated with leguminous plants; and (*d*) favoring the growth of those soil organisms that utilize atmospheric nitrogen independently of other plants (p. 147).

(3) *Calcium compounds in relation to some plant diseases.*—Calcium compounds are specific as a preventive of the disease known as "club-root" or "finger-and-toe" disease found in cabbage, turnips and similar crops, this disease occurring only in acid soils.

(4) *Harmful biological effects of calcium compounds.*—A strongly alkaline condition of soil, caused by the use of too much caustic lime (quicklime or slaked lime), is likely to affect injuriously, for a while, bacterial and other processes while excess of calcium sulfate may cause unfavorable conditions by producing soil acidity. These effects may be specified more fully, as follows:

(*a*) Injury to nitrifying and decay organisms. A strongly alkaline condition of soil causes suspension of the normal processes of decomposition and nitrification. After a time, when the caustic calcium is changed into carbonate, normal biological activities are resumed.

(*b*) Production of potato-scab. The application of calcium compounds in amounts sufficient to make a soil neutral or alkaline usually favors potato-scab disease.

(*c*) Unfavorable action on crops. Some crops, like watermelon, cranberry, serradella, blue lupine, etc., give best results on acid soils and are unfavorably affected by considerable amounts of calcium compounds.

Soil testing. *Left,* Taking sample. *Center,* filling bottle one-third full of soil and enough solution to fill it two-thirds full. *Right,* Bottle rack covered, ready for vigorous shaking for two or three minutes, after which the bottles must stand ten minutes before being "read."

Physical effects of calcium compounds.—The chemical changes produced in soil by calcium compounds are accompanied by marked physical changes.

(1) *Beneficial effects.*—Clay soils are made less sticky, more crumbly and more easily cultivated; water passes through them more readily as the result of increased porosity, and such soils furnish more congenial conditions, in general, for the growth of plant-roots. When used on some sandy soils, in small amounts, preferably in the form of carbonate, calcium tends to act like a cement in holding soil particles together, thus enabling the soil to hold water more tenaciously.

(2) *Harmful effects.*—The most common harm done soils by application of calcium compounds is related to their structure. Certain sandy and gravelly soils are made too open and are, therefore, injured by large application of calcium compounds, especially in the caustic forms. Muck and peat soils may be injured when a calcium compound is applied in a thin, superficial layer; the surface

humus is more quickly used up with the result that quick injury by drouth may follow due to lack of water-holding power. To prevent this, the calcium compound should be worked more deeply into the soil.

PRACTICAL USE OF CALCIUM COMPOUNDS

Having learned their composition and various effects we must now consider the more important points connected with their practical use under the following divisions: (1) How to learn if a soil needs calcium, (2) how much and how often to apply, (3) at what time to apply, (4) what forms to use, (5) how to apply, (6) commercial classification of calcium compounds for agricultural purposes.

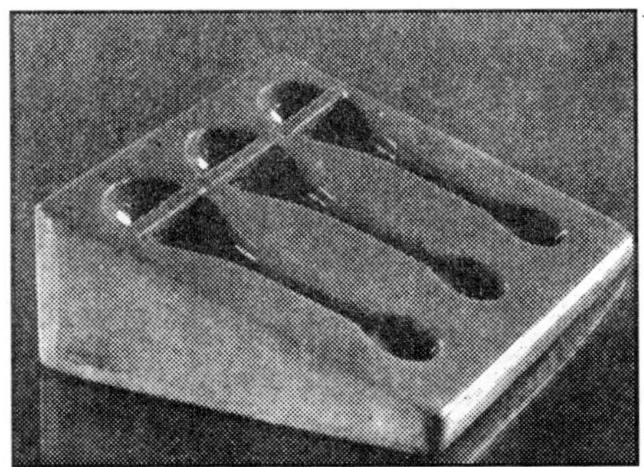

Morgan soil-test block used to obtain soil extracts for tests of soils.

How to learn whether soil needs calcium, either in the form of lime or of calcium carbonate. One or more of the following four tests may be used. The first two are merely tests for acidity. The third is more a specific test for calcium carbonate in the soil. But the fourth is specific in the sense that it shows the need of soil for either alkali or available calcium in order to produce conditions favorable for growth of the crop desired.

(1) *Litmus test for acidity.*—(See p. 21.)

(2) *pH determination on soil extract.*—Convenient apparatus and reagents for colorimetric pH determination in soil extracts, with directions for use, may be obtained from dealers.[1] The significance of the pH in soil extracts is as follows:

---

[1] Kits for the pH test can be obtained from the LaMotte Chemical Products Company, McCormich Bldg., Baltimore, Md., the Albert Dickson Company, Chicago, Ill., or Hellige, Inc., 179 E. 87th St., New York. The theory and technique of pH determinations by various methods are given in detail in "The Determination of Hydrogen Ions," W. Mansfield Clark. Published by William and Wilkins, Baltimore.

pH above 8.2, strongly alkaline.
pH 7.5-8.2, medium alkalinity.
pH 7.0-7.5, slightly alkaline.
pH 7.0, neutral.
pH 7.0-6.5, slightly acid.
pH 6.5-5.8, moderately acid.
pH 5.8-5.2, medium acidity.
pH 5.2-4.7, strongly acid.
pH less than 4.7, very strongly acid.

(3) *Dilute ammonia test.*

(4) *Growing of crops responsive to calcium carbonate.*—The foregoing tests have the advantage of rapidity and simplicity and, when both are used on the same soil, they give reliable indications, but there are cases in which the results may leave one in doubt or prove misleading. Though requiring time and care, the most reliable test is to raise on the soil some crop that requires calcium carbonate in generous amount, such as red clover or beets. This test is made by growing the crop on small portions of an acre, where the soil is uniform, one portion receiving some calcium compound like slaked lime or the carbonate and the other portion receiving none. In other respects the two portions of soil are treated alike; it is well to supply both with plant-food. In the case of red clover, the need of calcium carbonate is usually indicated when the crop after starting vigorously in the spring appears later to stand still in growth and finally disappears in part or wholly. Beets make poor growth on soils deficient in calcium carbonate. It has also been frequently noticed that horse-sorrel and some other weeds make vigorous growth on soils deficient in calcium carbonate, though such growth does not always show calcium carbonate deficiency.

Amount and frequency of application.—The amount of calcium compounds to apply is largely governed by three factors: (1) the character of the soil, (2) the kind of crops grown and (3) the form of calcium used, but in nearly all cases the general rule may be followed of making applications in smaller quantities at more frequent intervals rather than large amounts once in several years.

(1) *Character of soil.*—On poor soils and in case of light, dry soils, the amount of calcium compounds applied should be comparatively small, varying from 900 to 2,700 pounds of pulverized limestone (carbonate) an acre (equivalent to about 700 to 2,000 pounds of slaked lime and 500 to 1,500 pounds of quicklime). On heavy clay soils and on soils that contain large amounts of decaying, acid, organic material, the application may vary from 1,880 to 7,000 pounds of pulverized limestone an acre (equivalent to about 1,300 to 5,000 pounds of slaked lime and 1,000 to 4,000 pounds of quicklime), according to frequency of application, degree of soil acidity, etc. Soils rich in organic matter can utilize calcium compounds, in large amounts, more fully and with less danger of injury to soil or crops than can soils deficient in organic matter.

In 5 or 6 year rotations, one liming will suffice under usual conditions. It is preferably done at the time when the largest amount of organic matter is in the soil, as for example when a green crop or sod is plowed under. In general the importance of supplying the soil with abundance of vegetable matter along with the use of calcium compounds cannot be emphasized too strongly.

(2) *Kind of crops grown.*—Some crops are little affected by the amount of calcium carbonate present within wide limits. Therefore the amounts applied may vary widely without materially affecting plant growth, other conditions being favorable. However, most of the crops commonly grown are usually benefited by occasional use of calcium compounds in generous amount. Some few crops appear to be indifferent to the presence or absence of calcium carbonate in the soil, while a few plants are generally injured when much is present—watermelon, radish, potato, blue lupine, serradella, blackberry, raspberry and a few others of little importance. It should be stated that calcium carbonate may be applied in much larger amounts than quicklime or slaked lime without risk of injury to crops.

(3) *Form of calcium compounds used.*—In applying calcium compounds, we must keep in mind that the constituent of value supplied is *calcium.* When we put on a more dilute form, as the carbonate or the hydroxide (slaked lime), we must use more than when we apply the more concentrated form, quicklime or calcium oxide. From the data given (p. 197) we can make the following simple rule to find out how to make the necessary calculations, assuming that the various forms are fairly alike in purity of composition: *To find out how much slaked lime is equal to a given amount of quicklime, multiply the number of pounds of quicklime by 1.3; to find out how much calcium carbonate is equal to a given amount of quicklime, multiply the number of pounds of quicklime by 1.8.*

When to apply calcium compounds.—Generally speaking, the matter of convenience may guide one as to what particular time of year to apply calcium compounds. Usually spring is more convenient than fall. The character of rotation will decide in many cases. In general the application should be made when it can be done to best advantage in the rotation, for example, before seeding to clover or in connection with the use of a green-crop manure. When practicable, autumn is the best time to apply the caustic forms, *quicklime* or *slaked lime,* on land to be used for spring sown crops and on permanent grass lands. The power of these compounds to injure seeds is gradually lost by lying in the soil, since they are changed into harmless carbonate. *Calcium carbonate can be applied at any time* without risk of injuring seeds or crops. In case of fall seeding, quicklime or slaked lime can be scattered after plowing and then harrowed in thoroughly. For many crops, the caustic forms of calcium may be applied in spring with little risk of injury to seeds, provided it is worked into the soil completely and uniformly. In the case of sour soils, application in spring is often beneficial. Lime always works downward in the soil. When applied before plowing or digging it is buried several inches below the surface so only the soil below this inverted layer is benefited, the surface layer not at all. Therefore the important rule: Always apply lime after plowing or digging, then harrow or rake it in the surface.

When fertilizers that contain ammonia compounds are to be applied to a soil receiving quicklime or slaked lime, it is well to apply the latter at least two weeks before the fertilizer in order to avoid risk of loss of ammonia. Stable manure should be thoroughly incorporated in the soil when the caustic compounds of calcium are to be used; on soils that possess little power of retaining ammonia, like light, sandy soils, farm manure should not be applied until some weeks after caustic calcium compounds are used.

What form of calcium to use.—The particular calcium compound to be used must be determined by the conditions attending each farmer, among which the following are the most important: (1) The cost of the calcium, (2) the character of soil, (3) the kind of crop, (4) the rapidity of action desired, (5) fineness of division, and (6) convenience of handling.

(1) *Cost.*—When the material must be freighted or drawn a long distance, quicklime in the form of commercial stone-lime or granulated lime should be found cheapest, since it is the most concentrated form, containing per 100 pounds more of the desired element, calcium, than any other form suited to agricultural use.

Table 24A shows at what price one should be able to buy lime. The prices assume high grade of purity and equal fineness of division.

TABLE 24A—COMPARISON OF PRICES FOR VARIOUS CALCIUM COMPOUNDS

	Equivalent prices at which should be purchased one ton of		
Cost of one ton of quicklime (CaO)	*Slaked lime* (CaH_2O_2)	*Carbonate* ($CaCO_3$)	*Gypsum* ($CaSO_4.2H_2O$)
$8.00	$6.05	$4.50	$2.60
7.00	5.30	3.95	2.30
6.00	4.55	3.40	1.95
5.00	3.80	2.80	1.65
4.00	3.05	2.25	1.30
3.00	2.30	1.70	1.00

This table shows, for instance, that when quicklime costs $5. a ton delivered the cost of calcium in other forms should not exceed

\$3.80 for slaked lime, \$2.80 for carbonate and \$1.65 for gypsum. In the writer's neighborhood high grade carbonate (fine-ground marl) costs \$4. to \$5. a ton, delivery extra, thus making the calcium in it cost nearly twice as much as in quicklime.

Calcium-containing materials are sold as *agricultural lime, prepared lime, hydrated lime,* etc. Some of these are slaked lime, some carbonate and some mixtures of both. In some cases their coarseness reduces their efficiency. As their prices, \$8. to \$10. a ton, are higher than the best grade of lump lime, farmers are advised not to buy them. Some of them may be used without slaking but this convenience costs too much. When its cost and limitations are considered, gypsum can hardly be used profitably except, perhaps when applied directly to manure (p. 236).

(2) *Character of soil.*—On light, sandy or porous soils and on any soils low in amount of organic matter, application of caustic forms of calcium compounds tends to hasten the decomposition of organic matter, unless used in amounts of less than 1,000 pounds an acre. It is safer, therefore, to use on such soils the carbonate which can be applied in large amounts without risk of injury. On heavy, clay soils and on soils rich in organic matter, caustic forms of calcium can be used in large amounts (1 to 2 tons of quicklime) without danger of decomposing humus too rapidly. The effect that quicklime and slaked lime have on organic matter should always be met by keeping an abundant supply of vegetable matter in the soil by means either of farm manure or of green-crop manures. On some soils that contain large amounts of magnesium compounds, it is desirable to avoid the use of lime rich in such compounds. However, on most soils, no trouble occurs from the presence of such compounds in the material used for liming.

(3) *Kind of crop.*—In the case of crops which are so sensitive to an alkaline condition of soil as to be injured, the use of carbonate is preferable. For crops that flourish best in neutral or slightly alkaline condition of soil, quicklime or slaked lime can be used with the precaution of avoiding application of excessive amounts (p. 207).

(4) *Rapidity of action.*—Quicklime is changed into slaked lime before it becomes chemically active under ordinary soil conditions. Properly slaked, quicklime forms a finer powder than it is possible to obtain with ground limestone or marl; it can therefore be more thoroughly and uniformly distributed through the soil, a condition that enables it to act more quickly. Moreover, freshly-slaked lime consists mostly of calcium hydroxide, which is more soluble in water than is calcium carbonate, and the usual difference of solubility is made greater by the extreme fineness of the slaked lime. This solubility is an important factor when a freshly limed soil receives

abundance of rain, because a considerable amount of hydroxide goes into solution and on this account is more quickly and uniformly distributed through the soil. Then, again, calcium hydroxide, on account of its somewhat greater solubility, is more active than the carbonate in producing chemical changes in the soil. It is true that sooner or later the slaked lime changes into carbonate, but even under favorable conditions some weeks are required to complete the change and during the intermediate period the slaked lime must be regarded as more active chemically than the carbonate.

(5) *Fineness of division* has been referred to in the preceding paragraph. Quicklime in the form of commercial lump-lime or stone-lime must be slaked in order to permit even distribution in the soil. In properly slaked lime we have as fine a powder as it is possible to get; practically all of it passed through a 200-mesh sieve. Ground limestone and marl are not nearly as fine as well-slaked lime. For the more prompt action of ground limestone, not less than 75 per cent. should pass through a 100-mesh sieve. Fineness is of added importance in case of magnesian limestone. Where quick action is not desired, some farmers have used coarser limestone with apparent success, 60 per cent. passing a 50-mesh sieve, but none coarser than required to pass a 10-mesh sieve. When it will answer the purpose, limestone less finely ground has the advantage of lower cost.

(6) *Convenience of handling.*—The slaking of lump-lime and the distribution of the slaked lime cause some inconvenience and unpleasantness in handling. This disagreeable feature can be lessened by mixing the slaked lime with earth before distributing or by the use of ground burned lime, which is used for building purposes, and which is sold at reasonable prices. This can be handled with comparatively little trouble because it is usually put up in tight bags and does not require slaking before application. Much of the disagreeableness of distributing fine, slaked lime can be avoided by using a lime-spreader. Ground limestone or marl is also convenient to handle and is less disagreeable than slaked lime.

How to apply calcium compounds to soils.—In applying calcium compounds to soils, three facts should be kept in mind: (1) The lime should be distributed for the most part in the upper layer of the soil, say within 3 or 4 inches of the surface; (2) it should be distributed as uniformly and thoroughly as possible through the soil; (3) in the case of quicklime, it should be put into the soil while mostly in the form of slaked lime or hydroxide and before being changed more than slightly into carbonate.

The most convenient form of lime to apply is probably the ground quicklime, furnished for building purposes, which can usually be distributed with a fertilizer-drill. There are, however, special lime-spreaders which are more convenient.

Freshly burned lump-lime may be slaked in a large pile near the field and then distributed; but when one has no special means for distributing the slaked lime, it is more convenient and comfortable under ordinary conditions to distribute the lump-lime in small piles over the field after the land is prepared for a crop. A little water (equal to about one-third the weight of the lime or about 4 gallons of water for one bushel of lime), is slowly sprinkled on each heap and the lime covered with fine earth. To secure the finest powder possible, the slaking should not take place rapidly; the use of too much water should be avoided. When thoroughly slaked, the powder is mixed with more fine earth to make it less disagreeable to handle and distributed with a shovel as evenly as possible, after which it is promptly harrowed into the soil thoroughly. The size and frequency of the heaps of lime may be regulated by remembering that 20-pound heaps placed 20 feet apart make about one ton to the acre; or 25 heaps of quicklime with one bushel in each, make about one ton. After being spread, slaked lime should not be allowed to lie upon the surface overnight or during showers, owing to its tendency to cake and form lumps or crust, thus changing it from a fine powder into coarse particles or lumps.

Air-slaking is not advised, because it takes too long and the product is more largely carbonate than slaked lime proper.

MISCELLANEOUS MATERIALS USED AS INDIRECT FERTILIZERS

In addition to calcium and magnesium compounds and green-manure crops, many materials have been used as indirect fertilizers, some of undoubted value under special conditions and others of doubtful value or capable of positive harm. It is not desirable that we should discuss all these, but we will select for brief notice a few about which there is occasional inquiry for information. The materials to be discussed are the following: (1) Muck, peat, etc., (2) common salt, (3) magnesium compounds, (4) sulphate of iron, (5) soot, (6) coal ashes.

Muck, peat, etc.—The properties and composition of these materials have been already considered (p. 93). Their chief function as indirect fertilizers is due to their organic matter. The nitrogen they contain is slowly available. The effect of these materials is mainly physical, lightening clay soils and making them more porous, while rendering sandy soils more compact and more tenacious in holding water. In order to produce much effect on soils large amounts of muck or peat need to be added, 20 to 40 tons or more. Muck is more quickly effective than peat and easier to incorporate evenly into the soil. These materials cannot usually be economically employed unless close at hand and easy to handle. They should be removed and allowed to air-dry before distribution. Green crops will generally be found the most efficient source of organic matter. Muck and peat are best employed to mix with stable manure as an absorbent or as part of a compost mixture. While performing a useful service in saving urine in the stable, the insoluble nitrogen of the muck and similar materials is made available more quickly when mixed with stable manure than when the muck is put into the soil directly.

A good bed of muck or peat on a farm should be regarded as an important source of soil fertility, though largely indirect. It should be utilized systematically in one's plans for improving the soil. The value of these materials, when they are rightly utilized, can hardly be overestimated.

Common salt.—Common salt, consisting largely of sodium chloride (p. 44), has been used from time immemorial in agriculture, but its method of action has been little understood until recently and probably not fully even now. The form mostly used at present is known as "agricultural salt," which is refuse salt, too impure to utilize in regular commercial ways. The chief action of salt is believed to be that of changing the potassium already in the soil from an insoluble to a soluble form of combination. It also appears to economize the amount of potash used by crops in the case of root crops. Its extensive use on asparagus is of doubtful value aside from the setting free of potassium. In some cases experiments appear to show that salt renders insoluble phosphates soluble, while other experiments show the reverse. If used at all, it should be employed at first only experimentally on a small scale. The amount to use may vary from 200 to 600 pounds an acre. Larger applications at one time should be avoided.

Another effect of salt is to hold soil moisture and regulate to some extent the upward movement of water. The extent of this action is of theoretical interest more than of practical importance. Salt injures the quality of such crops as tobacco, sugar-beets, potatoes on account of the chlorine in it. One-third of kainite is common salt.

Sulfate of iron.—Sulfate of iron is mentioned chiefly because its agricultural use has not been justified by reliable experimental evidence. Claims, based on inference rather than on the results of rigid experimental investigation, have been made to the effect that the bright coloring of apples and certain flowers is due to the presence of iron compounds. When viewed in the present state of

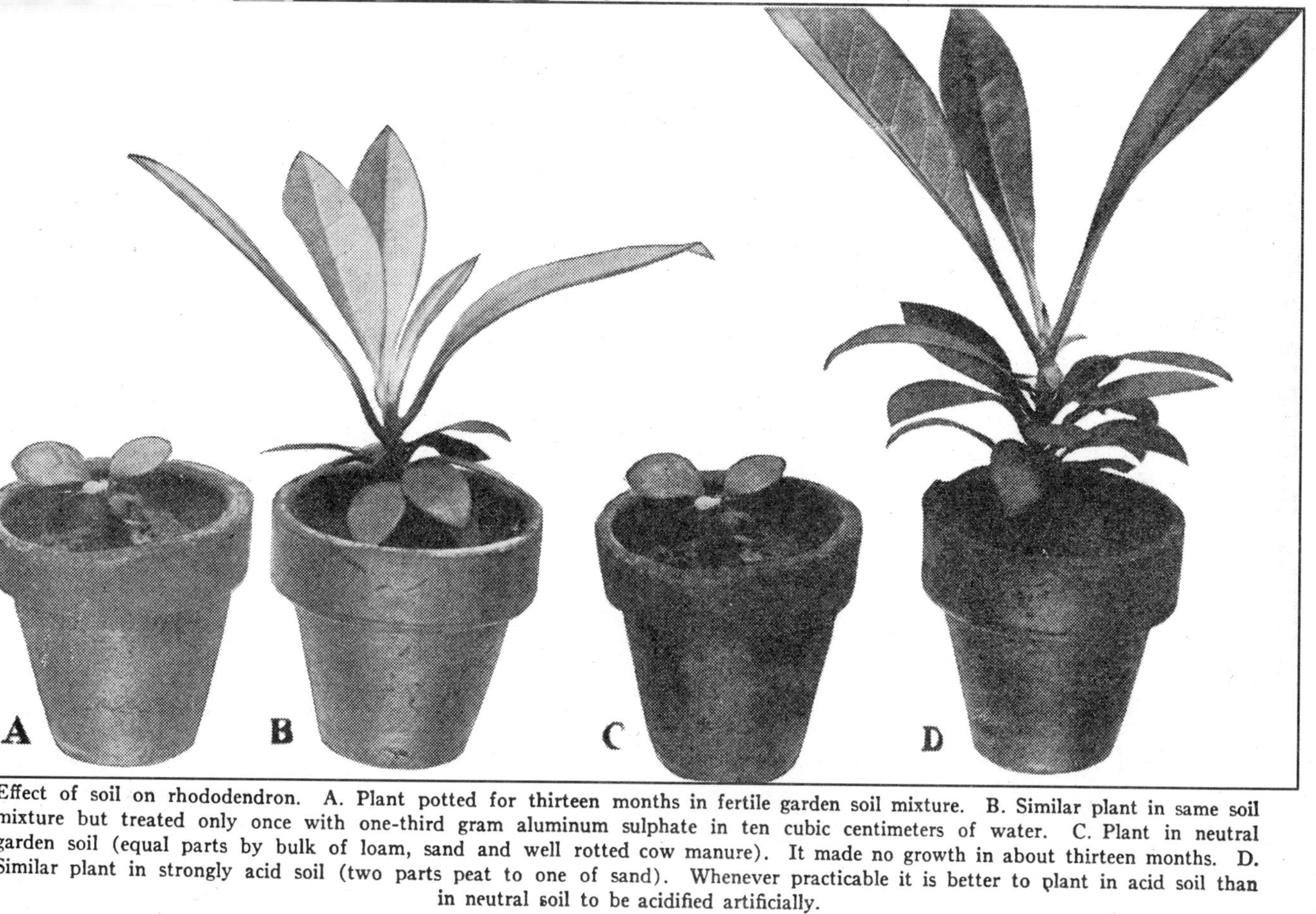

Effect of soil on rhododendron. A. Plant potted for thirteen months in fertile garden soil mixture. B. Similar plant in same soil mixture but treated only once with one-third gram aluminum sulphate in ten cubic centimeters of water. C. Plant in neutral garden soil (equal parts by bulk of loam, sand and well rotted cow manure). It made no growth in about thirteen months. D. Similar plant in strongly acid soil (two parts peat to one of sand). Whenever practicable it is better to plant in acid soil than in neutral soil to be acidified artificially.

our knowledge, the use of these compounds to supply iron to crops can be regarded as wasted effort. And when we consider the extremely minute amounts used by plants, all agricultural soils contain relatively immense amounts of iron compounds.

Soot.—While soot is to be regarded primarily as a nitrogenous fertilizer, containing on an average about 3 per cent. of nitrogen in the form of ammonia compounds, though varying from 0.5 to 11 per cent., it also possesses important properties as an indirect fertilizer. On account of its dark color, it absorbs heat when sprinkled on the surface of soil and appreciably raises the tempera-

In testing the effects of soils and fertilizers on plants of various kinds, metal and tile cylinders sunk in the ground but with their rims high enough above the surface to prevent splashing by rain have been found to provide more natural conditions than flower pots and crocks exposed to the air.

ture of the surface soil. Soot is effective in lightening heavy soils. Another property, not a part of its action on soil, is its killing value in case of slugs and small snails, which sometimes injure grain and other crops in their early stage of growth. Its use is of necessity limited on account of lack of supply. It can be used, when obtainable, as a top-dressing in spring on cereal crops.

Coal ashes.—Coal ashes possess marked value for their power to lighten heavy soils, and may be advantageously used for this purpose when near at hand and obtainable without further cost than that of hauling and spreading.

Aluminum sulfate as a soil amendment.—Probably the most important horticultural discovery of the present century is the treatment that assures the successful cultivation of blueberry, rhododendron, azalea, mountain laurel, trailing arbutus, gentian and various other plants which languish and die when planted in neutral or alkaline soils. Starting with blueberry, F. V. Coville, botanist of the United States department of Agriculture, conducted experiments which proved that these subjects require soils with positive acid reaction. He also proved that neutral and alkaline soils may be made congenial to them by being acidified. The chemical he recommends is aluminum sulfate, at the rate of "up to a pound" to the square yard for loams of ordinary fertility. The sulfate is spread on the ground, evenly mixed with the surface soil by raking and made to soak a foot deeper by rain or copious watering with a hose. See *Directions for Blueberry Culture,* United States Department of Agriculture Bulletins 334 (1916) and 974 (1921); also *The Effect of Aluminum Sulfate on Rhododendrons and Other Acid Soil Plants,* Smithsonian Report for 1926 pp. 369-382. (Publication 2897, 1927).—*Editor.*

Test for lime deficiency.—The dark color of the water in muck and peat swamps, and of forest ponds, lakes and streams is usually caused by decomposed organic compounds which have leached out of the soil. Somewhat the same kinds of leachings impart the still darker color to water that seeps from manure piles exposed to the weather. In cultivated soils they are less evident because they are either more dilute or are lacking through having been neutralized by alkalies. When calcium carbonate occurs in the soil or is supplied in sufficient quantities it unites with these organic acid compounds to form new compounds. When calcium compounds, especially the carbonate, are lacking or in small supply the organic acid compounds are not neutralized and so tend to make the soil acid.

This acidity may be readily determined by testing the soil for calcium compounds as follows: Half fill a glass bottle with water, preferably distilled or rain water because such water not having had contact with the soil is free of calcium. To this add a tablespoonful of soil and one or two teaspoonfuls of strong ammonia that has been diluted with four or five times as much water. Shake thoroughly and allow to stand over night or for several hours. If organic acid compounds are present the liquid will be dark brown or black or if not it will be more or less clear. The dark color is caused by the solution of combined ammonia and acid compounds. When calcium carbonate is present in the soil it combines with the organic acids to form new compounds which are insoluble in ammonia, but when calcium is lacking in the soil the organic acid compounds continue uncombined and reveal their presence by means of the ammonia test for calcium.

When manganese-deficient soil is treated with manganese sulfate it produces the striking contrast here shown. *Left*, String beans; *Right*, soybeans (20) without sulfate (1) sulfate added.

XVIII

FARM MANURE: COMPOSITION AND CHANGES

FARM MANURE consists of a mixture of three general classes of constituents: (1) Domestic animal dung or feces, (2) urine, and (3) materials used for bedding and often, in addition, miscellaneous waste farm products.

Farm manure is known under a variety of names, such as manure, mixed manure, barn manure, stable manure, farmyard or barnyard manure, etc. Usage of terms is not uniform. The term, mixed manure, is generally applied to a mixture of manures of various animals with addition of bedding. The manure from any particular class of animals is designated by the name of the animal, as horse manure, sheep manure, etc. We shall use the general terms, farm manure, manure, mixed manure, barn manure, etc., with a single meaning, namely, the mixed excrements, solid and liquid, of farm animals in general, containing material added for bedding. We shall use the term excrements, as applying to the feces and urine not mixed with other materials. The manurial products from individual classes of animals will be indicated by the name characteristic of the animal producing the excrements.

In connection with the use of farm manure as a source of nutrient materials, we are interested in the following points: (1) Composition of mixed farm manure, (2) conditions affecting composition, (3) care of farm manure, (4) fertilizing value, (5) methods of use.

PLANT-FOOD COMPOSITION OF MIXED FARM MANURE

We shall appreciate later how difficult it is to give a representative statement of the composition of a material which must of necessity vary greatly. The mixture may contain the excrements of one kind of animal or of several kinds. The bedding may be of several different materials. It seems desirable to start with some idea of the composition of mixed farm manure, including the approximately extreme limits of variation and what we may regard as a fair average. We shall limit our study of the composition of farm manure mainly to the nitrogen, phosphorus and potassium compounds.

TABLE 25—PERCENTAGE OF PLANT-FOOD CONSTITUENTS IN MIXED FARM MANURE

	Nitrogen		*Phosphoric acid (P_2O_5)*		*Potash (K_2O)*	
	Per cent.	*Pounds in one ton*	*Per cent.*	*Pounds in one ton*	*Per cent.*	*Pounds in one ton*
Lowest ..	0.4	8	0.2 (0.09P)	4 (1.8P)	0.4 (0.33K)	8 (6.6K)
Highest ..	0.8	16	0.4 (0.18P)	8 (3.5P)	0.8 (0.66K)	16 (13.3K)
Average .	0.5	10	0.25(0.11P)	5 (2.2P)	0.5 (0.42K)	10 (8.3K)

P, phosphorus. K, potassium.

It will be convenient to remember that one ton of average mixed stable manure contains approximately

10 pounds of nitrogen (12 lbs. ammonia).
5 pounds of phosphoric acid (2.2 lbs. phosphorus).
10 pounds of potash (8.3 lbs. potassium).

CONDITIONS AFFECTING COMPOSITION OF FARM MANURE

Variation in the plant-food composition of farm manure is produced by numerous causes, chief among which are the following: (1) Composition of fresh excrements, (2) kind and amount of bedding material in manure, (3) losses of plant-food constituents.

Variation in composition of fresh excrements of animals.—A study of the plant-food composition of the fresh excrements of various kinds of farm animals is desirable because this is the first factor in determining the nutrient value of farm manures. The composition of the excrements of farm animals varies greatly according to several conditions, among which the following are of special importance: (1) *The liquid and solid portions,* (2) *the kind of animals,* and (3) *the composition of the food consumed.*

Composition of solid and liquid excrements of various farm animals.—While the excrements of animals of the same kind vary greatly, and even of the same animal at different times, it is desirable that we have an approximate idea of their general plant-food composition. The following data (Table 26), taken from the most reliable sources available, will serve to show the approximate amounts of nitrogen, phosphoric acid and potash in the fresh excrements of farm animals. We state in round numbers (1) the usual limits of variation of each constituent in the solid and liquid portions re-

Effect of manure in continuous corn culture. Plot at the left received five tons of manure; that at the right, nothing.

spectively and (2) what may be regarded as the approximate average composition of solid and liquid portions, (*a*) separate and (*b*) mixed. In the case of water in fresh excrements, we give only the approximate average. In one column we state the approximate average percentage of the total excrements which the liquid and solid excrements respectively form.

TABLE 26—PERCENTAGE OF PLANT-FOOD CONSTITUENTS IN FRESH ANIMAL EXCREMENTS

Kind of Animal	*Excrement*		*Water*	*Nitrogen*	*Phosphoric acid*	*Potash*
	Portion	*Per cent.*	*Per cent.*	*Per cent.*	*Per cent.*	*Per cent.*
	Solid	80	75	0.55 (0.50-0.60)	0.30 (0.25-0.35)	0.40 (0.30-0.50)
Horse..	Liquid	20	90	1.35 (1.20-1.50)	trace	1.25 (1.00-1.50)
	Mixed	—	78	0.70	0.25	0.55
	Solid	70	85	0.40 (0.30-0.45)	0.20 (0.15-0.25)	0.10 (0.05-0.15)
Cow...	Liquid	30	92	1.00 (0.80-1.20)	trace	1.35 (1.30-1.40)
	Mixed	—	86	0.60	0.15	0.45
	Solid	60	80	0.55 (0.50-0.60)	0.50 (0.45-0.60)	0.40 (0.35-0.50)
Pig.....	Liquid	40	97	0.40 (0.30-0.50)	0.10 (0.07-0.15)	0.45 (0.20-0.70)
	Mixed	—	87	0.50	0.35	0.40
	Solid	67	60	0.75 (0.70-0.80)	0.50 (0.45-0.60)	0.45 (0.30-0.60)
Sheep...	Liquid	33	85	1.35 (1.30-1.40)	0.05 (0.02-0.08)	2.10 (2.00-2.25)
	Mixed	—	68	0.95	0.35	1.00
Hen....	Mixed	—	55	1.00 (0.55-1.40)	0.80 (0.35-1.00)	0.40 (0.25-0.50)

Attention is called to the following points, brought out in a study of the data embodied in Table 26:

(1) *Distribution of excrements between solid and liquid portions.*—The relative amounts of solid and liquid portions in fresh excrements vary greatly in the various classes of animals. In the case of horses, the solid portion is about four times as much as the liquid; in the case of pigs, the solid is only one and one-half times as great as the liquid portion. These amounts are, of course, only approximate averages; the proportions of solid and liquid excrements vary widely with the same animal under different conditions.

(2) *Percentage of water.*—There is a marked difference in the average amount of water in the excrements of various kinds of farm animals. In the solid portion, the excrement of sheep contains least moisture (60 per cent.); that of cows, most (85 per cent.). In the liquid portion, the amount of water is least (85 per cent.) in the case of sheep and greatest (97 per cent.) in the

case of pigs. In the case of poultry, there is no separation of solid and liquid excrements, but the amount of moisture in the combined excrement is much less than even in the solid portion of the excrement of the other animals under consideration.

(3) *Comparison of composition of solid and liquid portions.*—In comparing the composition of solid and liquid portions of animal excrements, we notice the following differences: (*a*) In respect to nitrogen, the liquid portion is much richer except in case of pigs; (*b*) in respect to phosphoric acid, the solid portion contains practically all, and the liquid portion only small amounts; (*c*) in respect to potash, the liquid portion is much richer except in case of pigs. The richness of the liquid excrement in plant-food constituents is made more striking if we compare them in respect to their composition in the water-free or dried condition, as illustrated in Table 27, in which only general average results are given:

TABLE 27—COMPOSITION OF DRIED OR WATER-FREE EXCREMENTS

Kind of Animal	*Portion of Excrement*	*Nitrogen*	*Phosphoric acid*	*Potash*
		Per cent.	*Per cent.*	*Per cent.*
Horse	Solid	2.20	1.20 (0.53P)	1.60 (1.33K)
	Liquid	13.50	——	12.50 (10.40K)
Cow	Solid	2.65	1.35 (0.60P)	0.65 (0.54K)
	Liquid	12.50	——	17.00 (14.10K)
Pig	Solid	2.75	2.50 (1.10P)	2.00 (1.65K)
	Liquid	13.00	4.00 (1.75P)	15.00 (12.45K)
Sheep	Solid	1.90	1.25 (0.55P)	1.15 (0.95K)
	Liquid	9.00	0.35 (0.15P)	14.00 (11.60K)

P, phosphorus. K, potassium.

Amounts of excrements produced by farm animals.—In order to make our discussion more complete, it is essential to consider not only the nutrient composition of the excrements of farm animals but that we should also know the total amounts and value of plant-food constituents annually produced by farm animals.

The amounts of fresh excrements produced by farm animals are subject to wide variations, being controlled by the kind of animal, age, amount of food, activity, etc. It will, however, be of some service to give approximate averages which can be used as a comparative basis of reference in estimating amounts of manure that can be reasonably expected from animals. The relative amounts of dung and urine vary with many conditions, so the proportions given below can be regarded as only rough approximations under average conditions. In order to be fairly specific and comparable, the results are stated on the basis of 1,000 pounds of animal live weight during a period of one year.

TABLE 28—AMOUNTS OF EXCREMENTS PRODUCED FOR 1,000 POUNDS OF LIVE WEIGHT BY FARM ANIMALS IN ONE YEAR

Kind of animal	*Total excrements*	*Solid*	*Liquid*	*Percentage of excrements in solid*	*Percentage of excrements in liquid*
	Pounds	*Pounds*	*Pounds*		
Horse	18,000	14,400	3,600	80	20
Cow	27,000	19,000	8,000	70	30
Pig	30,500	18,300	12,200	60	40
Sheep	12,500	8,300	4,200	67	33
Hen	8,500	——	——	—	—

Amounts and value of plant-food constituents produced in excrements.—From the data (Table 28) and from the average percentages of plant-food constituents in fresh excrements, as given in Table 26, we are able to obtain the results contained in Table 29, which shows the number of pounds of nitrogen, phosphorus and potassium produced on an average in the fresh excrements of various farm animals for one year, based on 1,000 pounds of live weight. The values in the last column are based on the following schedule pound prices: Nitrogen, 12 cents; phosphoric acid, 5 cents; potash, 4.5 cents.

Sheep are excellent grazers and spreaders of manure. Not only on range but on large lawns, in parks and golf courses they are often used instead of lawn mowers.

TABLE 29—PLANT-FOOD CONSTITUENTS PRODUCED ANNUALLY IN EXCREMENTS BY FARM ANIMALS PER 1,000 POUNDS OF LIVE WEIGHT

Kind of animal	Nitrogen		Phosphoric acid		Potash		Value of plant-food constituents
	Solid	Liquid	Solid	Liquid	Solid	Liquid	
	Pounds	Pounds	Pounds	Pounds	Pounds	Pounds	
Horse ...	79	49	43 (19 P)	——	58 (48K)	45 (37K)	$22.15
Cow	76	80	38 (17 P)	——	19 (16K)	108 (90K)	26.30
Pig	101	49	92 (40 P)	12 (5.3P)	73 (61K)	55 (46K)	29.00
Sheep ...	62	57	42 (18.5P)	2 (0.9P)	38 (32K)	88 (73K)	22.15
Hen	85	—	68 (30 P)	——	32 (27K)	——	15.00

P, phosphorus. K, potassium.

Table 29 can be made of additional interest by the supplementary table (30), in which the above data are presented in the form of percentages; for example, we state what percentage of the total nitrogen in the combined excrements is contained in the solid portion and what percentage in the liquid.

TABLE 30—RELATIVE AMOUNTS OF PLANT-FOOD CONSTITUENTS IN SOLID AND LIQUID EXCREMENTS

	Percentage of total nitrogen		Percentage of total phosphorus in		Percentage of total potassium in	
	Solid	Liquid	Solid	Liquid	Solid	Liquid
Horse	62	38	100	0	56	44
Cow	49	51	100	0	15	85
Pig	67	33	88	12	57	43
Sheep	52	48	95	5	30	70
Average	57	43	95	5	40	60

Somewhat striking differences exist in the case of various classes of animals. In general, most of the phosphorus appears in the solid excrement. In the case of horses and pigs, the yield of nitrogen is much greater in the solid excrement; in case of cows and sheep, the yield is approximately equally distributed between the solid and liquid portions. In the case of potassium, the excrements of horses and pigs show a larger amount in solid than in liquid, while the reverse is true of the excrements of cows and sheep, large proportions appearing in the urine as compared with the dung.

From the data in Table 31 the approximate amounts of plant-food constituents produced annually in the fresh excrements by a single animal of a given weight can be readily calculated. For illustration, a table such as this one can be prepared and easily extended to include an animal of any special weight. We repeat the precaution that undue dependence must not be placed upon data thus obtained, since they are not intended to apply accurately in detail to all cases. They should be regarded only as serving for a convenient, rough basis, to be used when other desirable details are lacking.

TABLE 31—APPROXIMATE PLANT-FOOD CONSTITUENTS PRODUCED ANNUALLY IN EXCREMENTS BY FARM ANIMALS OF GIVEN WEIGHT

Kind of animals	*Weight*	*Nitrogen*	*Phosphoric acid*	*Potash*	*Value of plant-food constituents*
	Pounds	*Pounds*	*Pounds*	*Pounds*	
Horse	900	115	38 (17P)	93 (77 K)	$19.90
	1200	154	52 (23P)	124 (103K)	26.65
Cow	800	125	30 (13P)	102 (85 K)	21.10
	1100	172	42 (19P)	140 (116K)	29.00
Pig	100	15	11 (5 P)	13 (11 K)	2.95
	200	30	21 (9 P)	26 (29 K)	5.90
Sheep	100	12	5 (2 P)	13 (11 K)	2.30
	150	18	7 (3 P)	19 (16 K)	3.45
Hens (100)	400	34	27 (12P)	13 (11 K)	6.00
	500	43	34 (15P)	16 (13 K)	7.50

P, phosphorus. K, potassium.

Influence of food upon composition of excrements.—Having studied the relation of the liquid and solid portions of excrements to the plant-food composition of fresh excrements in the case of various kinds of farm animals, we come now to take up the relation of the food consumed to the composition of the excrements. In the case of farm animals in general, the plant-food composition and value of the fresh excrements are, as a rule, more dependent upon the composition of the food consumed than upon any other single factor. The nitrogen, phosphorus and potassium compounds taken in the food into the body pass from the body more or less largely in the form of excrements. It is therefore obvious why the amounts of these constituents present in the excrements depend chiefly upon the amounts present in the food consumed. There can be in excrements no more nitrogen, phosphorus or potassium than is contained in the food materials taken into the body; but these constituents, as they are found in excrements, are on the whole in more readily available forms as plant-food than in the original food before it is digested.

Plant-food constituents in animal foods.—The amounts of nitrogen, phosphorus and potassium compounds in excrements primarily depend, as already stated, upon the amounts of these constituents in the food eaten. Concentrated foods, such as cottonseed-meal, linseed-meal, gluten-meal, etc., yield the largest quantity of one or more plant-food constituents; then follow leguminous plants, such as alfalfa, clover, etc.; next come cereals, and lastly root crops.

Of two foods costing the same price and having equal feeding value, it is economy for farmers to use that one which contains the largest amount of plant-food materials.

Proportion of fertilizing materials in foods recovered in fresh excrements.—In the case of farm animals in general, without special reference to kind or age, it is a fair estimate that, on an average, about 80 per cent. of the nitrogen and phosphorus and 90 per cent. of the potassium and 50 per cent. of the organic matter of the food consumed is recovered in the manure. Generally speaking, the excrements of working and fattening animals contain from 85 to 95 per cent. of the plant-food constituents present in the food eaten; in the case of cows giving milk and of young, growing animals, 50 to 75 per cent. When animals are not giving milk nor increasing in weight, the plant-food constituents in food and excrement are equal.

Digestibility of food in relation to plant-food value.—The solids of animal excrements consist largely of the undigested portions of food eaten. These undigested portions have not been acted on in the digestive process. This is a point of practical importance, because the portions of food not dissolved in animal digestion are likewise not easily made soluble in the soil by the ordinary processes of bacterial action. The plant-food constituents in the solid excrements are therefore not quickly available for supplying crops. In the process of digestion, the portions that appear in the solid excrement, though remaining mostly undissolved, are mechanically changed and improved, because they are converted into a finer state of division, having gone through a process of softening and grinding. On this account the plant-food constituents in solid excrements undergo decomposition in the soil and become available more quickly than the same constituents in the food before it is used by animals.

Another point worthy of notice in this connection is that in passing through animals the food becomes mixed with intestinal bacteria, which are responsible, in large measure, for the processes of decay that take place in the excrements, and make the plant-food available. From what has preceded, it is obvious that the larger the amount of indigestible matter in food, the larger is the amount of solid excrement and, therefore, of slowly available plant-food. Coarse, bulky fodder materials produce large amounts of solid excrement; concentrated, ground foods produce smaller amounts of solid excrement.

The urine contains those portions of plant-food which have been digested and utilized in the animal body; its constituents are all in soluble form and are either directly available as plant-food or readily become so. The nitrogen, phosphorus and potassium compounds in the urine are, therefore, more valuable as plant-foods than the insoluble, slowly available constituents contained in solid excrement. It therefore follows that the more digestible a food is, the larger is the proportion of its plant-food constituents that appear in the urine and the greater will be the plant-food value of the manure produced. Generally speaking, the richer the food is in nitrogen, the more digestible is the nitrogen and the larger the amounts of nitrogen in urine. The nitrogen compounds of the food are usually more extensively and completely transformed in digestion than are the phosphorus compounds and appear in the urine in larger amounts than in the solid excrements.

Relation of bedding to composition of farm manure.— Having considered the various conditions that influence the plant-food composition of the fresh excrements of animals, we are ready to take up the first factor of importance outside of the excrements that modifies the composition of manure; namely, the bedding or litter. This is used in stables primarily to furnish clean and comfortable places for animals to lie down. In relation to manure, bedding is used chiefly for the following purposes: (1) To absorb and retain the urine, thus preventing loss by drainage, (2) to increase the amount of slowly decomposable matter in manure in order to check and control the decomposition, (3) to increase the amount of organic matter and plant-food, (4) to retain ammonia and prevent its escape into the air, (5) to make manure easier to handle, and (6) to influence in some cases both the physical and the chemical action of manure.

The materials most commonly used are some kind of straw, such as wheat, oat, rye or barley. When easily procurable, muck and peat-moss are often used. Sawdust and shavings are used in cities owing to the relatively high cost of straw. The straws commonly used as bedding contain, on an average, about 0.5 per cent. of nitrogen, 0.3 per cent. of phosphoric acid (0.13 phosphorus) and 0.7 per cent. of potash (0.6 potassium). In general, bedding materials are not rich in plant-food constituents.

In respect to *absorbing* power, ordinary straw can take up two or three times its weight of water. Fine-cut straw has much greater absorbing power than coarse straw; thus, straw in pieces an inch long absorbs about three times as much liquid as uncut straw. *Peat-moss* of good quality can absorb and retain as much as ten times its weight of water. It also has the advantage of being able to absorb and retain ammonia. Muck is also an excellent absorbent. Manure containing liberal amounts of peat-moss or muck can be applied fresh, even to light soils, without disadvantage. Peat, or muck, adds nitrogen to manure. Though not in readily available form, it is made more available by fermentation (p. 198). One slight disadvantage connected with the use of peat or muck is that the manure is a little more difficult to handle. The absorbing power of *sawdust* and *shavings* is two or three times that of ordinary straw; they add to manure about half as much nitrogen as straw and approximately equal amounts of phosphorus and potassium. Fine dry earth, especially clay or loam with plenty of humus, is found a convenient and efficient absorbent.

TABLE 32—PLANT-FOOD IN MIXED EXCREMENTS AND BEDDING FOR ONE YEAR PER 1,000 POUNDS OF LIVE WEIGHT

Kind of animal	*Weight of mixed excrements*	*Weight of bedding*	*Weight of total manure*	*Nitrogen*	*Phosphoric acid*	*Potash*	*Value of plant-food*
	Pounds	*Pounds*	*Pounds*	*Pounds*	*Pounds*	*Pounds*	
Horse	18,000	6,000	24,000	158	61 (27P)	145 (120K)	$28.50
Cow	27,000	3,000	30,000	171	47 (21P)	148 (123K)	29.50
Pig	30,500	6,000	36,500	180	122 (54P)	170 (141K)	35.35
Sheep	12,500	7,000	19,500	154	65 (29P)	175 (145K)	29.60
Steer	15,000	3,000	18,000	135	54 (24P)	72 (60K)	22.15
Hen	8,500	——	8,500	85	68 (30P)	34 (28K)	15.15

P, Phosphorus. K, potassium.

The *amount* of bedding used in farm manure varies greatly according to various conditions. The chief point to keep in mind is to use enough to absorb the urine completely. The use of too much bedding material, especially of low grade, decreases the percentage of plant-food constituents, increases the cost of handling and may unfavorably affect the fermenting process (p. 145). Table 32 shows the average amounts of bedding used in connection with the excrements of various classes of farm animals. The data are based upon the average amounts of excrements produced annually per 1,000 pounds of live weight. Assuming that the bedding is good average straw and contains 0.5 per cent. of nitrogen, 0.3 per cent. of phosphoric acid and 0.7 per cent. of potash, we indicate the amount of plant-food constituents furnished by the bedding. With the exception of the data for steers, the figures in the last column are obtained from the table on page 225 by adding the value of plant-food in the bedding and in the fresh excrements. The data in the table are based on the assumption that all the excrements, both liquid and solid, are saved without loss. The values given represent, therefore, conditions more favorable than those prevailing in actual practice.

Approximate Plant-Food Value of Average Manure.—From the data in the preceding table, we obtain the following average values for *one ton* of manure in the case of different farm animals: Horse, $2.40; cow, $2.00; pig, $1.95; sheep, $2.75; steer, $2.45; hen, $3.55.

COMPOSITION OF FARM MANURE AS AFFECTED BY LOSS OF PLANT-FOOD CONSTITUENTS

Up to this point we have considered how the composition of farm manure is influenced by the composition of the fresh excrements and the bedding materials, the two factors which control the composition of the manure so far as its usual constituent materials are concerned. We should not need to go further in our study of composition of farm manure, if all excrements were saved without loss and if no chemical changes took place in the manure; but losses of plant-food do take place under ordinary conditions and necessarily affect the composition of the manure. Such losses may be due to one or all of three causes: (1) Loss of urine in stable, (2) loss of soluble plant-food by leaching, (3) loss of nitrogen by decomposition.

Loss of urine.—It has long been known that, taking the average for farm animals in general, one-half of the value of the nitrogen and two-thirds of that of the potash excreted are found in the urine; or, expressed on the basis of the total plant-food in urine and dung about 50 per cent. of the value is in the urine. When inefficient efforts, or none at all, are made to save the urine by use of absorbent materials (p. 235), it can be readily understood that large proportions of plant-food are lost and, as already pointed out above, *the portions thus lost are the most readily available forms of plant-food in the manure.* Too often these valuable constituents are allowed to run to waste through cracks of the stable floor.

Loss by leaching.—As a result of leaching by exposure to rain, the losses of plant-food in farm manure, have been studied at several experiment stations in America and Europe. All results indicate marked loss of valuable constituents, including, not nitrogen compounds alone, but phosphorus and potassium compounds as well. The method too commonly employed by many farmers is to throw the manure every day from the stable into the adjoining barnyard, where it lies for months unprotected from rain, sun and wind, often directly under the eaves of the barn. Investigations show, in brief, that in the course of six months stable manure may lose 60 to 70 per cent. of its plant-food constituents; in two months, when rain-

When rain and snow are allowed to seep through manure piles the escaping brown liquid contains over half the potash and nitrogen.

fall is fairly abundant, half or more of the plant-food value may be washed out. Leached manure necessarily shows greatly decreased effect on crops as compared with the unleached.

In general, it is safe to say that, under the usual careless manner of storing farm manure out of doors, *over one-half of the plant-food value is lost*. An important fact in connection with this loss is that *the portions dissolved out in leaching are the most readily soluble and therefore the most quickly available as plant-food*. The losses by leaching are not confined to the liquid portion of manure, but the solid excrement may in the course of a few months lose nearly one-half of its nitrogen and more than half of its phosphorus and potassium compounds. Taking into consideration both the amount and availability of the plant-food leached from stable-manure, it is not an exaggeration to say that two-thirds of the plant-food value is leached from much of the stable manure used on American farms, which means an average loss of at least $1.25 a ton.

Process of decomposition and loss of constituents.—It is a familiar farm experience to observe that stable manure rapidly undergoes characteristic changes, such as increase of temperature, development of marked odors, shrinkage in bulk and darkening in color. These changes are brought about by the action of various micro-organisms (p. 130) *especially bacteria, yeasts and molds.* The processes by which they are produced are complicated, especially since several different kinds of action may be going on at the same time. We cannot attempt to do more than to call attention briefly to some of the more important changes that take place during the decomposition of manure in so far as they relate to losses of material.

The constituents in which we are most interested are the following: (1) Phosphorus compounds, (2) potassium compounds, (3) nitrogen compounds and (4) nitrogen-free organic compounds (carbohydrates, starch, sugar, etc.).

(1) *Phosphorus compounds* are rarely, if ever, in danger of loss as the result of the decomposition of manure by micro-organisms. The insoluble phosphorus compounds contained in the solid excrement are more or less largely changed into available forms of plant-food. Therefore, decomposition affects the phosphorus compounds in manure mainly in a favorable way.

(2) *Potassium compounds* in manure undergo no loss as the result of decomposition changes. The potassium in the liquid portion of manure is usually converted into alkaline potassium carbonate sooner or later and in this form is useful in neutralizing acids formed by the decomposition of starch, sugar, etc. Insoluble potassium compounds in solid portions of excrements are changed during decomposition into more easily available forms of plant-food.

(3) *Nitrogen compounds.*—The most serious loss that takes place when manure undergoes decomposition is in relation to the nitrogen compounds. Results of experimental work indicate that under most favorable conditions about one-sixth of the nitrogen in manure is lost during decomposition and that, under ordinary conditions, not more than one-half of the nitrogen contained in fresh manure is likely to reach the soil. We can consider separately the nitrogen in the urine, which is soluble, and that in the solid excrement and bedding, which is largely insoluble.

(*a*) Decomposition of urinary nitrogen.—The smell of ammonia about horse stables is an experience familiar to every farmer; it is especially noticeable in warm weather when one first enters in the morning a stable which has been shut close during the night. The urine of animals contains certain nitrogen compounds which are the first to undergo bacterial change, resulting in the formation of ammonium carbonate, $(NH_4)_2CO_3$. This compound easily decomposes, forming ammonia gas, water, and carbon dioxide, as shown by the following equation:

$$\underset{\text{Ammonium carbonate}}{(NH_4)_2CO_3} = \underset{\text{Ammonia gas}}{2NH_3} + \underset{\text{Water}}{H_2O} + \underset{\text{Carbon dioxide}}{CO_2}$$

Ammonium carbonate decomposes easily at higher temperatures and rapidly passes into the air; it also goes into the air more easily when the liquid containing the ammonia is exposed to evaporation by distribution over more extended surfaces.

The bacteria that convert urinary nitrogen into ammonia are always abundant in stables, and their action is extremely rapid, especially in warm weather. Urine mixed with loose dung, like that of horses, undergoes decomposition even more rapidly than when kept separate. Urine that is richer in nitrogen compounds, as in case of horses, produces ammonia more rapidly than that containing less nitrogen, as in case of pigs, for example.

Urinary nitrogen is subject to serious loss on account of its rapid conversion into ammonium carbonate, which easily decomposes into products that escape into the air unless precautions are taken to prevent. The loss of such ammonia can be prevented or largely reduced by the presence of sufficient moisture and by the use of absorbents, (p. 235).

(*b*) Decomposition of nitrogen in solid excrement and bedding.—The nitrogen compounds present in the solid excrements of animals are largely insoluble, since they have already resisted the attack of the digestive fluids and of intestinal bacteria in animals and are, therefore, not easily changed further by the process of decomposition. The nitrogen compounds contained in bedding, such as straw or similar material, are, in part, at least, somewhat more easily decomposed than the nitrogen of solid excrement, but only slowly when compared with urinary nitrogen.

The rapidity of change, the kinds of products formed, and the amounts of nitrogen lost by the insoluble nitrogen compounds of manure during decomposition, depend mainly upon the amount of air in the manure. We have already had occasion to point out the effect of admitting or excluding air upon the decomposition of organic matter (p. 136) and to call attention to the differences in the process of decomposition in air (aërobic, p. 136) and decomposition without air (anaërobic, p. 137). Essentially the same results follow when these decomposition processes are applied to the insoluble organic nitrogen compounds of manure. The effect of presence or absence of air we will consider in connection with the following changes that may take place in the insoluble nitrogen of the solid portion of manure: (1) Formation of ammonia, (2) formation of free nitrogen, (3) conversion of soluble into insoluble nitrogen compounds, (4) production of nitrate nitrogen.

(1) *Formation of ammonia from insoluble nitrogen in manure.* —The organic nitrogen contained largely in insoluble and rather inert forms in the solid portions of animal excrements and of bedding (straw, shavings, peat, etc.), is acted upon by numerous kinds of micro-organisms.

When abundance of air is admitted to the manure, as happens in a loose heap or on the outside of an ordinary pile of manure, aërobic bacteria are active, and the insoluble nitrogen compounds are broken down with comparative rapidity, especially in case of favorably warm temperature; under these conditions ammonia is formed rapidly and, at the temperature developed in the manure heap, is likely to be lost in considerable amounts by escape into the air. Moreover, the ammonia that does not thus escape into the atmosphere may be decomposed into free nitrogen and water by other bacteria (p. 132) and thus lost to the manure as plant-food.

When air is excluded from manure, which is the case on the inside of a well-compacted, moist manure pile, the insoluble nitrogen is attacked by micro-organisms that slowly change it into various other nitrogen compounds and sooner or later into ammonia. When formed under these conditions, the loss of ammonia by escape into the air is slight, unless the manure pile is allowed to become dry or is forked over and exposed directly to the air.

(2) *Formation of free nitrogen* from ammonia may occur when air is freely admitted to manure, as already stated. Free nitrogen may also be produced by denitrification when nitrate nitrogen formed on the outside of a manure pile has been washed into the inside by rain.

(3) *Conversion of soluble into insoluble nitrogen compounds.* —Throughout all parts of a manure pile, immense numbers of bacteria are vigorously at work as long as conditions are favorable, and, it should be remembered, these bacteria must have nitrogen as food for their own use. When they can obtain it in soluble forms, they use these by preference. Therefore, in the process of decomposition of manure, bacteria appropriate to their own use ammonia or other soluble nitrogen compounds, which they convert into insoluble forms by storing it in their own bodily substance. The nitrogen thus used is, however, not lost, because it is again changed into soluble forms when the bacteria die and themselves undergo process of decomposition. The amount of insoluble nitrogen thus formed varies according to various conditions, but it may be stated that, in

general, the longer that manure is kept under conditions preventing access of air, the larger is the proportion of insoluble nitrogen and the smaller the proportion of soluble nitrogen. For this reason, old or well-rotted, is less active than fresh manure, which is rich in ammonia.

(4) *Production of nitrate nitrogen.*—Since nitrate nitrogen is formed only in the presence of oxygen, it can be produced only in those portions of manure that contain abundant air supply. In a moist, well-compacted pile of manure, nitrate is not produced except in the outer layers. If exposed to rain this is washed into the interior and then, in the absence of air, it may be converted into free nitrogen and lost or may be changed into insoluble nitrogen in the bodies of the active bacteria. Farm manure is not likely to contain much nitrogen in nitrate form, unless it has undergone a long process of fermentation under special conditions not commonly prevalent in manure piles. In the use of farm manure it is expected that the production of nitrate will take place chiefly after the manure has been incorporated in the soil.

(5) *Changes in nitrogen-free organic compounds in manure.*—Thus far we have not considered what happens during the process of decomposition to the large amount of nitrogen-free material, which constitutes so large a part of the dry matter in solid excrement and in bedding. The changes and results are essentially the same as those that take place in the decomposition of organic matter in soils.

The constituent forming the larger part of the nitrogen-free organic compounds is the ligno-cellulose or fiber present in relatively large amounts in the undigested, coarse portion of solid excrements and in the straw or similar material used for bedding. When air has free access, the fibrous material is largely changed into carbon dioxide and water; it is, in fact, practically burned up and largely disappears (p. 136). This occurs to some extent on the outside of a loose manure pile. The chief kind of change, however, is that which takes place on the inside of the pile in the absence of air. It is caused by anaërobic organisms. The material is changed into a dark-brown or black mass and all traces of original structure sooner or later disappear, the fibrous substances disintegrating into a dark, fine, soft mass. Owing to these characteristic changes, the process of manurial decomposition is commonly called "rotting." Fermented or decomposed manure is often known as "rotted" or "well-rotted" manure.

This change is accompanied by shrinkage in bulk, owing to actual loss of material. More or less of the carbon in the original carbohydrate material is changed into carbon dioxide gas and its hydrogen into free hydrogen or marsh gas (CH_4). It may be stated that, in general, from one-quarter to one-half of the dry matter originally present in manure is lost during the process of decomposition, the amount varying according to the conditions of decomposition and the length of time the process continues.

(6) *"Fire-fanging."*—In this connection an abnormal process occurs in loose and dry manure, as for example, a mixture of horse manure and straw. There is throughout the manure a white appearance, as if it had been sprinkled with flour, due to the growth of certain fungi. Manure in which such fungous growth has taken place is said to be "fire-fanged." Its value has been impaired.

Summary of decomposition changes of manure.—Enough has been said to show that the changes undergone by manure are varied and complex; many of them are taking place at the same time and, in some cases, changes that are reverse in character. To give a more comprehensive view of all these changes, we will briefly summarize the main facts.

(1) *Decomposition of urinary nitrogen.*—The first change to occur is the formation of ammonia in urine; loss of ammonia by volatilization is likely to occur unless the process is carefully managed by keeping the manure moist and compact.

(2) *Decomposition of insoluble nitrogen.*—The insoluble organic nitrogen compounds contained in solid animal excrements and in bedding undergo anaërobic decomposition with formation, sooner or later, of ammonia.

(3) *Conversion of soluble into insoluble nitrogen.*—Bacteria use ammonia and other soluble compounds of nitrogen in considerable amounts as food. This nitrogen is stored in their bodily substance in insoluble forms; nitrogen transformed in this way is not lost, but becomes available when the bacteria die and undergo decomposition.

(4) *Formation of free nitrogen.*—Under some conditions, the nitrogen of ammonia and of nitrates is decomposed with formation of free nitrogen and is consequently lost as plant-food.

(5) *Decomposition of nitrogen-free organic compounds.*—The portions of manure represented by fibrous materials like ligno-cellulose are disintegrated into a dark-brown or black mass of fine, soft humus material, in which all traces of original fibrous structure disappear sooner or later. During this change, carbon and hydrogen disappear in the form of gaseous compounds to such an extent that from one-fourth to one-half of the original dry matter in the manure is lost. For this reason, manure shrinks in bulk during decomposition.

Differences in fresh and decomposed manure.—In speaking of partially decomposed or rotted or fermented manure, we usually mean, as nearly as it is possible to describe, manure in which the fibrous material has been largely decomposed into a somewhat crumbly, dark-colored mass, the original structure of the materials having more or less largely disappeared. Assuming that the fresh manure is a normal mixture of urine and solid excrement and that the conditions of decomposition have been under good control, fresh and decomposed manure differ in composition as follows:

(1) *Decomposed manure richer in plant-food constituents.*—We have seen that fresh manure shrinks much in bulk, one ton of fresh manure forming only one-half to three-fourths of a ton of decomposed manure. Since this loss falls upon the organic nitrogen-free constituents which contain no plant-food, the plant-food constituents are, as a result, concentrated in the portion remaining.

(2) *Nitrogen in fresh manure more largely soluble.*—The nitrogen in decomposed manure is more largely in insoluble form than that in fresh manure. The urinary nitrogen in fresh manure is entirely in solution; but, during the decomposition of manure, bacteria use more or less of it as food and convert it into insoluble form as an integral part of their bodies.

(3) *Phosphorus and potassium more soluble in decomposed manure.*—The soluble phosphorus and potassium in fresh manure largely remain so during decomposition, while the insoluble forms of these two constituents present in solid excrement and bedding become more or less soluble.

Relation of air to decomposition of manure.—Before leaving this portion of our discussion, we wish to emphasize the importance of controlling the air supply of manure during the process of decomposition. We have already seen that serious losses of nitrogen in the form of ammonia and of free nitrogen are due to forms of micro-organisms that require abundance of oxygen, and that larger amounts of carbohydrate or nitrogen-free materials are converted into gaseous form in the presence of abundant air.

The bacterial processes that use atmospheric oxygen take place with great rapidity and intensity when the air supply is large. The process is one of oxygen consumption, a form of burning, and results in the production of heat, the temperature increasing with the air supply. Farmers commonly speak of this process in manure as "heating" or "hot" fermentation. The temperature may rise as high as 150° F. (65.5° C.) in the outer layers of a manure pile just under the surface or on the inside of a loose pile. At temperatures above 130° F. (54.5° C.), ammonium carbonate is decomposed and ammonia escapes into the air. When the air supply is limited or excluded, as on the inside of a compact heap, the temperature is usually below 100° F. (37.8° C.).

The presence of moisture in large amounts tends to exclude air and therefore favors anaërobic instead of aërobic decomposition.

It is largely the difference in moisture and, consequently, the amount of air present, that makes the manure of different animals decompose with different degrees of ease or rapidity. The manure of horses and sheep, on account of dryness, porosity and ready accessibility to air, decomposes easily; such manures are called "hot." The manure of cattle and pigs contains a high percentage of water, ferments slowly and is called "cold." Moist manure, like that of cattle, forms a hard, impervious coat on the surface when it dries. This protects the interior from rapid loss of moisture and also makes it less accessible to external bacteria.

XIX

FARM MANURE: CARE, VALUE AND USE

THE fundamental object in caring for farm manures is to prevent loss of plant-food constituents as far as possible. Even under favorable conditions, it is practically impossible to prevent some loss of nitrogen, probably, 15 to 20 per cent. However, there is no difficulty in conserving all the phosphorus and potassium compounds. In considering methods of caring for farm manures, certain conditions must be observed in all methods; so before taking up details of special methods, we will consider some of the more obvious means of preventing loss of plant-food, mainly suggested by the discussion in the preceding chapter.

PREVENTION OF LOSS OF PLANT-FOOD IN STABLE MANURE

The usual losses of plant-food in farm manures can be largely avoided by observing the following conditions: (1) Use of mechanical and chemical absorbents, (2) regulation of the process of fermentation, (3) protection from leaching.

Use of absorbents.—Urine must be saved from mechanical loss, and, after amomnia is formed in it, this must be kept from escaping into the air. Such losses are prevented or lessened (1) by means of water-tight floors and stable gutters, and (2) by means of mechanical and chemical absorbents. For mechanical absorption, materials must be used which will take up the liquid and hold it securely without danger of loss from dripping when the manure is handled, (see discussion of bedding p. 224). Several chemical materials are also used, which have a minor value as mechanical absorbents, but are more useful because they can form, with ammonia, chemical compounds which do not escape into the air. Of such materials, those most commonly used are land-plaster (gypsum), kainite and superphosphate. Such materials as wood-ashes or lime, either slaked or unslaked, should *in no case* be mixed with stable manure, because they produce rapid decomposition and loss of nitrogenous compounds. The use of ground rock-phosphate ("floats") with manure will also be discussed in this connection.

(1) *Land-plaster* or gypsum ($CaSO_4.2H_2O$, p. 236) has been used for a long time. Its chief value is not that it possesses unusual power as a mechanical absorbent of liquid or of ammonia, a common belief, but that by chemical action it changes the easily volatile ammonia into a more stable form. This change can be represented as follows:

$CaSO_4$	+	$(NH_4)_2CO_3$	=	$(NH_4)_2SO_4$	+	$CaCO_3$
Calcium sulfate		Ammonium carbonate		Ammonium sulfate		Calcium carbonate

Ammonium sulfate is a compound in which the ammonia is held fast and is not in danger of loss by escape into the air. The calcium carbonate thus formed is beneficial to the soil (p. 166). There is, however, one objection to the use of land-plaster; during the anaërobic fermentation in the manure, it is possible that the sulfate may become converted into *sulfide* of calcium, which acts as a poison to plants. This possible danger is probably one that does not deserve serious consideration.

In order to be effective in manure, the land-plaster *must be finely ground and moist.* One method is to sprinkle it on the urine or moist dung. Some men prefer to apply it on the stable floor after cleaning and before spreading the bedding. To be of most use, the amount applied should be fairly large, not less than 100 pounds for one ton of manure, which is rather more than is usually recommended. The amounts used daily for single animals would be approximately as follows: Horses, 2 to 3 pounds; cattle, 3 to 4 pounds; pigs, 8 ounces to 1 pound; sheep, 2 to 4 ounces; 100 hens, 8 ounces to 1 pound. The comparatively high cost of gypsum and its smaller efficiency, compared with superphosphate, make its use somewhat expensive.

(2) *Kainite* sprinkled on manure tends to attract and hold moisture. As in case of calcium in land-plaster, the potassium and magnesium compounds in kainite tend to change ammonium carbonate into more stable compounds, such as ammonium chloride and sulfate. Kainite has the advantage of being easily soluble. It is therefore more readily distributed through the manure and more completely utilized than land-plaster. The application of 50 pounds of kainite is recommended for one ton of manure. One precaution should be observed in the use of kainite or any similar material; it should be kept from direct contact with the feet of animals, since it is irritating. It is therefore best applied to fresh manure and covered with bedding or it may be applied and carefully mixed with the manure daily when the stable is cleaned.

(3) *Superphosphate* (acid phosphate, p. 177) contains gypsum

(land-plaster), and, to this extent, its action is the same as that of land-plaster. The soluble calcium phosphate in superphosphate combines chemically with and holds ammonia firmly. Superphosphate is more effective than either land-plaster alone or kainite in holding ammonia. To be effective, about 50 pounds of plain superphosphate (containing 16 to 20 per cent. of phosphoric acid) or 25 pounds of a concentrated superphosphate (containing 40 to 45 per cent.) should be used for one ton of manure. Superphosphate, like kainite, should be kept from coming into direct contact with the feet of animals. The objection may be made that the soluble phosphoric acid of the superphosphate is changed into an insoluble form by combination with ammonia. This objection is not serious, since the phosphoric acid again becomes easily available in the soil. The use of superphosphate with manure should for reasons of efficiency and economy supplant land-plaster.

A mixture of acid phosphate and kainite may be used at the rate of 25 to 50 pounds of each for one ton of manure. This mixture has the advantage of increasing the amounts of phosphorus and potassium, so that the treated manure contains plant-food constituents in proportions nearer to those commonly used on crops instead of being over-rich in nitrogen and poor in phosphorus. An ideal compound both to stabilize ammonia and add potassium, is monopotassium phosphate (p. 37) when this is obtainable at a reasonable price. Fifty pounds are used to a ton of manure.

(4) *Ground rock-phosphate* or *"floats"* has been extensively used in connection with stable manure at the rate of 100 pounds for one ton of manure. This, however, has no chemical action in combining with ammonia, therefore, its behavior is that of a mechanical absorbent, like any other inert, finely divided substance. The chief advantage appears to be that insoluble phosphate applied to fresh manure becomes slowly available to crops to which the manure is applied.

(5) *Sulfurous acid.*—Another compound recommended for use on manure to prevent losses during the decomposition process is a solution of sulfurous acid sprinkled on manure freely. It combines with ammonia and soon changes to ammonium sulfate, which is in no danger of loss by escape into the air.

Regulation of the decomposition process.—We have seen that the plant-food constituent lost in bacterial decomposition of manure is nitrogen and that this is lost by escape into the air in the form of either ammonia or of free nitrogen. We have also seen that the largest nitrogen losses in decomposition take place when the manure is dry and loose, furnishing ready access of air to all parts. While the loss of 15 to 20 per cent. of the nitrogen originally present in the manure is practically unavoidable, additional loss can be largely prevented by controlling the conditions of decomposition; (1) by keeping the manure compact, (2) by keeping it moist, (3) by use of preservatives, and (4) by using old manure as a foundation of the manure pile.

(1) *Keeping manure compact.*—The rapid decomposition that results in loss of nitrogen from manure is caused by organisms that require an abundant supply of oxygen. In cutting down the supply of air in manure, this kind of destructive decomposition is prevented or kept under control. It is essential to keep the mass of manure, whether stored in a pile or some other way, as compact as possible; this can be done by trampling after every addition is made to the mass. In the case of manure piles, the sides and tops should be kept as smooth as practicable in order to exclude air. Under conditions of proper compactness, the temperature does not become high enough during fermentation to cause serious loss of nitrogen.

(2) *Keeping manure moist.*—The effect of an abundance of moisture in the manure pile is twofold: (1) It tends to keep the temperature down and thus prevents destructive decomposition of nitrogen compounds, and (2) it shuts out the air and thus prevents rapid decomposition. Water should be added to a manure heap whenever signs of hot decomposition are apparent. One must, however, avoid adding an excess of water, by which is meant more than can be absorbed and held without draining away from the pile.

By keeping manure sufficiently compact and moist, one prevents serious loss of nitrogen without checking the slower forms of fermentation.

In connection with this treatment, attention should be called to the decided advantage of mixing various kinds of fresh farm manures when they are available. Thus, horse or sheep manure, which is dry and porous and quick to decompose, should be mixed with cow or pig manure, which is moist and compact, and slow to decompose.

(3) *Use of preservatives.*—We have already (p. 235) discussed the use of some substances which, when added to manure, combine with and firmly hold ammonia, which prevents it escaping into the air. Some of these materials, especially *superphosphate* and *kainite,* have the additional power of retarding decomposition; they are poisonous to micro-organisms. When used in moderate amounts, rapid decomposition is prevented. On the whole, however, it appears advisable to depend largely upon keeping the manure compact and moist to prevent rapid decomposition.

(4) *Use of old manure as foundation of pile.*—Experiments have shown that losses of nitrogen by decomposition can be largely controlled by using as a foundation of a new manure pile a layer

of well-rotted farm manure, a foot deep or more. The probable explanation of the beneficial results of this treatment lies in the fact that the old manure gives off abundant amounts of carbon dioxide gas and expels, or, at least, dilutes the air in the fresh manure, thus diminishing the rate of decomposition by reducing the air supply.

Protection from leaching.—When farm manure is to be stored long, it should be kept under shelter, if possible. When conditions compel its storage outdoors, precautions can be taken to lessen greatly the loss by leaching. The pile should be made high enough to absorb and hold all the rain that falls on it without leakage. The sides of the pile should be kept as straight and smooth as possible, the top should slope toward the center a little, and the heap should be made as compact as possible. It is not the rain that falls on the pile which causes loss, but the water which runs through the pile and drains away at the bottom.

SPECIAL METHODS OF CARING FOR FARM MANURE

Among special methods of storing and handling manure we select the following as the most practicable: (1) Covered yards, (2) deep-stall system, (3) composting, (4) special treatment of poultry manure.

Covered manure yards are well-roofed sheds, with or without sides, and provided with water-tight floors, which may be used for storing stable manure during decomposition when it cannot otherwise be advantageously disposed of. The manure can be piled and kept under control by properly compacting and moistening when necessary. Another use of covered manure yards amounts practically to putting the barnyard under cover, allowing the animals to run there. The manure from the stables is spread uniformly over this yard daily. The trampling of the manure by the animals keeps it compact and moist, and the decomposition is easily kept under proper control. Under this system the losses of nitrogen during decomposition are comparatively small, the manure is convenient for handling and the plant-food is in a good condition of availability. Excellent results, have been obtained also by using the covered barnyard as a place for keeping cows except during milking-time. Mangers for feeding are provided and enough bedding distributed to keep the animals clean. A simple modification of this system is to put horse and cattle manure under shelter, as for example in the basement of the barn, where swine can work it over, adding their excrements and keeping the mass compact and moist.

Another modification of covered manure yard or shed is an inclosed structure, located conveniently near the stables, with a watertight floor sloping slightly from the sides to the center. Thick walls and a shaded location will contribute much to control of the conditions of decomposition. A convenient plan for such a building is to have it wide enough to form piles of manure on each side and leave space enough in the center to drive through with a wagon. Upon the floor is daily placed the manure taken from the stables and a pile built up five or six feet high. In starting the pile, it will be found advantageous to place as the foundation a six-inch layer of well-decomposed manure. The pile is kept moistened if necessary. Should liquid drain from underneath it should be put back on the pile or absorbed by muck, dry earth, or some other good absorbent.

After a pile has been completed, it is allowed to decompose about a month in warm weather or two months in cold, when it is well to fork it over in order to make the decomposed product more uniform in character and also to hasten the process somewhat by a moderate air supply. Water is added after the turning if the pile shows signs of being dry. It is also well after turning manure to cover the pile with land-plaster, dry earth, muck or similar material to the depth of one or two inches in order to retain ammonia that might otherwise escape into the air. Slightly modified, this method may be used for making compost. Along with the stable manure may be thrown in any waste organic material from the house or the farm. After the layer is about a foot deep, it is covered one or two inches deep with dry earth, muck, inverted sod or similar material, and then another layer of manure, etc. The same precautions are taken to keep the pile moist and cool, and the process is the same in other respects.

Deep-stall system.—A method of storing manure prevalent in Europe is to allow the manure to accumulate under animals in something like box-stalls, the floor of the stall being below the general level of the barn floor. The fresh manure is daily distributed in a uniform layer over the stall floor and covered with abundance of litter. The results are good in respect to the character of the manure produced, but the method is objectionable on sanitary grounds, at least for use with dairy animals.

Composting.—This method consists, in general, of forming a heap of fresh manure in alternate layers with absorbent materials, keeping as moist as necessary and turning over occasionally. The details vary in individual cases in respect to the absorbent and other materials used, the thickness of layers, size of heap, frequency of turning over, location outdoors or under shelter, etc.

One method of making a compost heap is to use as a foundation four to six inches of fine dry earth, or, better, peat or muck, or well-decomposed manure. On this foundation a layer of fresh manure is spread of equal or greater thickness and then alternating layers of absorbent material (peat, muck or earth) and manure. Care is taken to see that the manure is sufficiently moist when first placed and that it be kept so all the while, in order to prevent too rapid fermentation. The addition of so much water as to cause drainage from the bottom of the pile must be avoided. The earth or muck is to hold moisture and absorb ammonia. The entire outer surface of the mass should be kept covered all the time with a

Tobacco seedlings on manganese-deficient soil. *Left,* untreated; *right,* treated with manganese sulfate.

coating of earth. In six weeks or two months the completed pile is forked over, uniformly mixed and moistened if necessary and, unless used at once, again covered with a layer of absorbent. This is to increase somewhat the air supply and make conditions more uniform through the mass. It also favors the formation of nitrates. In place of earth or muck one can use grass sod, a mixture popular for forcing-house composts. Composted along with manure or separately can be used any waste organic materials of household or farm, such as kitchen garbage, dead animals, weeds, plant wastes, human excrements, etc. One advantage of composting such waste materials in addition to utilizing their plant-food, is that they can be made to decompose readily without the offensive odors characteristic of decay of many waste materials.

The composted product can be made more valuable as a source of plant-food by mixing with either fresh or decomposed manure bone-meal, ground rock-phosphate ("floats"), acid phosphate, bone-flour, kainite, muriate of potash, or other materals rich in phosphorus and potassium.

Bones can be composted and reduced to a condition that enables one to grind them easily. This is done by mixing them with quicklime or unleached hard-wood ashes, covering with a thick layer of earth or muck to absorb ammonia and allowing to remain for some weeks.

Attention is briefly called to a method treating manure, straw or any kind of organic material with a mixture of quicklime, unleached wood-ashes, salt, saltpeter, land-plaster and night-soil. Carefully applied this treatment would undoubtedly cause rapid changes, but special care, by a generous use of absorbent materials, would be needed to prevent loss of nitrogen. Quicklime or slaked lime can be used in small amounts in any compost heap where rapid action is desired, but extreme caution must be used to keep the heap compact and moist, and absorbent materials must be employed in abundance. The use of lime in this way will insure effective action of the compost when applied to soil. We should, however, advise the use of finely ground carbonate of lime in larger amounts rather than the use of small amounts of quicklime or slaked lime.

Special treatment of poultry manure.—Poultry manure is of sufficient importance to deserve a paragraph by itself. On account of its high percentage of nitrogen, poultry manure easily loses nitrogen by decomposition, especially in warm weather. It is therefore important to mix the manure while fresh with materials that will retard decomposition and act as absorbents. For this purpose, the following mixture is recommended: For 10 pounds of poultry manure, use 4 pounds of sawdust or dry muck, 4 pounds of acid phosphate, and 2 pounds of kainite, and at proportional rates for larger amounts of manure. One ton of such a mixture contains about 10 pounds of nitrogen, 65 pounds of phosphoric acid and 30 pounds of potash, furnishing a complete fertilizer of good quality adapted to general farm use.

FERTILIZING VALUE OF FARM MANURES

Until the development of the fertilizer industry and the utilization of leguminous crops, animal manures have always been the main source of supply of plant-food. In many countries this is still true. In the light of our modern knowledge relating to soil fertility and the feeding of crops, the long-continued use of farm manures has been well justified. Some of the main reasons why stable manure has special value are the following: (1) As a source of plant-food, (2) as a source of organic matter, (3) as a promoter of useful organisms, (4) as a promoter of crop growth, (5) long-continued effects, (6) conservation of fertility, (7) commercial and agricultural value, (8) unbalanced character of farm manure.

As a source of plant-food.—Stable manure contains nitrogen, phosphorus and potassium compounds, which are the constituents usually most needed in cultivated soils. When farm manure is properly cared for and applied, the plant-food constituents possess varying degrees of availability, portions of each being ready for immediate use and other portions becoming gradually available during the growing period of crops and, to some extent, during two or three or more seasons in succession. Generally speaking, farm manures are deficient in phosphorus in comparison with nitrogen and potassium and, in addition, nearly all of the phosphorus is in the solid portion of the manure and more or less largely insoluble, at least in the case of fresh manure.

As a source of organic matter.—The organic matter is an important factor in giving added value to the action of stable manure in the soil. When manure is applied fresh, the organic matter is gradually decomposed in the soil and its decomposition products tend to make available some of the insoluble plant-food constituents in those portions of the soil that come into contact with the decomposing substance and its products. In well-rotted manure, the organic matter is largely decomposed already. In addition to the value of its plant-food constituents, stable manure has, therefore, special value in helping to maintain the soil's supply of organic matter with all the attendant results in relation to furnishing colloidal material, improving soil texture, water-holding power, warmth, resistance to effects of wind, etc.

As a promoter of useful organisms.—Farm manures favorably influence the germ life of soils. Certain organisms thus added to the soil convert organic nitrogen into ammonia, others change ammonia into nitrate nitrogen, and still others decompose carbohydrate materials like cellulose. Manures also supply food material for the various organisms desirable in soils.

As a promoter of crop growth.—Numerous experiments, in which stable manure and commercial fertilizers have been compared, have shown beyond question that the plant-food in properly-handled manure, in the long run, has a crop-growing value equal to the same amount of plant-food in the form of commercial fertilizers, while the organic matter gives farm manure a decided advantage in the many cases where organic matter is deficient in soils. The crop-growing ability of most soils may be maintained indefinitely by the exclusive and abundant use of stable manure; but whether this is the most economical source of plant-food depends upon special conditions, which must be taken into consideration by each individual farmer.

Long-continued effects of stable manure.—A generous application (20 tons) of good stable manure shows a favorable influence on crops for several years following. Even more striking results in showing long-continued effects have been obtained by abundant applications of manure for several years in succession and then discontinuing or largely reducing the amount applied. The lasting effect is due, in part, to the fact that some farm manure plant-food constituents become soluble slowly and, in part, to the fact that some of the organic matter furnished may remain for several years and aid in gradually making available some of the insoluble plant-food in the soil.

Stable manure and conservation of fertility.—When it is kept in mind that about 80 per cent. of the plant-food constituents of the food of animals can be found in the fresh manure and 50 per cent. in decomposed manure, it is readily appreciated that the crops raised on a farm and fed to the animals belonging to the same farm do not constitute such an exhausting drain on fertility as when crops are removed from the farm. The loss of plant-food in the feeding of animals and in caring for the manure falls most heavily on the nitrogen compounds, which can be readily and inexpensively replaced by means of leguminous crops (p. 259), or by purchase of foods high in nitrogen to be fed on the farm.

Commercial and agricultural value.—The value of a ton of farm manure expressed in dollars and cents has been the subject of much discussion. From our knowledge of the extensive variations found in the composition of manure, it is obvious that, even when we know the amounts of plant-food constituents present, we are far from being able to place a precise value upon the manure which shall express its agricultural value, that is, what it is worth to the farmer in the growing of crops.

Commercial values assigned to manures are based upon the total amounts of plant-food contained and do not generally take into consideration the availability of the plant-food constituents or the agricultural value of the organic matter present. These factors depend so much for their value to the farmer upon the kinds of crops grown, the condition of his soil, etc., that it varies according to the circumstances of each case. On page 221 we have given conservative commercial estimates of value based on the plant-food constituents of fresh manures. In general, it should be true that the agricultural value is somewhat greater than is expressed in a statement based solely on plant-food constituents, since organic matter has some value. In regard to stable manure, the composition and history of which we have no means of knowing, any estimate of value nearer than 25 or 50 per cent. must be regarded largely as guesswork.

Unbalanced character of farm manure.—Farm manure in generous amounts, especially when used fresh, generally stimulates the growth of stems and leaves in a marked way, a condition characteristic of a high proportion of available nitrogen relative to available phosphorus and potassium. Under ordinary conditions, farm manure is regarded as a nitrogenous plant-food and as unbalanced with reference to the needs of most crops, relatively lacking in available phosphorus and potassium. This behavior of fresh manure of good quality is explained in large part when we consider that much of its nitrogen is in the form of ammonia, or soon becomes so. Under favorable soil conditions the ammonia is rapidly changed into nitrate, thus providing relatively large amounts of nitrate nitrogen long before the phosphorus, most of which is present in slowly available condition, becomes useful to plants. When we take into consideration the relatively small amounts of available phosphorus in most soils, this condition aids also in explaining why fresh manure when applied in large amounts, so often acts as a one-sided, highly nitrogenous fertilizer.

The question naturally arises as to why well-decomposed manure does not produce the effects of overfeeding with nitrogen as much as, or even more than, fresh manure. We have been in the habit of thinking of well-decomposed manure as containing much nitrate nitrogen; as a matter of fact, its nitrogen is likely to be, on the whole, more largely in forms much less quickly available than nitrogen in the urine contained in fresh manure. We have already pointed out the fact (p. 228) that, during the process of decomposition in a manure pile, bacteria are using large amounts of available nitrogen for their own food and are storing it away in their own body substance as insoluble protein, which becomes available only when the bacteria die and undergo decomposition. It is for this reason that well-decomposed farm manure is not likely to contain large amounts of nitrate nitrogen and that its nitrogen, on the whole, is much less quickly available than the urinary nitrogen of fresh manures. However, the case is somewhat different with available phosphorus and potassium, which are present in well-decomposed manure in larger amounts than in fresh manure. For these reasons, well-decomposed manure is to be regarded as more nearly approaching what we call a balanced combination of plant-foods than is fresh manure.

METHODS OF USING FARM MANURE

The effectiveness of farm manure as a fertilizer depends upon a number of conditions. No fixed rules for greatest efficiency can be given which will apply uniformly under all circumstances. We can, however, state certain known facts so the individual farmer may apply them in specific cases. He must adapt his practice to the particular conditions under which he is working. Among the points of greatest practical importance are the following: (1) Advantages and disadvantages of use of fresh manure, (2) use of decomposed manure, (3) kind of soil, (4) kind of crops, (5) amounts to use, (6) methods of distribution, (7) when to apply, (8) depth to cover, (9) effective methods of general use.

Advantages of use of fresh manure.—Among the advantages of applying farm manures to the soil in as fresh condition as possible are prominent:

(1) *Plant-food utilized.*—Under most conditions, the application of manure when fresh utilizes the largest amount of plant-food. Losses of plant-food through leaching or destructive forms of decomposition are prevented. When used as a top-dressing, the portions leached by rain go directly into the soil; or, in case of rain, the manure may dry with little loss from rapid decomposition. Mixed with the soil, fresh manure decomposes readily in warm or mild weather, its constituents being made gradually available as plant-food without risk of loss of nitrogen in the form of ammonia, except in some cases of light, porous soils exposed to high temperature.

(2) *Plant-food of soil made available.*—Not only are the plant-food constituents of farm manure, when applied fresh to soil, made more available, but the insoluble plant-food compounds of the soil itself are made more soluble wherever the decomposing manure comes into contact with the soil particles.

(3) *Desirable organisms supplied.*—Fresh manure contains various kinds of organisms effective in causing desirable chemical changes in the organic matter of the soil, and which the manure carries to the soil.

(4) *Lightening heavy soils.*—Fresh manure is most efficient for improving the texture of heavy clay soils, since it contains organic matter utilized in the soil with all the consequent benefits.

(5) *Prolonged effects.*—The plant-food constituents in fresh manure include both those quickly and those slowly available. The soluble portions are immediately available or soon become so in the soil; the insoluble portions become available gradually, furnishing plant-food supplies continuously, not alone during one growing season, but through several.

(6) *Favors growth of foliage.*—Fresh manure used alone and in large amounts (p. 243) favors vigorous growth of stems and leaves. It is, therefore, useful in the case of many crops that furnish as their marketable product the stems and leaves, which are especially dependent upon rapid growth of foliage.

Disadvantages of use of fresh manure.—The principal disadvantages that may attend the use of fresh farm manures under certain conditions are the following:

(1) *Unfavorable effects on soil.*—Excessive application of fresh stable manure, especially on light, porous soils during hot, dry seasons, may either not ferment at all or may ferment too rapidly according to special conditions. If the manure contains large amounts of coarse vegetable matter from bedding, which persistently resists decay, a condition common in leached manure, the soil may be made so loose and open that the manure and the soil both lose moisture to such an extent that decomposition will stop and the soil be left in a condition unsuited to crops. On the other hand, if the soil contains enough moisture to favor decomposition, the action will take place so rapidly under the conditions of air supply and temperature that the soil will be heated and dried out to such an extent as to be left with greatly diminished water-holding power, owing to increased openness; the coarser the manure, the greater the injurious effect on the soil.

(2) *Unfavorable effects on crops.*—Under the conditions described in the latter portion of the preceding paragraph, when fresh manure containing abundance of urine is applied to light, open soils and undergoes extremely rapid decomposition, direct injury, known as "burning out," may be done to crops. This results from the rapid fermentation of urinary nitrogen in the presence of a large air supply and high temperature. Large amounts of ammonium carbonate and its nitric acid compounds are quickly produced and become so concentrated by the rapid evaporation of water as to "burn" plants (p. 125), causing them to wilt and die. When the soil solution becomes so concentrated that it contains one part of soluble compounds, especially such as ammonia and nitrate, in about 400 parts of water, it is likely to damage tender rootlets. When applied as a mulch about young plants, fresh manure may give off enough ammonia to injure foliage

(3) *Carrier of weed seeds and disease germs.*—Fresh manure may carry to a soil weed seeds and germs of plant diseases. This disadvantage is usually overcome by making application of such manure to crops not affected by the disease germs present or which are least troubled by weed seeds or which furnish an opportunity for destroying the weeds.

(4) *Not adapted to certain crops.*—Direct applications of fresh manure, especially in large amounts and on any but heavy soils, are undesirable for cereal and some other crops, since the quickly available nitrogen furnished by the urinary nitrogen of fresh manure may injure the crop by causing too rank a growth of stems and leaves to the detriment of the quality or quantity of the marketable part of the crop.

Advantages of use of decomposed manure.—We have already called attention to the principal differences between fresh and partially fermented manure in respect to chemical composition and physical characteristics. Generally speaking, the phosphorus and potassium in partially decomposed manure are available to a larger extent than in fresh manure and are present in somewhat higher percentages. The nitrogen as a whole is in a generally less available condition. Such manure possesses the following obvious advantages:

(1) *More even action.*—Whenever action of plant-foods is desired calling for available nitrogen in moderate amounts and for readily available phosphorus and potassium, rather than for the action of quickly available nitrogen in relatively large amounts, partially decomposed manure is well adapted to meet such needs. Instead of being a highly nitrogenous fertilizer in its effects, partially decomposed manure may properly be regarded as a more evenly balanced combination of plant-foods.

(2) *Advantage on light soils.*—Decomposed manures are useful to furnish organic matter to improve with its colloidal material the mechanical condition of light soils without risk of producing increased openness and consequent "burning out," which may result from the use of fresh manure. The decomposed organic matter in old manure increases the water-holding power of light soils.

(3) *Convenience in handling.*—Decomposed manure, being more concentrated than fresh, is less bulky for the same amount of plant-food and in this respect offers some advantage in convenience of handling.

(4) *Desirable micro-organisms supplied.*—Decomposed manures supply to soils beneficial organisms and also aid in making conditions favorable for increased activity of some desirable soil organisms.

(5) *Decomposition of weed seeds.*—When manure undergoes decomposition, the weed seeds are largely rendered harmless, their power of germination being destroyed.

Kind of soil in relation to use of manure.—Generally speaking, farm manure, whether fresh or partly decomposed, can be used to advantage on soils deficient in organic matter. When it is desirable to introduce well-decomposed organic matter into the soil at once, decomposed manure can be used on any soil. When it is desired to have the decomposition of organic matter take place in the soil for the double purpose of making the insoluble plant-food available and leaving instead a residue of humus in the soil, then fresh manure is preferable, especially on clayey soils. Fresh manure must be used with caution and preferably in comparatively small applications at one time on light, open soils, especially when there is likelihood of hot, dry spells of weather. In relation to plant-food action, fresh manure is safer on heavy than on light soils, especially in large applications.

Kind of crops in relation to manure.—Farm manures, while beneficial for the growth of practically all common farm crops, requires for best results the observance of certain precautions in their use, especially with a few special crops.

It may be stated, as a preliminary, that in all cases where the direct application of fresh stable manure may do harm to the crop on account of the quickly available urinary nitrogen, the undesirable effects can be avoided by making generous application of the manure to a suitable crop preceding the one which is more sensitive. The first crop profits by the direct application, the unused residue being left for the sensitive crop, as will be illustrated later. Another method is to apply the manure in the fall previous to growing the sensitive crop in the spring following. Still another method is to apply generous amounts of acid phosphate and potassium chloride or sulfate.

(1) *Grass-lands* are generally much benefited by top-dressing with farm manure, either fresh or fermented. The manure serves, in addition to furnishing plant-food, to form a mulch, often favoring somewhat earlier growth.

(2) *Root crops* usually respond most satisfactorily to generous applications of stable manure. However, some precautions must be observed in the case of suger-beets and potatoes. Fresh farm manure used in excessive amounts especially on light soils and loams, may cause beets to grow large but with a low percentage of sugar; but on clay soils fairly large amounts (20 tons) of partially rotted manure may be used directly without harm. In case of potatoes, direct use of stable manure favors the development of scab, and excessive amounts of fresh manure produce overgrowth of tops at the expense of the yield of tubers.

(3) *Corn, millet* and leafy crops, in general, which produce large amounts of foliage and stems, utilize to advantage generous amounts of stable manure when directly applied, whether fresh or rotted.

(4) *Cereals* are injured by large direct applications of fresh manure, the straw growing large at the expense of the grain and lodging easily. Wheat and barley are particularly sensitive in this respect.

(5) *Tobacco,* when directly treated with large amounts of stable manure, produces large leaves of coarse texture and poor quality for use.

(6) *Flax* may also be unfavorably affected in quality by direct application of too much farm manure.

(7) *Garden crops* in general respond satisfactorily to generous applications of manure.

(8) *Young trees, shrubbery, rose-bushes, etc.*, are greatly benefited by applications of stable manure, moderate in case of fresh manure, generous in case of well-rotted.

(9) *As a mulch* in connection with any growing plant, good stable manure is highly effective and useful. Rotted manure is, of course, superior to fresh manure for mulching purposes on account of its greater water-holding power. Moreover, fresh manure, rich in urinary ammonia, when applied as a mulch about green plants, may injure the foliage, especially in warm weather. The ammonia volatilizes and burns the foliage.

Amount of farm manure to use.—No statement can be made to meet all conditions in respect to the amount of stable manure to use in the growing of crops. Obviously it will depend upon a variety of conditions, such as physical character of the soil, amount of available plant-food in the soil, condition and composition of the manure, kind of crop, frequency of application, etc. In general, 5 tons an acre may be regarded as a light application; 10 tons, moderate; 20 tons or more, heavy. It should be kept in mind that the larger amounts are better utilized when the soil is kept well supplied with calcium carbonate. In some cases, as on grass-lands, or where manure is used in the first part of a well-planned rotation, large amounts are applied once in 4 or 5 years. Generally speaking, the most economical results are obtained by making smaller applications at more frequent intervals. The frequent use of light applications keeps the soil in a more uniform condition of fertility, and the nitrogen is utilized more completely, with less risk of loss by leaching.

TABLE 33—AMOUNT OF PLANT-FOOD IN STABLE MANURE

	5 tons	*10 tons*	*15 tons*	*20 tons*
	Pounds	*Pounds*	*Pounds*	*Pounds*
Nitrogen	50	100	150	200
Phosphoric acid	25 (11P)	50 (22P)	75 (33P)	100 (44P)
Potash	50 (42K)	100 (83K)	150 (125K)	200 (166K)

P, phosphorus. K, potassium.

In this connection one should take into consideration the amount of plant-food contained in manure. Taking good farm manure of average composition, one ton contains about 10 pounds of nitrogen, 5 pounds of phosphoric acid and 10 pounds of potash. On this basis we prepare the Table 33.

In applying 5 tons of good farm manure to an acre of ground,

Spreading phosphated manure on clover sod in early fall. In spring, after plowing, two tons of pulverized limestone to the acre will be harrowed in.

one is putting on the same number of pounds of nitrogen contained in 300 pounds of sodium nitrate, or in 1,000 pounds of fertilizer containing 5 per cent. of nitrogen; the same number of pounds of phosphorus as in 150 pounds of 16 per cent superphosphate, or in 300 pounds of a fertilizer containing 8 per cent. of phosphoric acid (3.5 phosphorus); and the same amount of potassium as in 100 pounds of high-grade potassium chloride or sulfate, or in 400 pounds of kainite, or in 500 pounds of a fertilizer which contains say 10 per cent. of potash (8.3 potassium). When we consider the application of 20 tons of manure an acre, we are putting on large amounts of plant-food as compared with the amounts usually applied in the form of commercial fertilizers. But a mere statement of the number of pounds of plant-food applied to an acre in the form of farm manure is misleading, if we fail to consider at the same time the availability of the plant-food thus applied. Even under favorable conditions, it is doubtful whether we can reasonably expect more than one-third to one-half of the total plant-food in manure to be used by crops during the first season. Therefore, in considering the amounts of plant-food to supply crops in the form of farm manure during the growing season, it is usually safe to count on using two or three times as many pounds of nitrogen, phosphorus and potassium as are used with success in the form of commercial fertilizers of the best quality under similar conditions of soil and crop.

Methods of distribution.—In distributing farm manure on fields, three methods are commonly employed: (1) Distributing in heaps; (2) broadcasting from wagon or manure spreader; (3) applying in hill or row with seed.

(1) *Distributing manure in heaps,* small or large, over the field and permitting it to lie until convenient to spread, whether the time be weeks or months, while possessing the one advantage of temporary convenience, is open to several pronounced objections: (*a*) The labor of handling is increased; (*b*) the spots of ground under the heaps receive the leachings—the quickly available part of the manure—and are thus differently fertilized, as shown by results in unevenness of crops in respect to growth and maturing; (*c*) the manure may undergo rapid decomposition and lose much of its easily available nitrogen. Storing manure in large heaps on the field is less objectionable in respect to loss of nitrogen, provided that the heaps are kept moist, compacted, carefully covered with a thin layer of fine earth, so managed as to prevent loss by drainage, and are not allowed to lie too long before final distribution. In this connection the droppings of animals at pasture—if undisturbed, invariably produce uneven results of fertilization. In order to distribute these heaps of excrements it is well occasionally to go

over the field with a drag or some implement which will scatter the droppings more uniformly.

(2) *Applying manure broadcast* may be done by hand directly from a wagon, but on farms where large amounts are to be handled a manure spreader may be regarded as an economic necessity, since it distributes manure evenly and saves time and labor. Broadcasting is preferably practiced on fairly level fields where the manure is allowed to lie some time before plowing. If so applied on steep hillsides and allowed to lie before covering there is danger of leaching and washing down the slope, especially when the soil is frozen. Similarly, serious losses may be experienced in case of fields subject to flooding.

The method of broadcasting manure directly from wagons has two marked advantages: (1) There is economy of time and labor in handling as compared with distributing in heaps before broadcasting; and (2) any leachings from the manure except on slopes and when the ground is frozen before being plowed under pass directly into the soil in uniform amounts over the field, so no unevenness of crops results as in the case of leachings from distribution in heaps.

(3) *Applying in hill or row* is practiced when it is desired to have the effect of the manure concentrated where it will benefit the growing plants most quickly. It is especially applicable for forcing some kinds of garden crops. For example, when a shovelful or more of manure is put in each hill in planting cucumbers, melons, squashes, etc., the same amount of manure can be made to go further than when broadcasted. Rotted manure gives excellent results when used in this way.

When to apply manure.—Farm manure can be applied within wide limits as to time, depending on various conditions. Generally speaking, least risk of loss of ammonia occurs when manure is applied to soil only a short time before plowing. One plan in considerable favor is to draw and distribute on the field in a uniform layer of any desired depth the fresh manure daily or at intervals as frequent as convenience permits, care being taken to treat the manure in the stable in some one of the ways already described (pp. 235 to 240).

Direct distribution from stable to field has several pronounced advantages: (1) It economizes time and labor in handling. (2) It relieves the barn and its vicinity from the presence of accumulating manure. (3) It is generally more convenient, especially during fall and winter, since farmers have more time then than in the spring when all kinds of work are pressing. (4) On the whole, this method brings to the soil the largest amount of plant-food, since risk of loss due to destructive fermentation, leaching, etc., is avoided.

(5) The soluble portions of manure are gradually, completely and uniformly mixed with the soil. (6) The amount of nitrogen lost as ammonia is small if the manure has been properly treated in the stable.

Experiments indicate that manure spread in thin layers over fields loses little nitrogen either in the form of ammonia or of free nitrogen, provided it is not in a state of decomposition so active as to contain much ammonia when applied. In dry weather the manure quickly dries and decomposition stops; in wet weather the soluble portions are washed into the soil beyond danger of loss by escape into the air. There is, however, one distinct disadvantage in allowing manure to lie on the surface of the soil until its soluble nitrogen compounds have been largely dissolved and removed even though not lost to the soil. The soluble nitrogen compounds, as contained in fresh manure, are essential factors in the decomposition of the insoluble constituents of manure, so when they are once separated, as in leaching, the insoluble portion does not ferment as readily in the soil as when these are present. On this account it is desirable, especially in case of light, porous soils, not to allow the manure to lie on the surface long, but to apply only a short time before being plowed in. It is possible also on light soils which leach easily that some small amount of the nitrogen may be lost to the soil by drainage if the manure is spread some months before it is plowed. In the case of fields subject to flooding and of steep hillsides, the manure is best applied just before plowing.

Depth to cover manure in soil.—In heavy soils manure should be plowed under about 4 inches; if covered too deeply in such soils, decomposition is delayed or prevented by exclusion of air. In light, porous soils the manure may be covered more deeply. Generally speaking, well-rotted fine manure gives best results near the surface, while coarse manure gives better results more deeply covered.

Effective methods of using farm manure.—Among the best methods of utilizing farm manure in feeding crops and making soil conditions favorable to plant growth, are the following:

(1) *Use in rotations with commercial fertilizers.*—Farm manure is often applied directly to the crop in a rotation that utilizes it to best advantage; with the succeeding crops of the rotation commercial fertilizers are used to supplement the plant-food needs according to the indications of experience. For example, farm manure is often applied directly to a corn crop, which is followed by wheat or any crop that might be injured by direct use of farm manure (p. 247) and for which an application of quickly available commercial fertilizer is used. It is generally used once in large appli-

cation in 3 to 5-year rotations, and commercial fertilizer, in some form, is applied during the intervening period.

(2) *Fortifying farm manure with concentrated plant-food materials.*—The manure is treated in the stable or in the pile immediately after removal from stable with some such material as fine-ground rock-phosphate ("floats"), bone-flour, acid phosphate, potassium salts (especially kainite), etc. (p. 235). This treatment increases the amount of phosphorus or potassium or both and, properly regulated, has the effect of making the proportions of plant-food constituents of the treated manure more nearly like those present in commercial fertilizers used with success.

(3) *Farm manure and calcium carbonate.*—The action of farm manure is made more effective by keeping the soil well supplied with calcium carbonate. The application of manure to a "sour" soil increases rather than remedies the condition; with insufficient calcium carbonate in soils, there is loss of time and material in applying stable manure.

XX

GREEN-CROP MANURES

Green-crop manures or *green-manures* are crops grown to be plowed under and thus make soil conditions more favorable for the growth of succeeding crops. They are known as *cover-crops* when grown to occupy the ground only part of the growing season, during winter, or when used in orchards, the object being, in part, to protect the soil during the latter part of summer and over winter, and, in part, to furnish organic matter for the soil. Green crops are known also as *catch-crops* when following other regular crops or when used to take the place of a crop that has failed. The plowing under of sod and of stubble and roots is a form of green-crop manuring.

The beneficial effects of green-crop manures were observed long before it was known in what particular ways they act to produce better crops and numerous experiments have served only to confirm and emphasize as well as explain more fully the fact that green-crop manures favorably affect succeeding crops. The subject will be treated under the following divisions:

1. Effects.
2. Various crops used.
3. Conditions and methods of use.

EFFECTS OF GREEN-CROP MANURES

The beneficial effects obtained by the use of green-crop manures are briefly summarized under the following heads: (1) Organic matter, (2) conservation of soluble plant-food, (3) addition of nitrogen, (4) transfer of plant-food from subsoil to surface, (5) concentration of plant-food, (6) effect on bacterial life in soil, (7) increase of available plant-food, (8) improvement of subsoil, (9) relation to surface of soil, (10) effect on growth of orchard trees.

A few crops used as green-manures are able to produce all of these effects; all are able to furnish material for humus, to diminish the loss of soluble plant-food and while growing to influence the surface by their presence while many, in addition, increase the actual supply of soil nitrogen.

Organic matter.—The most extensive action of any green-crop manure is to furnish a supply of organic matter, the amount varying according to the crop and the extent of growth. The usual range of yield of green vegetable matter, including roots, may be placed between 5 and 10 tons an acre, though higher and lower yields may occur; this consists approximately of 1 to 2 tons of dry matter, and 4 to 8 tons of water. There is no other way in which such large amounts of organic matter can be added to the soil so cheaply under agricultural conditions commonly prevailing. The decomposition of the vegetable matter is attended by all the beneficial effects characteristic of this important soil constituent (p. 95). The roots are first to decompose, then the more succulent portions of the green tops and finally the more woody parts.

Conservation of soluble plant-food.—On land not occupied with crops for a considerable time, as in case of land lying fallow after the removal of a crop during summer or early fall, there is danger of loss of nitrate nitrogen in one or both of two ways: (1) Nitrates are easily lost in drainage water and (2) under some conditions they are decomposed into free nitrogen (p. 144). Experiments have fully established the fact that when the soil is occupied by a crop, the soluble plant-food goes into drainage water in much smaller amounts, if at all, but is taken into the plants and becomes a part of the crop growth. Moreover, under such conditions, nitrate nitrogen is in little danger of being lost as free nitrogen. The soluble plant-food is transferred into the crop, in which form there is little danger of loss under ordinary conditions. Crops having extensive root systems are especially efficient in thus collecting, concentrating and holding plant-food. Nitrogen as nitrate may be lost annually to the extent of 40 pounds or more an acre on fallow ground. The losses are greatest on light soils. Other conditions being the same, during the warm summer months, especially July and August, when nitrification is taking place most rapidly in the soil, nitrate nitrogen is lost more rapidly than at other times of year. When wheat is grown continuously on the same soil, without any cover-crops between, there may be lost from the soil 4 to 6 pounds of nitrogen for each pound used by the crop.

Addition of nitrogen.—Owing to their ability to use atmospheric nitrogen in their growth when their roots are properly supplied with nodules (p. 147) leguminous crops used as green-crop manures add to the amount of nitrogen in the soil. They contain 50 to 100 pounds and more of nitrogen an acre, not including stubble and roots, the amount depending upon the kind, yield, maturity, etc., of crop. Including all the vegetable growth of the crop, roots as well as tops, the total amount of nitrogen usually runs from 100 to over 150 pounds an acre. Not all of this nitrogen is taken from the air, but a considerable portion is known to be, the exact amount varying according to the crop and conditions of growth. The nitrogen compounds in the vegetable matter of green crops undergo decomposition in the soil, with final conversion into nitrate nitrogen. For example, if the amount of nitrogen added is 50 pounds, this is equal to 300 pounds of sodium nitrate.

Concentration of plant-food.—In the case of plants whose roots penetrate the subsoil, nitrogen, phosphorus and potassium compounds are collected from all parts of the soil; then when the crop decomposes in the upper layers of the soil, this food is concentrated within a more limited area of the soil than previously. In the case of leguminous crops, the concentration of nitrogen becomes greater than in other cases, since we have not only the nitrogen collected from the subsoil but also that taken from the air.

Increase of available plant-food.—The action of growing roots appears to be effective in making some unavailable plant-food (phosphorus and potassium) readily available. This is taken into the plant and, in case of green-crop manures, returned to the soil in forms more available than before use by the crop.

Transfer of plant-food from subsoil.—Crops with long roots gather from the subsoil considerable amounts of plant-food, thus lifting it to the surface and making it available for the use of succeeding crops. Alfalfa is famous for the length of its roots, sending them down under favorable conditions anywhere from 12 to 30 feet or more.

Effect on bacterial life in soil.—The various kinds of bacteria that convert organic nitrogen into ammonia and nitrate nitrogen are found to flourish in the presence of the organic matter furnished by green crops. In soils left fallow for part of the year, as in case of continuous wheat growing, bacteria prevail which actively decompose humus and set nitrogen free. These bacteria are found to disappear mostly in soils in which green crops, and especially leguminous crops, are incorporated as a part of a system of crop-rotation.

Improvement of subsoil.—Those plants having long tap-roots penetrate the subsoil deeply and when the roots decay, extensive passageways are opened for the percolation of water and circulation of air.

Relation to surface of soil.—When used as cover-crops green crops serve to protect the surface of the soil from washing, especially on hilly lands, and, in case of light soils, from being blown about by high winds. On soil during winter they hold drifting snow, keep the soil warmer and prevent injury to roots caused by alternately freezing and thawing. In some cases where the foliage is broad-leaved and extensive, the soil is shaded and protected from needless loss of moisture by direct evaporation.

Effect on growth of orchard trees.—Cover-crops in orchards compete with trees for plant-food, including water. In the case of soils that contain nitrate nitrogen in such proportions as to make a prolonged, succulent growth of new wood, liable to injury by freezing in winter, cover-crops serve to use the surplus nitrogen, check the growth of trees late in summer and induce early maturity in wood.

CROPS USED AS GREEN-MANURES

From the preceding statements, it is obvious that the purposes of green-crop manures are most completely fulfilled by those crops which, in addition to heavy foliage and deep roots, have the power to utilize and store atmospheric nitrogen. The particular kind of crop used as green-manure must depend upon the conditions present in each case, such as climate, soil, kind of rotation, special objects to be attained, the value of the green-crop for forage, etc.

The crops used as green-manures can be conveniently divided into two groups: (1) The first includes those plants which contain more or less nitrogen obtained from the air, and which, therefore, when incorporated in the soil, actually add nitrogen to the soil's previous supply. These are commonly known as *nitrogen-gatherers* of which those in actual use belong to the leguminous family of plants. (2) The second group includes non-leguminous plants, the nitrogen of which is taken directly from the soil, and which give back to the soil only the nitrogen taken from it. These, commonly known as *nitrogen-consumers,* include such crops as rye, buckwheat, oats, rape, mustard, weeds, etc.

Nitrogen-gathering or leguminous green-crop manures.—In the United States and Canada several leguminous crops are grown for partial or complete use as green-manures, among which the following are of most practical interest: Red clover, alfalfa, cowpeas, alsike, crimson and white clover, Canada peas, soy-beans, peanuts, vetch, velvet beans, Japan clover, sweet-clover, beggar-weed, horse-beans, field-beans, lupine, sainfoin, serradella, and lentil. Of these legumes, those most commonly grown as green-manures are field peas, cowpeas, soy-beans and vetch. The clovers and alfalfa serve the purpose of green-manure when the entire crop or the stubble and roots are plowed under. The various crops mentioned can be grown under best conditions only in more or less definitely limited areas. In the northern portions of the United States, red clover, mammoth clover, alsike clover, field-peas and common vetch are most commonly used; in regions having a moderate to warm temperate climate, cowpeas and crimson clover are extensively grown; in many parts of the West, alfalfa, clovers, cow-peas and soy-beans largely furnish green-crop manures.

For details in regard to the value and cultivation of leguminous crops for green-manure, see page 259.

CONDITIONS AND METHODS OF USE OF GREEN-CROP MANURES

In addition to the statements already made, the following details should be briefly considered before leaving the subject of green-crop manures:

(1) When green-crop manures should be used.
(2) What crops should be used.
(3) Preparation for growing green-manure crops.
(4) Relations to different soils.
(5) Time of plowing under.
(6) Crops to follow green-manure crops.
(7) Effect on suceeding crops.

When green-crop manures should be used.—The indiscriminate use of green-crop manures cannot be recommended. Used properly when conditions call for them, they are extremely effective; but, when used under some conditions, they may be either useless or else harmful. Green-crop manures can be used profitably to improve soils poor in nitrogen and organic matter, since such soils are always defective in physical conditions. They are especially efficient on many light, sandy soils and also on heavy clay soils. In orchards, they may be advantageously used as cover-crops.

On soils in good physical condition that grow good yields of crops, green-crop manures should be used only at intervals sufficient to maintain the supply of organic matter and nitrogen. This is best done ordinarily by using some leguminous crop as part of a regular rotation.

When farm manure can be had in abundance, as on dairy or stock farms, it is found most profitable to raise as green-manures only those crops which, like red clover and alfalfa, can be largely utilized for feeding animals, depending upon the roots and stubble, but supplementing with farm manure, as sources of organic matter and nitrogen supply.

The use of green-crop manures in dry regions is not practicable without irrigation, since with insufficient moisture, the vegetable matter decays with extreme slowness, leaving the soil filled with air spaces and promoting loss of water by evaporation.

What crops should be used as green-manures.—It is obvious that those crops should be used best adapted to the climate, soil, system of farming and the special purpose one has in view. Leguminous crops should be given preference when they can be grown satisfactorily. The particular legume to use must be determined mainly by the choice of the one that can be grown most advantageously under the conditions present in each case. In this connection, the prominent characteristics of the various plants will be briefly reviewed.

Some are perennials, some biennials or annuals; some are summer or warm-weather crops and others are cool-weather crops, adapted to late fall and early spring growth. Thus, mammoth, alsike and white clovers and alfalfa are *perennial;* sweet clover is *biennial,* while red clover, though it may persist for three years, is more often grown for only two years continuously and may therefore be regarded as a practical biennial. These crops make slow growth at first—a serious objection to their use in cases where quick results are needed, as with soils deficient in humus.

Of the annual legumes adapted to use as cover-crops, the following are *summer* annuals: Cowpeas, soy-beans, peanuts, beans, velvet beans, common vetch and field-peas; while the following are winter annuals: Crimson clover, hairy vetch, and, in the South, common vetch and field-peas.

In regard to the extent of root systems, red clover, mammoth clover, alfalfa and sweet clover are deep-rooted, and the roots alone may form from one-third to one-half of the plant; the other leguminous plants have comparatively short root systems, constituting approximately one-sixth to one-tenth of the plant. All these as well as other individual characteristics must be taken into consideration in making a choice of the crop that will best adapt itself to the needs and conditions of the farmer.

Preparation for growing green-manure crops.—While most of the green-manure crops are not difficult to grow under proper climatic conditions, and several of them even on poor soils, yet some precautions should be observed if one is to expect success. The points to which special attention should be given are summarized as follows:

(1) *Preparation of soil.*—The soil should be put into fine tilth before seeding, especially in the case of leguminous crops.

(2) *Use of calcium compounds.*—Application of ground limestone or of slaked lime should be made if there is marked indication of acidity (p. 95).

(3) *Addition of plant-food.*—In case of poor soils, it is desirable to use a small application of fertilizer (pp. 360 to 363). In case of

non-leguminous crops, a complete fertilizer should be used; in case of leguminous crops, phosphorus and potassium are particularly desirable, with a small amount of nitrogen to insure the crop a good start (p. 346).

(4) *Inoculation of soil.*—For leguminous crops, especially when grown for the first time, it is often necessary to inoculate the soil for the special crop grown in order to promote the development of root-nodules and the formation of compounds from atmospheric nitrogen (p. 150).

(5) *Relations to various soils.*—Light, sandy soils are generally most benefited by the use of green-crop manures, because they are usually poor in plant-food, lacking in humus, porous, and unable to hold moisture sufficiently for production of good crops. Green crops that will supply nitrogen, moisture and organic matter should be used. The bacterial decomposition of organic matter is rapid, owing to the porous condition and consequent supply of air, and in such soils the organic matter is used up rapidly. Therefore, green crops must be used more frequently.

In growing green crops on light soils, one serious difficulty often encountered is the limited amount of moisture available for crop growth. In case of cover-crops, the preceding crop may take so much water from the soil as to prevent the germination of seed and thus make it difficult to start the green corp. Then, again, green crops may use up so much soil moisture for their growth that, while the green crop grows well, the following crop may suffer from lack of moisture. The excessive drying out of a soil generally results also in marked decrease of the soil bacteria necessary for the decomposition of organic matter and the conversion of its nitrogen into nitrate. It follows, therefore, that green crops should not be seeded in an excessively dry soil.

The incorporation of a large amount of organic matter in a light soil increases the porosity and therefore the rapidity of evaporation of water. Hence, care should be taken not to turn under too large an amount of vegetable matter at one time. In the case of bulky crops, especially if the crop is so mature as to be somewhat dry, it is well to remove a considerable portion of the crop, turning under only stubble and roots. Excessive evaporation of water may be lessened in light soils after turning under a green crop by compacting the surface with a roller.

While light soils are peculiarly benefited by a proper use of green-crop manures, heavier soils are also improved under many conditions. Some of the more striking differences we will briefly notice. Increase of water-holding power in heavy soils is less important than in light soils. Decomposition of green crops in heavy soils is slower, the organic matter lasts longer and green-crop

Uneven spreading of fertilizer can't be kept a secret—the grass will tell.

manures are therefore needed less often. Loss of nitrate nitrogen in drainage water is smaller in case of heavy soils; cover-crops are more effective in saving nitrate from loss by drainage in light soils. Loss of nitrate nitrogen as free nitrogen, except in case of water-logged soils, is of less frequent occurrence in heavy soils. Another point of difference is that in heavy soils spring growth is less rapid than in light soils, which must be taken into consideration in the selection of green crops. More care is required in selecting green-manure crops for heavy soils. The plowing under of green-manure should be less deep in heavy soils than in light soils, since too deep covering retards decomposition of the vegetable matter and promotes the accumulation of organic acids in the soil. The decomposition of green-manure may be hastened by spreading over the surface of the green crop 2 or 3 tons of partially decomposed farm manure an acre, just before plowing the crop under, thus inoculating the soil with vigorous bacteria that cause decomposition of organic matter.

Time of plowing green crops under is of much importance and is dependent upon several conditions, among which are the season of growth, the condition and character of the soil, the degree of maturity of the green crop, as well as weather and seasonal conditions. Generally speaking, green crops should be plowed in when they are still green and full of moisture, since they then decay most quickly, especially in light soils. If a green crop is allowed to become so mature as to contain large amounts of dry, coarse and woody stems, decomposition takes place slowly and the material remains long unchanged in the soil, resulting in drying the soil by increased evaporation caused by abnormal looseness of soil and, consequent upon this condition, a serious decrease of those soil bacteria that produce decomposition of organic matter. There is more danger of this condition in light than in heavy soils. A similar result may occur if a green crop is plowed under when a soil is dry.

Best results can come only when there is enough moisture in the soil to insure prompt decomposition of vegetable matter. Slow decomposition with injurious results is likewise experienced when a large amount of green matter is plowed under in warm, dry weather, or when a hot dry spell follows the plowing in. On the other hand, when a green crop is turned under in late summer or fall, and cool, wet weather follows, slow decomposition with production of acidity may result. The ideal condition for rapid decomposition of green-crop manures, without loss of nitrogen, is abundance of moisture and heat. When fall-sown crops are to follow the green-crop manure, it is desirable to turn the green crop under 4 to 6 weeks before planting time, harrowing frequently or compacting the soil in some other way. Fall plowing of green crops for spring planting may be of advantage in respect to the increase of nitrate nitrogen in the northern states.

In plowing under green-manure crops, the material is evenly distributed from the surface downward to the bottom of the furrow, instead of being covered so as to form a continuous layer between the top soil and subsoil. This is accomplished by turning the soil partly over instead of turning the furrow over flat.

Crops to follow green-manure crops.—In general, it is well to follow green-manures with cultivated crops, such as corn, potatoes, tobacco, cotton, etc. The cultivation of such crops produces conditions that favor the decomposition of the vegetable matter and the consequent increase of available plant-food. Rye and oats generally thrive on green-crop manures, while wheat and barley appear to give variable results, especially the first season after plowing under a green crop.

Effect on succeeding crops.—Numerous experiments with various green-crop manures on many crops, under a great variety of conditions of soil and climate, have demonstrated that when properly used, the succeeding crops are generally much increased. As illustrations of the results of some of these experiments, red-clover increased the yield of corn 20 bushels an acre, of oats 10 bushels or more, of potatoes over 30 bushels; cowpeas increased the yield of seed-cotton 700 pounds an acre, of oats 10 bushels, while cowpea stubble and roots increased the yield of sorghum 2 tons an acre and wheat 4 bushels for four years in succession; crimson clover increased the yield of potatoes 20 bushels an acre the first year and 27 bushels the second.

It is noticeable that the effects of leguminous green-crop manures last through several seasons. There is, however, a possibility that some crops immediately following leguminous green-manures may be unfavorably affected. For example, when a large amount of organic nitrogen is incorporated in the soil by a green crop followed by hot, moist weather, favoring rapid decomposition with formation of nitrate nitrogen, the crop may get too much available nitrogen for normal growth and develop too luxuriantly in stems and leaves, as in the case of excessive applications of farm manure; under these conditions the ripening of fruit is delayed, and there is generally a decreased yield; with grains like wheat and barley, the straw is excessive in growth, lodging easily, and the yield of grain is diminished.

PART IV

FACTORS THAT GUIDE SELECTION OF FERTILIZING MATERIALS

XXI

CONDITIONS THAT INFLUENCE EFFECTIVENESS IN THE USE OF FERTILIZERS

IN using fertilizers, whether commercial or farm-produced, whether direct or indirect, it is highly desirable to know as far as possible something of the conditions under which they can be used with most effective results. It is necessary to discuss this subject as a preliminary to the consideration of two fundamental questions of practical importance which sooner or later come to every farmer: (1) Under what conditions is it necessary to use fertilizers? (2) What fertilizers will give greatest profit? These questions are pertinent also to all fertilizing materials but are asked most frequently with reference to commercial fertilizers, so will be considered more particularly with reference to them. Farmers who purchase without reference to the conditions affecting their efficiency, on the sole basis of the hope of obtaining larger crops and greater net income, inevitably run the risk of throwing away money by buying what may not give good commercial returns, due either to applying constituents not needed or to the application of needed materials under conditions so abnormal as to make impossible proper utilization by crops. Many millions of dollars have been spent for fertilizers without any adequate return.

In Part I, the more prominent facts underlying our knowledge of plant nutrition, including the relation of plant-foods to soils have been studied. In presenting the facts, effort has been made also to keep in mind the fact that the effectiveness of plant-foods in relation to crop production, especially when applied in the form of purchased fertilizers, is unavoidably connected with and dependent upon numerous other factors. It is our purpose now to review and summarize the more important data, preliminary to taking up a detailed study of the practical use of fertilizers in crop-growing.

When considered in the most comprehensive way, several kinds of factors enter into the successful growing of crops, such as (1) the personal knowledge and ability of the grower himself, (2) the negative factors that appear in the form of insect pests, plant diseases, etc., (3) most important for our purpose, the natural factors that directly determine the conditions of plant growth, and (4) artificial methods of control, including (*a*) tillage, (*b*) crop rotations and (*c*) fertilizers. Tillage has been already discussed (p. 106) as much as the limits of this book permit.

External natural factors of crop production.—The natural factors that directly control plant growth proper may be conveniently divided into two sets, those external to the soil and those directly centered in the soil itself.

In the first set of factors, we have: (1) Good seed, (2) sunshine (light and heat), (3) physiological peculiarities of plant or crop, that require adaptation of crops to conditions of climate, etc., and (4) sufficient rainfall favorably distributed in time.

Without detailed discussion, it is obvious that fertilizers bring no profit when any of these factors is responsible for poor crops. In general, it is assumed, in the treatment of the subject, that farmers consider such fundamental conditions carefully before they attempt to grow any particular crop, though there are probably many exceptional cases, especially when new crops or new varieties are being grown experimentally for the first time. It is obvious, also, that some of these conditions are not under one's control and, when they are unfavorable, it is hopeless to work against them.

Good seed is one condition that can be controlled, a fact being increasingly emphasized. By careful selection and breeding, varieties of seed have been obtained especially adapted to certain climatic conditions, so that one variety will produce two or three times as large a crop as another, when both are grown under the same soil conditions. For example, the provincial department of agriculture developed varieties of early maturing wheat well adapted to the conditions of the short hot summers of the Canadian Northwest. Some of the experiment stations have produced variations in corn such that one variety contains much more protein, while another contains much more oil. Similarly, beets grown from some varieties of seed contain less than 5 per cent. of sugar, while those grown from other selected varieties commonly contain 12 per cent. and sometimes over 16 per cent.

It is well known that in various lots of the same variety of seed important differences are found. Generally speaking large, plump, well-ripened seeds give better crop yields than small seed or those immature or shrunken. The importance of seed selection has been especially well illustrated in the case of corn growing, the yield having been greatly increased as the result of care and intelligence.

Purity of seed is another important condition. The necessity of freedom from weed seeds needs no discussion. In case of doubt, farmers should purchase only from reliable parties under a guarantee of purity and should have their seeds tested in respect to purity.

This work of testing is generally performed gratuitously at the agricultural colleges or experiment stations.

Soil factors of crop production.—The conditions required in soils to promote the growth of crops are:

(1) Abundance of available plant-food.

(2) A suitable physical structure, (*a*) which combines mellowness and firmness, permitting plant-roots to extend their growth freely; (*b*) which enables the soil to receive water easily, distribute it uniformly, hold it with sufficient tenacity, and give it up as needed by plants; (*c*) which permits some circulation of air, furnishing needed supplies of oxygen; and (*d*) which enables the soil to absorb heat and maintain a degree of warmth suited to plant growth.

(3) The presence of beneficial micro-organisms and conditions favorable to their growth.

(4) The absence of substances injurious to plants, including acids, inorganic salts, and organic compounds.

A soil may contain an abundance of available plant-food and yet fail to grow satisfactory crops, if any serious defect exists in the soil structure or if toxic substances are present. When it comes to the economical use of certain forms of plant-food supplies and the conversion of unavailable into available forms, we must also take the micro-organisms present into consideration and the conditions affecting their existence and growth. We may briefly repeat here what has already been more fully stated in Part I, that the physical conditions most favorable to the growth of plant-roots are the precise conditions most favorable to the growth of desirable agricultural micro-organisms.

We have previously learned (p. 71) that the physical structure of a soil is largely governed by the texture or fineness of particles and by the arrangement of the soil particles, the best structure being that known as the granular or crumbly structure. To a large extent, soil structure is under control and is influenced by tillage (p. 108), decaying organic matter (p. 108), drainage (p. 108), root growth (p. 97), soluble salts (p. 78), etc. Any one of the numerous conditions unfavorably affecting soil structure may, therefore, make useless the application of fertilizers.

Selecting the special conditions or factors which most commonly affect the extent to which crops are influenced by plant-foods and which are largely controllable, the following should be emphasized in this connection as of special importance: (1) Soil organic matter. (2) Soil acidity. (3) Soil structure. (4) Soil moisture.

To have plant-food utilized by plants to best advantage and, for reasons presented in the earlier chapters, we must have an abundance of decaying organic matter (p. 97), a soil free from acid compounds in case of most crops (p. 94), an abundance of soil moisture (p. 77), and a soil structure that permits the best conditions for root growth (p. 71). Though it is obvious that these conditions are more or less interdependent, we have arbitrarily separated them for the purpose of emphasis. As a preliminary step in ascertaining when it is necessary to use fertilizers, one must first make sure that the soil conditions are satisfactory in these respects. If any defects in these respects exist, they must be corrected. This may be done, according to the special demands peculiar to each case, by supplying organic matter through the agency of farm manure (p. 243) or of green crops (p. 259), draining (p. 109), irrigating (p. 110), liming (p. 192), improved tillage (p. 106), etc.

Relation of soil conditions to available plant-food.—The fact cannot be too strongly emphasized that the conditions of the soil to which attention has been called in connection with efficiency of applied fertilizers are just the conditions most effective in changing the unavailable or slowly available forms of soil plant-food into readily available forms. As one of their chief purposes, the best methods of soil management seek to utilize the slowly available plant-food supplies present in the soil, to make them gradually available during the crop-growing season in amounts sufficient to meet the needs of crops in part or in whole.

With the foregoing statements in mind, it can readily be appreciated how hopeless it is to try to answer some of the questions asked by farmers, one of the most common of which is: "What fertilizer will give me the best results for this or that particular crop?" It is, of course, impossible to give any definite, reliable answer to such a general question, involving so many different factors, none of which is clearly known. As we have repeatedly pointed out, the composition, the physical properties and the biological conditions of the soil, the extent and manner in which it has been previously cropped, fertilized and managed, the kind of crop one wishes to grow, all of these conditions need to be known, and even then it will require some practical experimenting on the part of the individual farmer to determine what forms and amounts of fertilizers can be used most economically.

Our knowledge of the functions and proper use of fertilizers has been largely increased in the past as the result of experimental investigation, and even more rapid advances are now being made; yet it must be confessed that, with all these additions of knowledge and experience, there is yet no simple way of determining beyond doubt when and how best to use fertilizers in crop growing. The conditions involved are extremely complex and do not usually permit any reliable, shortcut methods of reaching satisfactory conclusions.

In selecting fertilizers for use, we need to consider numerous questions of fundamental importance, among which are the following:

1. Conditions under which fertilizers should be used.
2. The specific constituents of plant-food needed.
3. The amount of each fertilizing constituent needed.
4. The forms in which it is best to supply plant-food constituents.
5. Facts about commercial fertilizers.
6. Home-mixed fertilizers.
7. Methods and seasons of applying fertilizers.
8. Crop-rotation in relation to the use of fertilizers.
9. Plant-food mixtures for various crops.

XXII

CONDITIONS THAT REQUIRE FERTILIZERS—PLANT-FOOD CONSTITUENTS NEEDED

In the case of fertile virgin soils, abundant crops are raised without application of any form of fertilizing material, direct or indirect, but under such treatment, even the richest soils sooner or later begin to show decreased yield. Then the farmer begins to use farm manure and other resources of the farm to check the decrease.

Assuming that this is all that is needed at first, there comes a time in most systems of farming when all the resources of the farm capable of furnishing fertilizing materials are insufficient to maintain the former yield of crops. Assuming that the conditions affecting the efficiency of plant-food materials do not require attention, the next step necessary to maintain crop yield is the use of purchased plant-foods or commercial fertilizers. Stated in a broad way, a farmer must resort to commercial fertilizers when he has exhausted all the resources of the farm in producing his own fertilizing materials and finds that commercial fertilizers increase crops and profit. Under conditions of good soil management, when his crops cannot get from the soil as much nitrogen or phosphorus or potassium or other plant-food constituents as they need for profitable production, and when the resources of the farm in furnishing plant-food materials have been utilized to their fullest extent without maintaining desirable crop yield, he may resort successfully to the use of commercial fertilizers. In order to make such use a paying operation, one must secure an increase of crop more than sufficient to pay for the cost and the labor of applying the fertilizer used.

Before resorting to commercial fertilizers, if economy of production is to be realized, one must be sure that lack of crop yield is not due to some unfavorable soil conditions other than deficient supply of plant-food. (See Chapter XXI.)

It is, therefore, readily seen that it is not a simple matter to tell when one should use commercial fertilizers. The general rule will be to use them when their application is attended with increased profit due to larger crops. It is obvious that this condition can be ascertained only by actual trial or experiment. (p. 279.)

Indications of various kinds may suggest the need of applying purchased plant-food materials, so we will consider these in connection with our discussion of what plant-food constituents are needed.

How to ascertain what specific plant-food constituents are needed.—When it has been settled that a soil needs the addition of purchased plant-foods in order to grow crops more successfully, the question at once presents itself as to what kinds of plant-food are required. Is nitrogen needed or is it phosphorus, or is it potassium? In some special cases, may it not be sulfur, or manganese or some other constituent not present in the soil in sufficient amount? How can we ascertain what kind of plant-food is required? Several methods have been employed, some of which are suggestive, but there is in reality no simple method upon which we can rely for all soils and under all conditions. Experience and practical judgment often count for much but may lead one wholly astray. Among the ways of finding what particular plant-food constituent or constituents a soil needs, we will consider the following: (1) Analysis of soils, (2) appearance of crops, (3) special character of certain soils, (4) suggestions furnished by previous cropping, (5) fertilizer experiments.

Analysis of soil as indication of plant-food needs.—It was formerly thought that chemical analysis would readily furnish information from which the amounts of nitrogen, phosphorus and potassium necessary to assure an abundant crop could be calculated and applied to the soil. A trained analytical chemist can easily determine the amount of each plant-food constituent in a soil and show the total amounts held as a store for the future, but such results fail to furnish information sufficient for practical purposes as to how much of these total amounts is unavailable or slowly available or immediately available for crop growth. Usually, a soil analysis is made by treating a sample of soil with a dilute acid; it is supposed that the amounts of plant-food constituents thus dissolved approximate the proportions more readily available for the use of plants. When these results are compared with actual crop growths on the field from which the sample of soil analyzed is taken, wide discrepancies often occur. The actual value of soil analysis in determining positively and definitely the plant-food needs of a soil has been and is still a matter of dispute. All investigators agree that when analysis shows that the total amount of any plant-food constituent is present in small amount or wholly absent, the results of soil analysis are negatively helpful in enabling one to reach conclusions. Soil analysis is found useful in careful investigation, especially in connection with field experiments, but, as previously stated, the two methods do not necessarily furnish results that agree. Studies have been made to learn what relations might exist between water-solutions of soils and available plant-food, but the results have not been sufficiently reliable as a guide to applying fertilizers.

Another consideration will illustrate how misleading the result of a soil analysis by itself may be. A soil may be shown by analysis to have large amounts of available plant-food constituents and yet fail to grow satisfactory crops owing to defects in physical or other conditions of the soil such as discussed in the preceding chapter.

Soil analysis, however, is an essential part of any complete soil investigation, and, taken in connection with other data, furnishes valuable information. In the case of a region where soils have been extensively studied, it may furnish desired information in the case of individual farms characteristic of the region. But an ordinary soil analysis, taken by itself, is generally insufficient as a basis in deciding what kind and amounts of plant-food to use for crops.

Appearance of crops as indication of needs of plant-food.—To one experienced in watching the behavior of crops, much may often be learned regarding the plant-food conditions by closely observing the appearance of the plants. The precaution should be given that this method is not infallible, because an abnormal condition of appearance caused by deficient plant-food may likewise be caused by some other condition not necessarily so connected.

(1) *Nitrogen.*—(*a*) Deficiency. When the physical condition of the soil is favorable and when there is sufficient rain and sunshine, without excess, a pale-green or yellow color of leaves or stems, or a slow, small growth of leaf or stalk, is commonly an indication of nitrogen starvation, due usually to an insufficient supply of available nitrogen. The same effects may be caused by some forms of plant disease or by some insect pests, due probably to inability of plants to use their nitrogen supply even when present in abundance.

The yellow appearance of crops in time of drouth is largely due to the inability of the plants to obtain needed plant-food because there is insufficient moisture to dissolve and carry it or, in some cases, because the soil solution becomes so concentrated as to injure plant roots. In the case of heavy soils, a cold, wet spring delays nitrification (p. 139) to such an extent that the soil does not furnish available nitrogen rapidly enough to meet the demands of the young crop, a condition shown by the unhealthy, pale appearance of the plants. A change to warm and less wet weather or an application of sodium nitrate under such circumstances usually

Soil house where crop samples grown in these cylinders are dried, ground and stored. These experiments were instituted in 1898 and are continuous to date.

produces sudden and marked change in the color and growth of the crop.

Lack of nitrogen may be indicated also when trees drop their leaves early, or when plants ripen prematurely or fruit trees are stunted, bearing little or no fruit.

(*b*) Abundance of nitrogen.—A bright, deep-green color, with a vigorous growth of leaves and stalks, is, in case of most crops, a sign that available nitrogen is not lacking, but does not necessarily indicate that more nitrogen could not be used to advantage

(*c*) Excess of nitrogen.—An excessive growth of leaves and stalks, accompanied by an imperfect development of bud, flower or fruit, generally indicates too much nitrogen in relation to phosphorus and potassium. The application of large amounts of fresh farm manure furnishes relatively large quantities of urinary nitrogen, which, under favorable soil conditions of warmth, moisture, ventilation, etc., may be so rapidly changed into nitrate as to furnish the crop an excessive supply. This has been a common experience with farmers. The results have been manifested in the case of grain crops by large growth of straw and small yield of grain, the crop showing a tendency to "lodge" (p. 50); in the case of crops like potatoes, there is an enormous growth of tops with comparatively few and small tubers. Such excess of nitrogen has been reported also as the result of plowing under leguminous crops too frequently.

This condition can be remedied by avoiding large applications of nitrogen-containing materials or by accompanying such applications with the addition of phosphorus or potassium compounds or both. As will be pointed out later (p. 413), some crops require large amounts of nitrogen for best results in growth and quality; amounts which would injure cereal crops, for example, are not too much for the needs of such crops as celery, asparagus, lettuce, etc.

(2) *Phosphorus.*—Under favorable soil and climatic conditions when grain crops mature early and furnish smooth, plump, and heavy kernels, a good supply of available phosphorus in the soil is indicated. Slow maturing of grain crops, with light, shrunken kernels, is usually due to an excess of nitrogen in relation to phosphorus, as in case of fresh farm manure. Lack of phosphorus is also indicated when the growth of straw in cereals is normal but the yield of grain is below normal.

(3) *Potassium.*—When such crops as corn, cabbage, grass, potatoes, tomatoes, beets, etc., have a vigorous, strong, healthful growth, there cannot be a deficiency of potassium compounds in the soil. Another indication of the presence of abundance of potassium is the growth of fleshy fruits of fine flavor and texture. Deficiency of potas-

sium is often indicated by weakness and brittleness in stalks and leaves and in some cases by the sensitiveness of plants to disease.

As already stated, such special indications may often be helpful, so the conditions of crops should be carefully studied with all the facts in mind. In interpreting such signs too narrowly, and without taking all conditions into consideration, such as season, diseases, insects, etc., there is always danger of being misled. Then, too, unfavorable indications shown by crops may be due to conditions which are temporary and which will adjust themselves, as, for example, in case of unfavorable weather.

Plant-food needs in relation to character of soil.—In certain special kinds of soils, such as clay, sandy and peaty soils, the character of the soil often furnishes valuable information as to what kind of plant-food may be lacking.

(1) *Clay soils,* owing to their origin (p. 73), are usually well supplied with potassium compounds but are likely to be deficient in phosphorus and often in calcium compounds. Therefore, treatment of many clay soils with phosphorus will supply that element, while application of calcium compounds may be sufficient to insure a generous supply of potassium for crops.

(2) *Sandy soils,* as might be judged from their origin (p. 72), often lack an adequate supply of potassium and often require phosphorus also. The same is true of many gravelly soils, owing to their liability to lose plant-food by leaching.

(3) *Peaty soils* are rich in nitrogen in forms not quickly available and are often deficient in potassium and phosphorus. Therefore, the application of potassium and phosphorus compounds to drained swamp-lands is usually attended with marked advantage in crop yield.

In the numerous mixed types of soils, where no one character is pre-eminent, these suggestions find no application. Most soils are of the mixed types.

Kind of plant-food suggested by previous cropping.—The growing of the same kind of crop continuously on the same land may be exhausting in regard to one or more elements of plant-food. For example, when wheat is grown year after year on the same soil, the available nitrogen and phosphorus are used in larger proportions than potassium. When such deficiency actually occurs, application of nitrogen and phosphorus compounds is called for. On old wheat-growing soils, it is generally true that special benefit comes from applying phosphorus compounds. The continuous growth of grass crops takes large amounts of potassium compared with phosphorus.

Fertilizer experiments.—During the past the most common method employed in soil investigations to ascertain what constituents of plant-food a particular soil needs to produce increased crops of a given kind has been to apply various plant-food constituents to various portions of a soil and observe whether or not any increase of crop results. Theoretically, this is the simplest and most direct method and should give helpful results. While it frequently does so, it does not always, and in some cases, may give results that are contradictory, meaningless or misleading. When such experiments are carried on systematically for long periods by skilled investigators and the results checked by all possible precautions, useful results are obtained. Such experiments carried on for a single season may be wholly misleading, especially if some abnormal weather conditions happen to be present. Under ordinary conditions and in the hands of one willing to give the work reasonable attention, the method may be helpful for the purpose in view.

Another method of soil-testing is to grow plants in boxes or pots filled with the soil to be experimented with, applying various fertilizers and combinations in varying amounts. Under some circumstances helpful results are often obtained. A modification of this method is known as the "paraffin-basket" method of soil testing. Another method is to grow plants in water solutions containing the various plant-food constituents. Without going into the details of these methods of experimenting, it is sufficient for our purpose to say that they are not suitable for use in the hands of farmers for the purpose of ascertaining with any satisfactory degree of certainty or accuracy the plant-food needs of a soil.

Amounts of plant-food constituents needed.—How can we ascertain how much nitrogen, phosphorus, potassium or other constituent it may be necessary to apply to a soil to insure available plant-food enough to produce a satisfactory crop? This matter of amounts of plant-food to use is one of great complexity; no simple way is known to answer the question. If we could know how much available nitrogen, potassium and phosphorus could be furnished a given crop by the soil during the growing season, we could estimate from data similar to those given in Table 20 (p. 113), the amounts of these plant-food constituents that would be needed for a certain yield of the crop, and by comparing these figures, we could ascertain whether the soil supply would meet the demands of the crop.

The factors that largely determine the amounts of plant-food constituents needed are: (1) The amount of available plant-food present in the soil, (2) the amounts of plant-food constituents required by the crop grown, and (3) the special kind of farming practiced. We will briefly consider these points.

(1) *The amount of available plant-food in the soil.*—The methods of experiment mentioned above are not adapted to use except for investigational work in the hands, or under the direction,

of trained scientists. Though field experiments in the hands of many farmers often prove disappointing because definite conclusions cannot be drawn, such experimental work cannot be too highly recommended for the general training and educational benefit it affords. This line of work, properly conducted, may furnish the progressive farmer information of great practical and economic value, but it will also exert a profound educational influence upon him, making him personally familiar with plant-food materials and their specific effects, causing him to take a deeper interest in the growing of crops and leading him to practice better methods.

(2) *The amounts of plant-food constituents required by crops.* —We have already shown (p. 111) that various crops use varying amounts of nitrogen, phosphorus and potassium. When we know approximately how much of each plant-food constituent an average yield of any particular crop will remove from the soil, we have fairly definite knowledge of the amount of each necessary to have

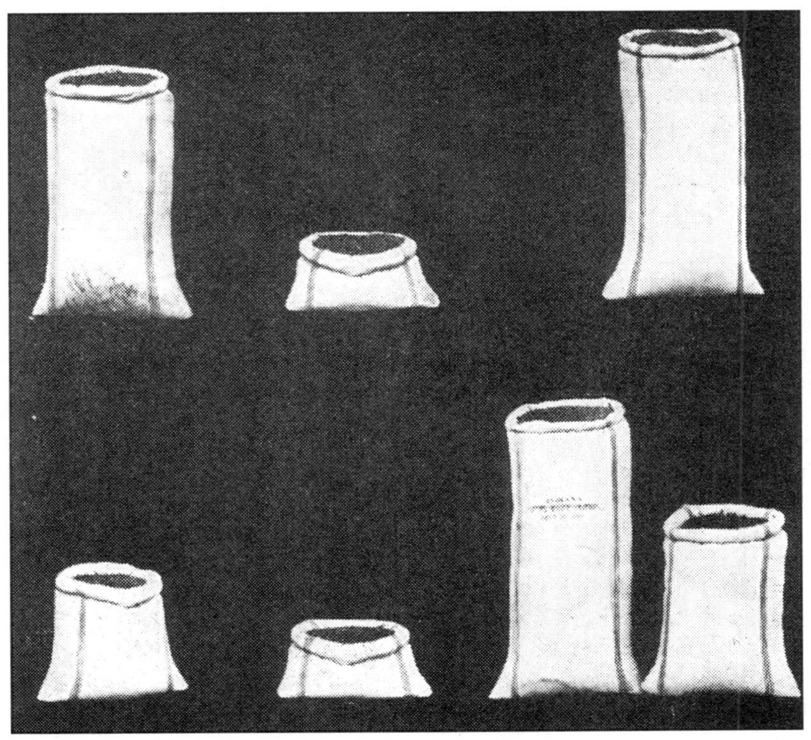

Effect of fertilizers on wheat. Crops in center bags received no fertilizer. Upper left had phosphorus and nitrogen; upper right, phosphorus and potash; lower left, potash and nitrogen; lower right, potash, phosphorus and nitrogen.

within reach of the plant-roots during the growing season of the crop. However, such figures do not in every case represent the quantities of plant-food removed from the soil. Thus, in the case of leguminous crops, a portion of the nitrogen is generally obtained from the air, so for such crops less nitrogen need be applied than the composition of the plant indicates. While such figures are useful as valuable suggestions, we cannot rely upon them as strictly accurate guides in telling us how much nitrogen or phosphorus or potassium to furnish a crop.

(3) *The special kind of farming practiced.*—The amount of plant-foods required for the economical growing of crops must depend to some extent upon the specific character and methods of farming employed. In the case of market-garden crops, where it is desired to secure maximum crops at the earliest possible time of season, it is the practice to apply amounts of fertilizers that would be regarded as wastefully excessive in general farm practice. Again, the farmer who utilizes an intelligent form of crop-rotation can raise crops with less expenditure for fertilizers than can the one who does not use a desirable system of rotation, as we shall point out later more in detail (p. 336).

From what has preceded, it is clear that we have no reliable method for determining accurately beyond question the plant-food needs of soils. The application of fertilizers in crop growing is still, and will probably always be, more or less largely empirical, resting on experience and observation. The matter of soil fertilization is really more than the feeding of crops, since other conditions affect plant growth. When we apply plant-foods to soils, we not only place the material within reach of the growing plant, but the presence of the added material in the soil usually produces modifications of a physical, chemical or biological character in the soil itself, and these modifications influence plant growth to an extent we cannot easily measure. It is this multiplicity of effects produced by a fertilizer that often makes difficult the correct interpretation of soil experiments.

While we cannot tell with any degree of accuracy in each special case how much plant-food to apply to a soil for best results, we are not left in hopeless confusion when it comes to the practical growing of crops. We know many facts of importance and we have the results of long-continued experience and observation that afford a basis for agricultural practice. While specific rules of general value cannot be given we can study the facts that have been accumulated and use them as suggestions for application to individual cases.

XXIII

FACTORS THAT GOVERN SELECTION OF PLANT-FOOD MATERIALS

We have seen in Part III that we can obtain nitrogen, phosphorus and potassium, each in several forms. When we come to make use of fertilizing materials as sources of plant-food, we must decide which specific forms we shall purchase. What principles, all things considered, will guide us in making the best choice? We can be guided by: (1) Availability of plant-food, (2) effect upon soil, (3) adaptation to crops, (4) comparative cost.

AVAILABILITY OF VARIOUS FORMS OF PLANT-FOOD

As a rule, commercial fertilizers are used for their chief effect upon crops during the single season immediately following application. Farmers generally give to each crop by itself the plant-food supplies they think may be needed, and avoid furnishing any crop more material than it can use to best advantage in one year. This is especially true of those who do not own the farms they work and who plan each year's work by itself without reference to the future. On the other hand, the farmer who owns his land frequently desires to use materials the plant-food of which will be utilized gradually by crops and furnish a continuous supply through several successive seasons. For some crops it is a convenience and an advantage to use slowly acting forms of fertilizers. It will thus be seen that if one desires a fertilizer which will act at once and be largely used by the immediate crop, he will need to purchase his plant-food in forms different from those used by the man who desires longer-continued action.

Availability of different forms of nitrogen.—We have discussed in other pages the composition and general properties of the nitrogen-containing materials used as sources of plant-food (pp. 29, 162). We will now consider them particularly with reference to the quickness with which plants can use their nitrogen placed within reach of growing roots.

(1) *Nitrate nitrogen* (sodium nitrate, ammonium nitrate, calcium nitrate, potassium nitrate) is easily soluble in water and its nitrogen is directly used (p. 284). Under ordinary conditions one part of sodium nitrate dissolves in only one part of water; that is equal to saying that one pound of sodium nitrate is completely dissolved in only one pint of pure water; other nitrates are equally or more soluble. The ease with which a nitrate compound dissolves enables it, in the presence of soil moisture, to be distributed quickly through the soil and reach the plant-roots. Therefore, because of its ease of solution and because the solution can be directly absorbed and used at once by plants, nitrate is the most easily avail-

Experiment plots at Tennessee Experiment Station.

able form of nitrogen we can furnish plants. Under favorable conditions, the effect of applying a nitrate to soils can be observed in the appearance of plants in one or two days.

The extremely easy solubility of a compound nitrate in water is a distinct advantage in respect to availability, but is a disadvantage under conditions where a large amount of rainfall dissolves and carries the solution down below the reach of plant roots, with likelihood of loss in the drainage water. (p. 118.) On this account, nitrates should be applied only in moderate amounts at a time, and preferably when crops are growing on the soil. Because of their rapid action on plant growth, nitrates are extensively used by market-gardeners and in forcing-house operations and because of their easy solubility and rapid action they are of special value as top-dressing in early spring on meadow-lands and on small grains before nitrification has begun, when soil nitrates are usually least abundant.

(2) *Ammonia nitrogen in ammonium sulfate* (p. 29), ammonium carbonate and ammonium compounds in general follow ni-

trate nitrogen in point of availability as plant-food. The compound that most commonly occurs in commercial fertilizers, ammonium sulfate, is easily soluble in water, about as much so as sodium nitrate (p. 166), and has, therefore, the advantage of quick distribution in the soil. Some plants can absorb and directly use ammonia nitrogen somewhat as they do nitrate; but, to whatever extent this may be true, ammonia nitrogen is changed in the soil rapidly into nitrate (pp. 138, 144) and under normal soil conditions the ammonia is largely changed into nitrate before it reaches the plant. Ammonium compounds are not leached from soils as readily as nitrates (p. 118) and are therefore likely to give better results in wet seasons. The action of the ammonia nitrogen is somewhat more gradual and prolonged than that of nitrate and is, therefore, preferred when the effect of the plant-food is to be extended through a longer period of crop growth.

(3) *Ammonium nitrate* (p. 31) combines the properties of both the nitrate and the ammonia nitrogen, the quickness of one and the moderation of the other.

(4) *Calcium nitrate* (p. 165) closely resembles sodium nitrate in solubility and availability in furnishing plants with nitrogen.

(5) *Urea* (p. 167) is readily soluble in water and completely available. It does not leach from soil as easily as nitrates. While urea nitrogen may be directly assimilated by some plants to some extent, it undergoes bacterial decomposition into ammonia and the conversion into nitrate under usual soil conditions, before utilization by crops.

(6) *Cyanamid nitrogen* (p. 285) in calcium cyanamid has been shown to have about the same degree of availability when compared for a growing season as that shown by ammonia nitrogen. This would be expected from the fact that, after it is applied to soils, its nitrogen changes into ammonia. Calcium cyanamid itself is not used as food by plants, but must decompose first into ammonia and then change to nitrate.

(8) *Nitrogen contained in animal and vegetable materials* is not readily soluble in water and not in condition for immediate use by plants. As explained (pp. 138, 144), these materials first undergo bacterial decomposition, changing through a series of complex compounds into simpler forms, and finally into ammonia, and this in turn, by the nitrification process, into nitrate nitrogen. We thus see that the availability of organic nitrogen depends upon the rapidity with which it undergoes bacterial decomposition, and this in turn depends upon (1) the source of the organic nitrogen, (2) the fineness of division of the organic material, and (3) the soil conditions that affect bacterial decomposition, such as the presence of the

right kinds of bacteria, proper degree of warmth, moisture, air supply, calcium carbonate, etc.

Generally speaking, the better forms of these materials can be used to advantage on crops whose period of growth is prolonged during the season, since they furnish a gradual supply of nitrate nitrogen as they undergo decomposition and nitrification. It is generally a wise plan to have in fertilizers some nitrate nitrogen for the immediate use of the young plants and furnish the remainder of the nitrogen in the form of ammonium compounds and organic matter, the nitrogen of which becomes available after the nitrate is used.

Of the various materials that contain organic nitrogen, *dried blood* and "base-goods" nitrogen (p. 286) become most quickly available. Of nearly equal value to blood in availability are finely ground *fish*, *cottonseed-meal,* and *castor-pomace.* Acidulated fish is somewhat more available than untreated fish. *Tankage,* on account of its variable composition and fineness, varies in availability, but may be generally regarded as considerably slower than dried blood. The availability of the organic nitrogen in *bone* varies greatly, depending on the fineness and previous treatment. In raw, coarse bone, the nitrogen is slow to become available. The removal of fat by steaming or boiling makes the nitrogen more available for two reasons: In the first place, the presence of the fat interferes with the decomposition of the organic nitrogen, and, second, the bone can be much more finely ground after removal of fat. *Horn* and *hoof meal* (p. 286), when fine, appear to undergo decomposition about as rapidly as ground fish. Their availability can be increased by treatment with sulfuric acid. *Leather* (p. 286) in raw condition is practically valueless as a source of nitrogen on account of the extreme slowness with which it decomposes. Even when roasted or steamed and ground fine, it has little value. However, in such form when carefully treated with strong sulfuric acid, it acquires a degree of availability about equal to that of ground fish. The nitrogen in unprocessed *hair* and *wool-waste* is of much the same character of availability as that in raw leather.

Farm manure, as already shown (pp. 215-234), varies in availability according to a variety of conditions. The nitrogen in urine becomes quickly available, almost as readily as ammonia; under favorable conditions, it requires only a few hours to change the nitrogen of urine into ammonium carbonate. The nitrogen in the solid portion of animal manure is not quickly available, not as quickly as the nitrogen in the food before consumption, because in the process of animal digestion the less-resistant portions of nitrogen compounds are retained in the body, while the more-resistant portions remain undigested and pass from the body as solid excrement. The nitrogen in *green crops* varies with diverse conditions, but, in general, is more quickly available than that in the solid portion of animal manure; the decomposition may be hastened by treatment, previous to plowing under, with good quality of mixed farm manure or with liquid manure, which furnishes the micro-organisms that produce decomposition. *Muck, peat* and similar materials usually contain little nitrogen that becomes readily available (p. 93). When such materials are used on soils as a source of nitrogen, they should be applied in rather large amounts. Their decomposition may be hastened by treatment with freshly slaked lime at the rate of 1,000 to 1,500 pounds an acre.

Availability tests of nitrogen-containing materials.—Numerous experiments have been made to determine the comparative plant-food value of nitrogen in the more common animal and vegetable materials, both by comparative tests in growing crops and by chemical methods. While the results of vegetation experiments have not been uniform, they serve as a fair guide in regard to the plant-food availability. Taking results secured with nitrate nitrogen as 100, Table 34 can be made:

TABLE 34—COMPARATIVE PLANT-FOOD VALUE OF NITROGEN IN VARIOUS MATERIALS

Material containing nitrogen	*Availability or Crop-producing value*
Sodium nitrate	100
Ammonium sulfate	95
Dried blood	70
Cottonseed-meal	70
Castor-pomace	70
Fish, dried and ground	68
Horn and hoof meal	65
Leather, dissolved with acid	65
Garbage tankage	55
Leather, steamed	15
Leather, roasted	10
Leather, raw	2

It must be understood that these figures are only approximate. They indicate what is fairly true when the materials are used on the same crop under the same set of conditions.

Numerous attempts have been made to apply chemical methods to organic material in order to distinguish slowly from quickly available forms of nitrogen in such a way as to determine approximately the relative availability for crop use. The most promising method as yet tried is treatment of the water-insoluble portion of organic matter with potassium permanganate, with and without an alkali. For example, a mixture of dried blood and cottonseed-meal is found by this method to contain about 72 per cent. of available nitrogen, while a mixture of treated leather, garbage tankage, and peat contains only 44 per cent. The method is useful, at least, in enabling chemists to distinguish readily available from unavailable nitrogen, even if it cannot show exactly how much nitrogen a crop can use during a growing season.

Availability of various forms of phosphorus or phosphate compounds.—The composition and general properties of phosphorus compounds have been previously treated (pp. 173-185); we will discuss here only those points that relate particularly to the availability of the various phosphorus compounds which form the chief source of plant-food supply. As a general statement, it can be said that the availability of phosphates, as determined by experiments, is in the following order: (1) Mono-calcium phosphate, (2) di-calcium phosphate, (3) precipitated tri-calcium phosphate, bone meal, basic slag, (4) bone ash and rock phosphate (floats).

(1) *Phosphates, soluble in water,* are the only phosphorus compounds directly absorbed by plant-roots. They include in fertilizer materials superphosphates (mono-calcium phosphate) and phosphates of ammonium and of potassium (pp. 37, 180), the compounds used to furnish quickly-available phosphorus for feeding crops. Mono-calcium phosphate dissolves under ordinary conditions at the rate of about 1 part in 100 parts of water. This is the chief phosphorus compound of superphosphates (acid phosphates).

(2) *Di-calcium phosphate* or "reverted" phosphate (pp. 32-36)—When applied to soils, soluble phosphate compounds generally undergo a change in composition, forming with the calcium carbonate of the soil di-calcium phosphate (p. 33), known also as "reverted" and "citrate-soluble" phosphate. "Precipitated" bone (p. 33) consists largely of citrate-soluble phosphate. This compound is less soluble in water than the superphosphate, but is fairly soluble in water containing carbon dioxide, as in the case of soil water. While one part of reverted phosphate requires 7,500 parts of pure water to dissolve it under usual conditions, it requires only 1,800 parts of carbonated water. In the case of soils that contain large amounts of iron or aluminum compounds and only small amounts of calcium carbonate, the reverted phosphate is an iron or aluminum compound, which is so little soluble in water as to be practically unavailable as plant-food for prompt use. This condition means loss of applied soluble phosphate, at least, in large part, so far as the season's crop is concerned; it can be prevented by keeping the soil well supplied with calcium carbonate (pp. 192-214).

The question naturally arises as to why it is not better to apply phosphorus directly in the form of "reverted" phosphate instead of applying it as soluble phosphate, which soon changes to the reverted form. Even though phosphate applied in soluble form becomes reverted in soils before it can be used by a crop, it is better to apply it in the soluble rather than the reverted form. When the soluble phosphate is applied to the soil, it goes into solution as quickly as the amount of water in the soil permits and is at once distributed uniformly among the soil particles; when it reverts and goes into the precipitated form, the soil particles are coated with this finely divided phosphate and under these circumstances it is in the best possible condition for undergoing solution in the soil water and coming into contact with plant roots. On the other hand, if the phosphate is applied in the reverted form, even if finely divided, it is impossible to distribute it as thoroughly and uniformly as in the case of a solution.

(3) *Tri-calcium phosphate* or insoluble phosphate (pp. 33 and 36) is less easily soluble in water than the preceding forms. One part requires about 50,000 parts of pure water for solution under ordinary conditions, but the solubility in water is considerably increased by the presence of carbon dioxide and also of some materials applied to soils as fertilizers. Among these are sodium nitrate, ammonium sulfate and potassium sulfate; aluminum and iron compounds make it less soluble. The presence of decaying organic matter, which furnishes carbon dioxide and free acids, creates conditions which make the insoluble phosphate dissolve more quickly, and so become more readily available as plant-food. Tri-calcium phosphate was formerly generally regarded as of little value as a practical source of phosphorus for plants, but recently it has come to be increasingly valued under certain conditions, especially where soils are well supplied with decaying organic matter and calcium carbonate. The presence of calcium carbonate is of special value in preventing the formation of the insoluble iron or aluminum phosphates. One form in which insoluble phosphate is extensively used is that of finely-ground "raw" or untreated phosphate rock, commonly known as "floats" (p. 235). Other things being equal, availability of such insoluble phosphate depends upon the fineness of its particles; extreme fineness is essential to promote more uniform distribution in soil and to offer more extensive surface for attack of dissolving agents. On definitely acid muck soils insoluble phosphate in the form of "floats" may give as good crop yields as the dissolved phosphate, other conditions being favorable. The availability of insoluble phosphate appears to vary for diverse crops. For example, turnips, rye, buckwheat, alfalfa and sweet clover appear often to use fine-ground insoluble rock-phosphate with beneficial results. Barley and corn appear generally to do better with superphosphate. Ground rock-phosphate on acid soils gives excel-

lent results in some cases. Results, however, often depend upon various soil conditions as much as upon the kind of crop.

(4) *Bone.*—The availability of the tri-calcium phosphate in animal bone is greater than in ordinary tri-calcium phosphate from mineral sources when the condition of fineness is the same. This is because the phosphate in bone is intimately associated with organic nitrogenous material that decomposes easily under favorable soil conditions and the insoluble phosphate is made gradually soluble during this process of decomposition. The phosphate in raw bone becomes available less easily than in bone from which the fat has been removed, because the fat serves to retard the decomposition of the bone and also to make impossible equally fine grinding. Increased fineness of bone increases the availability. Under favorable conditions, it is believed that one-third to one-half of the phosphate in bone becomes available for crops during the first growing season after application. The balance becomes available approximately in the course of two or three years. So much depends upon soil conditions that great variation inevitably occurs.

(5) *Basic-slag phosphate,* on the basis of experimental evidence, is believed, to be about one-half as quickly available as soluble calcium phosphate, but, of necessity, much depends on special soil conditions and crops. This material has shown good results on acid soils and especially on swampy and wet soils. Its degree of availability is dependent also upon its fineness of division.

Availability of potassium compounds.—All the potassium salts commonly used in agriculture are easily soluble in water (p. 38). They are, therefore, in condition to be distributed promptly and uniformly when the soil contains sufficient moisture. In this dissolved condition they are ready for direct use by plant roots. However, after potassium salts are applied to a soil, considerable time may elapse before the crops can use the solution. Meanwhile, they are undergoing reactions or exchanges with some of the soil constituents, by which the potassium changes into compounds less easily soluble (p. 117). This fixation prevents excessive loss of potassium by leaching. In the soil water the potassium again slowly changes back to soluble compounds as rapidly as called for by crop demands.

EFFECTS OF FERTILIZING MATERIALS UPON SOILS

The thought of farmers has naturally not gone further than to apply plant-food materials to soils to furnish nutriment for crops. But we are coming to understand that the application of a chemical compound to the soil may, and usually does, do much more than furnish plant-food. What we may call secondary effects often influence the chemical, physical and biological properties of the soil, and, therefore, may affect crops in ways not anticipated. The secondary effects thus produced by fertilizers deserve consideration, because they bear upon the selection of fertilizing materials. The effects of organic materials have been previously discussed (pp. 97-100). Here we will confine our attention more particularly to inorganic compounds.

Effect of sodium nitrate on soils.—In applying sodium nitrate as a fertilizer, we have had in mind only the effect of nitrate nitrogen upon plants; we have not usually considered the sodium as contributing anything to the welfare or injury of soil or crop.

(1) *Beneficial effect.*—Sodium in sodium nitrate may be valuable in crop growing, not because it is in any way necessary to the direct nutrition of plants, but because of its action upon the insoluble potassium compounds in the soil, in which sodium changes places with potassium and a soluble potassium compound results. To such an extent may this action take place in many soils, especially clays, that an application of sodium nitrate brings so much potassium into solution as to make unnecessary the application of soluble potassium compounds.

(2) *Injurious effect.*—When sodium nitrate is used alone continuously for long it may destroy the granular crumb structure of the soil (p. 76). Under such conditions the soil remains wet and puddles badly (p. 74), if disturbed before it dries. Such land becomes intolerably sticky after rains and dries into hard, unmanageable lumps, so much so that it is often difficult to secure a start for young plant roots except in very favorable seasons. These effects are caused by the sodium contained in the nitrate and left behind after the nitrate is taken into the plant. It may be explained that plants do not take the whole compound of sodium nitrate into their roots (p. 127), but take the nitrate and leave practically all of the sodium behind in the soil. This residue combines readily with carbon dioxide in the soil water and becomes sodium carbonate.

It is well known that alkali carbonates, such as sodium carbonate, produce deflocculation (p. 78), that is, cause the soil grains or crumbs to fall apart into the very fine particles composing the larger crumbs. In such cases the finest material of the soil is slowly removed in the course of years. Bad soil structure, thus produced, is difficult to correct. The use of calcium compounds, such as hydrate or carbonate, under these conditions only make matters worse. The most effective remedy is the application of superphosphate or ammonium sulfate or gypsum to furnish acid to neutralize the alkali of the sodium carbonate and convert it into a harmless compound. This condition is prevented by using ammonium sulfate along with sodium nitrate in furnishing nitrogen, or by substituting calcium nitrate for sodium nitrate.

(3) *Effect on soil acidity.*—In the preceding paragraph we have noted the production of sodium carbonate in the soil as a result of the use of sodium nitrate. This carbonate neutralizes acids and prevents a soil becoming acid. It therefore saves the calcium carbonate of the soil. Generally speaking, the effect of 100 pounds of sodium nitrate in the soil, when used by plants, saves over 100 pounds of calcium carbonate.

Effect of ammonium sulfate on soils.—Ammonium sulfate produces in soils a chemical condition which is the reverse of that formed by sodium nitrate. The ammonia is changed in the soil into nitric acid (p. 139), leaving free or uncombined sulfuric acid, which combines with calcium carbonate present in the soil. The result is that by long-continued use of ammonium sulfate as a fertilizer, we gradually use up the calcium carbonate of the soil and have an increasingly acid soil. Under such circumstances, additional application of ammonium sulfate or of organic nitrogen does no good, because the nitrogen is not changed into nitrate nitrogen by bacteria in the presence of acids (p. 133). Therefore, crops cannot thrive. In such cases the simplest remedy is to apply calcium (lime) compounds, either the carbonate or slaked lime (p. 142). This condition can be prevented by the application of calcium carbonate or by using along with ammonium sulfate some sodium or calcium nitrate. Obviously it is well never to use either sodium nitrate or ammonium sulfate exclusively for long periods, but to alternate them, or, better, always use them together, unless other materials are applied to prevent the injurious action of either when used alone.

Effect of use of ammonium chloride on soils.—Ammonium chloride leaves in the soil hydrochloric acid, which combines with calcium carbonate in normal soils to produce by prolonged or excessive use loss of bases, especially calcium.

Effect of calcium nitrate.—After the nitrate of calcium nitrate has been absorbed by plant-roots it leaves calcium in the soil. This becomes carbonate under normal conditions. On this account calcium nitrate is a more desirable compound to use than sodium nitrate, which leaves sodium carbonate.

Effect of ammonium nitrate and potassium nitrate on soils.—When ammonium nitrate is applied to soils, the nitrate is used as such and the ammonia is converted into nitrate and then absorbed. Under conditions which enable plants to use all the nitrogen of this compound, no residue is left in the soil, so that, theoretically at least, the soil is left in the same condition as before, so far as the effect of the ammonium nitrate is concerned. The ammonium nitrate itself undoubtedly has some direct chemical and physical effects upon some of the soil constituents before it is used by plants, but we are here considering more particularly the effects of compounds left as residues after the plant has absorbed the plant-food constituents.

Effect of previous crop on barley. 1. Following alfalfa on hill soil. 3. Following timothy on hill soil. 9. Following alfalfa on productive soil. 11. Following timothy on productive soil. 15. Following clover on productive soil. 7. Following alfalfa on hill soil, alfalfa fertilized with nitrogen. Other plots received no nitrogen.

In the case of potassium nitrate both constituents may be regarded as being used by the growing plant and, under ordinary conditions, leaving no residue.

Effect of calcium cyanamid on soils.—Since calcium cyanamid contains a considerable amount of calcium hydroxide and calcium carbonate, it adds to the soil a direct acid-neutralizing agent. The nitrogen of cyanamid is changed to ammonia and then to nitric acid, which combines with the calcium to form calcium nitrate.

Effect of urea on soils.—Urea is changed in normal soils to ammonia and then to nitric acid, which combines with calcium to form calcium nitrate. In *calurea,* the presence of calcium carbonate furnishes directly the calcium for the nitric acid to form calcium nitrate.

Effect of leunasalpeter, leunaphos ammo-phos and leunaphoska on soils.—These materials contain ammonium sulfate (p. 166) and have, therefore, the same effect on soils, tending with prolonged or excessive use to add to the acid content of the soil in the absence of a sufficient amount of calcium carbonate.

Effect of superphosphates on soils.—Whether we use a superphosphate (p. 173) made by dissolving rock-phosphate or bone, we have a mixture of mono-calcium phosphate and hydrated calcium sulfate (gypsum, land plaster). We have previously seen that the soluble phosphate is more or less quickly changed in normal soils into di-calcium or "reverted" phosphate through combination with calcium. When this compound again becomes soluble, it is probably absorbed as a whole without leaving any residue, or if any is left, it is calcium, which under normal soil conditions, becomes carbonate. The amount of gypsum applied to a soil in a superphosphate is insufficient to have a marked effect, in relation to soil acidity. Some popular prejudice has existed against superphosphate (acid phosphate), based on the belief that it produces or increases soil acidity. Careful experiments have shown that this prejudice is not justified by facts. Superphosphate has no tendency to make a soil acid.

Effect of ammonium and potassium phosphates on soils.—These compounds contain only available plant-food constituents and leave no residue in the soil.

Effect of rock phosphate, colloidal phosphate, etc. on soils.—These materials have little or no direct effect upon soils. They consist largely of tri-calcium phosphate (p. 33). When, however, tri-calcium phosphate is changed into mono-calcium phosphate and the phosphoric acid is taken up by plants, calcium is left in the soil to combine with any acid present.

Effect of bone meal on soil.—Bone meal contains tri-calcium phosphate and calcium carbonate. It has, therefore, the same effect on soils as these two compounds, neutralizing acids.

Effect of basic slag on soils.—Basic slag contains 40 to 50 per cent. of calcium and magnesium compounds, largely in the form of oxide, and has therefore the same effect as an equal amount of quicklime in neutralizing soil acidity or in making a neutral soil alkaline or in affecting the physical condition of a soil.

Effect of potassium compounds on soils.—It would naturally be thought that when we apply to soils potassium chloride or sulfate, the plant after absorbing the potassium and rejecting most of the hydrochloric or sulfuric acid would tend to bring about an acid condition in the soil as the result of continued application of such potassium compounds. We have previously stated (p. 117) that when these potassium salts are applied to soils they are usually converted into insoluble compounds, their acids combining with calcium or magnesium to form the chloride or the sulfate of these elements. These compounds are found extensively in drainage waters. In this way, the acid constituents of applied potassium chloride and sulfate are removed from soils to such an extent as to prevent accumulation and acidity. Of course the presence of an abundance of calcium carbonate in soils favors the rapid formation and removal of any excess of hydrochloric or sulfuric acid.

Effect of potassium carbonate on soils.—Potassium carbonate, the chief potassium compound in wood-ashes (p. 189), when applied to soils is a quick and powerful neutralizer of acids on account of its easy solubility and alkalinity. If used in excess, the effect is the same as in case of sodium carbonate (p. 78), to destroy the crumb structure of the soil.

ADAPTATION OF FORMS OF PLANT-FOODS TO CROPS

Each crop possesses individual characteristics, which vary from others in many ways. Among these differences are the habits of root growth, so-called "feeding-power" (p. 128), season of growth, etc. On account of such variations, some forms of plant-food constituents are better adapted to promote normal growth than others. Another difference found in some cases is that the character and commercial quality of the crop are affected favorably by one plant-food compound and unfavorably by another containing the same plant-food constituent. In furnishing fertilizers it is desirable, as a matter of their most effective use, to consider these individual peculiarities so far as we have learned them.

Adaptation of fertilizers to feeding-habits of crops, season of growth, etc.—In discussing those individual characteristics, we can conveniently divide crops into several different classes; the members of each class, while differing in many ways, have in common certain resemblances of importance in this connection. The classes are: (1) Cereal, (2) leguminous, (3) grass, (4) orchard, (5) root, and (6) garden crops.

(1) *Cereal crops* are comparatively shallow-rooted. Starting near the surface, the main roots send out feeding roots in every direction, occupying the upper layer of soil extensively before the end of the growing season and also reaching more deeply down into the soil. On account of location of the root system, fertilizers are most effectively applied in the upper layer of soil. The nearer very soluble forms of compounds are placed to the surface the better, provided rain can be relied upon to dissolve them. Such cereals as barley, oats, rye and wheat need a large proportion of their nitrogen during the early part of the growing season. At this time there is likely to be a lack of nitrate nitrogen because nitrification does not get well started until the weather has warmed the soil (p. 139). On this account, the application of nitrate in spring to these crops is usually found most beneficial.

In the case of corn, which has a long growing season, making its best growth after the other cereals have matured, nitrogen is better applied for the most part in the form of organic materials, since nitrification is most active by the time the corn crop makes its largest demands. For this reason farm manure gives good results. All cereals are generally benefited by application of superphosphate. Oats and corn appear to be able to utilize some forms of plant-food in soils to better advantage than wheat does. Barley should have comparatively small applications of soluble nitrogen compounds, since the quality of the grain for malting purposes is injured by too much nitrogen in the early stages of growth. Owing to the comparative shallowness of the root system, barley cannot forage for phosphorus as well as other cereals and therefore should be furnished with a supply.

(2) *Leguminous crops* for the most part, have comparatively long roots, which in some cases go deeply into the soil; they are able to absorb plant-food compounds from different portions of the soil. Their chief distinguishing characteristic is their power to make use of atmospheric nitrogen (p. 146), which reduces greatly the amount of nitrogen that needs to be furnished. In starting some of the leguminous crops on poor soils, alfalfa particularly, application of a nitrate is helpful. On most soils, the fertilizing materials found most effective on leguminous crops are calcium, potassium and phosphorus compounds.

(3) *Grass crops,* generally shallow-rooted, obtain their food supplies from a comparatively narrow range of soil. Grasses in permanent meadows and pastures are greatly benefited by fertilizers. Nitrogen in soluble forms is especially useful because the

chief part of the crop consists of stems and leaves, the growth of which is largely dependent on nitrogen (p. 49). In grass lands where the crop has occupied the soil for over a year, decomposition changes are less active because the air supply is reduced; nitrification is, therefore, slow and the crop is generally in condition to respond promptly to nitrogen feeding.

(4) *Orchard crops* are capable of collecting food from a wide extent of soil and at considerable depth. They should have a continuous supply of plant-food materials in moderate amounts and in fairly soluble form, but they should not be so fertilized as to make too rapid growth of new wood, or to prolong their growth too late in the fall.

(5) *Root crops* do not have the ability to range through wide reaches of soil; each plant must take its food from a comparatively limited area. It is important, therefore, to supplement the soil food supply with fertilizers in fairly soluble forms. Turnips appear to respond to phosphate compounds, while carrots, beets, parsnips, etc., are benefited by readily available forms of nitrogen.

(6) *Garden crops* include plants that differ greatly in their habits and seasons of growth. The chief purpose for which these crops are grown is the production of leaves and stalks, in most cases, of a tender, succulent character. The desired quality is dependent on rapidity of growth; therefore, abundance of soluble nitrogen is essential. Enough available phosphorus and potassium must be assured to enable the nitrogen to accomplish its work most effectively. Farm manure in large amount is often used.

Adaptation of fertilizers to quality of product.—In the preceding paragraphs we have shown that some crops thrive better on sodium nitrate at certain stages of growth than at others. But there are some cases in which the quality of the product is affected favorably by some constituent of plant-food in the form of one compound and unfavorably by another containing the same constituent. This is strikingly shown in the case of tobacco. The quality of tobacco is injured by chlorine. Potassium carbonate or sulfate, not chloride, is, therefore, used in tobacco fertilizers. It has been stated that the quality of sugar-beets is better with potassium sulfate than with chloride. This probably refers to the character of crystallization, which causes less sugar to remain in molasses. Much has been said about the effect of potassium salts upon potatoes. So far as facts are available at present, it is believed that on light soils, either chloride or sulfate of potassium can be used with equally satisfactory results; but, on clay soils and especially in wet seasons, the use of chloride appears to produce potatoes which contain less starch and which are heavy or watery when cooked, while sulfate produces potatoes that contain more starch. Probably many new facts in this line remain to be discovered. If so they will enable us to control more or less directly the quality of crops.

COMPARATIVE COST OF PLANT-FOOD MATERIALS

The materials used to supply plant-foods to crops vary in cost for the same plant-food constituent. Considerable saving in the purchase of fertilizing material can be made by a careful selection based upon a study of market values. In many cases the less costly form will answer the farmer's purpose as well as the more expensive. Other things being equal, it is economy to purchase the cheapest form that will meet one's needs. Prices are constantly varying so we can discuss the subject only in an illustrative way.

Nitrogen-containing materials.—In discussing cost of nitrogen, we can illustrate by making use of prices at which a consumer can at present (1931) purchase materials in Central New York State. The figures in the last column of Table 35 show the cost of one pound of nitrogen in various materials.

TABLE 35—PERCENTAGE AND RETAIL COST OF NITROGEN IN DIFFERENT MATERIALS

Material	*Retail price per ton*	*Per cent. of nitrogen*	*Pounds of nitrogen in one ton*	*Cost of one pound of nitrogen*
	dollars			*cents*
Ammonium sulfate	40.50	20.5	410	9.9
Calcium cyanamid	48.25	21.5	430	11.2
Leunasalpeter	60.00	26.0	520	11.5
Urea	125.00	46.0	920	13.6
Sodium nitrate	51.00	15.0	300	17.0
Calcium nitrate	51.00	15.0	300	17.0
Dried blood	85.00	12.0	240	35.4
Animal tankage	38.00	7.0	140	27.1

Phosphorus-containing materials.—The market materials that contain only phosphorus without nitrogen or potassium are superphosphate, basic-slag phosphate and ground phosphate-rock or "floats." These are, of course, not comparable in respect to cost. For soluble compounds of phosphoric acid, we have plain and concentrated superphosphates. At present market prices of $17.50 for 16 per cent. superphosphate, $22.50 for 20 per cent., and $44.25 for 40 per cent., the cost of one pound of phosphoric acid is about the same, in superphosphates and two grades of Ammo-Phos.

TABLE 36—PERCENTAGE AND COST OF PHOSPHORIC ACID IN VARIOUS MATERIALS

	Retail price per ton	Per cent. of phosphoric acid	Per cent. of nitrogen	Pounds of phosphoric acid in one ton	Pounds of nitrogen in one ton	Cost of one pound of Phosphoric acid	Cost of one pound of Nitrogen
						cents	
Superphosphate	$17.50	16	0	320	—	5.49	—
Superphosphate	22.50	20	0	400	—	5.62	—
Superphosphate	44.25	40	0	800	—	5.53	—
Ammo-Phos.	65.50	20	16.5	400	330	6.10	12.45
Ammo-Phos.	76.00	46	10.7	920	214	5.60	11.45

Materials that contain nitrogen and phosphorus.—The comparison of prices in cases of materials that contain more than one plant-food constituent is somewhat more complicated, but the results are readily comparable in materials like bone, tankage, fish, etc., as shown in Table 37.

TABLE 37—PERCENTAGE AND COST OF NITROGEN AND PHOSPHORUS IN BONE, TANKAGE, ETC.

Material	*Retail price per ton*	*Per cent. of nitrogen*	*Per cent. of phosphoric acid*	*Pounds of nitrogen per ton*	*Pounds of P_2O_5 per ton*	*Cost of one pound of nitrogen*	*Cost of one pound of P_2O_5*
	dollars		*(P_2O_5)*			*cents*	*cents*
Ground steamed bone	29	1.0	24.0	20	480	20.0	4.0
Raw bone	34	3.3	20.0	66	400	22.0	4.4
Bone-tankage ..	26	4.0	18.0	80	360	17.1	3.5
Bone-tankage ..	28	6.5	9.0	130	180	16.8	3.4
Fish scrap ground	48	8.0	5.0	160	100	26.7	5.4

The cheapest forms of organic nitrogen are found in nitrogenous tankage. Phosphorus is likewise cheapest in tankage.

Potassium-containing compounds.—Of all commercial plant-food materials, potassium compounds have varied least in price, owing to the German monopoly of the supply. Table 38 gives a fair idea of the relative prices of the various materials:

Their cost makes wood-ashes a luxury as a source of potassium, and the cost is still high even if we make generous allowance for the calcium and phosphorus compounds. For most crops, chloride answers every purpose and costs least. Commercial potassium carbonate is little used at present; this can be used by those who want a substitute for the potassium carbonate in wood-ashes.

Whether we consider unmixed fertilizing materials or mixed commercial fertilizers, the constant aim should be to purchase the largest amounts of nitrogen, phosphorus and potassium in available form that we can buy for one dollar, instead of as many pounds as possible of material without reference to what plant-food these represent.

TABLE 38—PERCENTAGE AND COST OF POTASSIUM IN VARIOUS MATERIALS

Material	*Retail price per ton dollars*	*Per cent. of K_2O*	*Per cent. of K*	*Pounds in one ton* K_2O	K	*Cost of one pound* K_2O *cents*	K *cents*
Potassium chloride (muriate) ..	53	50	41.5	1000	830	5.3	6.5
Potassium sulfate	66	48	40	960	800	6.9	8.2
Potassium carbonate ..	127	60	50	1200	1000	10.4	12.7
Kanite	22	12	10	240	200	9.1	11.0
Wood-ashes .	22	5	4	100	80	22	27

XXIV

FACTS ABOUT COMMERCIAL FERTILIZERS

COMMERCIAL FERTILIZERS are manufactured preparations, made by mixing plant-food materials of various kinds, and commonly sold under special trade-names. They are also known as *artificial, prepared* or *manufactured* fertilizers or manures. When they consist only of inorganic materials, they are often called *chemical* fertilizers or manures. The term "agricultural chemicals" is often applied to such materials as sodium nitrate, ammonium sulfate, calcium cyanamid, calcium nitrate, potassium chloride, potassium sulfate, superphosphates, etc.

The materials used to make commercial fertilizers include manufactured products, substances found in, or prepared from, natural deposits, and, in addition, by-products of numerous industries, obtainable only through the channels of trade (pp. 162-212).

While commercial fertilizers contain numerous constituent elements, their chief value depends, in most cases, on compounds of *nitrogen, phosphorus* and *potassium;* in exceptional conditions, some other constituent may be included, such as manganese or sulfur. The agricultural value varies in accordance with the forms, and relative proportions of the various plant-food compounds.

Commercial fertilizers are either *complete* or *incomplete.* Complete fertilizers, also called general fertilizers, contain the three plant-food constituents, nitrogen, phosphorus and potassium. Incomplete fertilizers contain only one or two of these three.

The quantity of commercial fertilizers used in the United States is estimated to be not less than 5,000,000 tons a year It therefore deserves thorough discussion in all of its relations to consumers under the following headings: (1) Meaning of guarantee-analysis, (2) high-grade and low-grade fertilizers, (3) brand names of fertilizers, (4) meaning and use of commercial valuations, (5) fertilizer laws—objects, methods, (6) advantages and disadvantages in use of commercial fertilizers.

MEANING OF GUARANTEE-ANALYSIS

In most states where fertilizers are used, manufacturers are required by law to state on each package the percentage of nitrogen, phosphorus (phosphoric acid, P_2O_5), and potassium (potash, K_2O) which the fertilizer contains; in some states they are also required to state the source or form of nitrogen and potassium.

Formerly many and various terms were used, sometimes eight to ten different ones being employed to specify the amount of nitrogen, phosphoric acid and potash guaranteed to be present in the fertilizer. The old method was confusing and, in some cases, misleading. At present, the usual form of guarantee-analysis states simply the percentage of nitrogen, phosphoric acid and potash, as for example: Nitrogen, 5 per cent.; available phosphoric acid, 10 per cent.; potash, 5 per cent. In some cases the equivalent percentage of ammonia may be added and after the term, potash, sulfate or muriate (chloride).

Formerly the figures were frequently given with fractions. The tendency now is to give the guarantee-analysis only in whole numbers.

Terms used in stating guarantee of nitrogen in commercial fertilizers: (1) Nitrogen, (2) ammonia, (3) nitrogen equal (or equivalent) to ammonia, (4) nitrogen from animal matter, (5) nitrogen from organic matter, (6) total available nitrogen.

(1) *Nitrogen,* as used in statements of guaranteed analysis, refers to the total amount of nitrogen present in the fertilizer without reference to the form of combination in which it exists. It does not state whether the nitrogen is present as nitrate, ammonia, or organic nitrogen, but simply that, whatever form it is in, the specified amount is as given.

(2) *Ammonia* was formerly used exclusively in stating the amount of nitrogen in fertilizers and is still used extensively in commercial market quotations, and among dealers in fertilizers, regardless of whether or not the material in question contains any ammonia compound. The use of the term in a guarantee-analysis is misleading because most farmers suppose that nitrogen and ammonia are the same. A pound of ammonia contains less than a pound of nitrogen because ammonia consists of nitrogen in combination, with hydrogen (p. 29); 100 pounds of ammonia contain 82 1/3 pounds of nitrogen and 17 2/3 pounds of hydrogen, or, stated another way, 100 pounds of nitrogen will make 121½ pounds of ammonia. The figure giving the percentage of nitrogen as ammonia is, therefore, higher than that for nitrogen. In purchasing fertilizers, farmers should insist on a guaranteed statement of the percentage of nitrogen and should pay no attention whatever to the ammonia percentage.

The National Fertilizer Association now recommends that the guarantee-analysis give the percentage of nitrogen, with or without adding the equivalent of ammonia.

(3) *Nitrogen equal or equivalent to ammonia* is an expression which means simply that, if the nitrogen were present as ammonia, there would be the percentage stated; it does not necessarily mean that any form of ammonia is present.

(4) *Nitrogen from animal matter,* occasionally found in guarantees, means that the fertilizer contains such materials as dried blood, ground meat, fish, bone, horn or hoof meal, etc., but does not necessarily exclude useless untreated leather and hair.

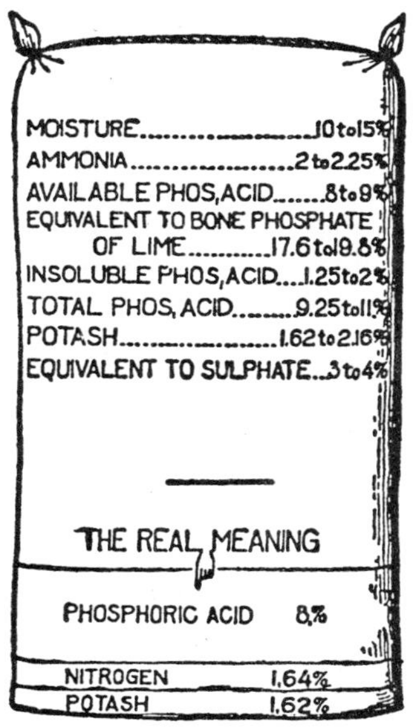

The bag and the plant-food in it.

(5) *Nitrogen from organic matter* furnishes little information. It implies that the fertilizer contains animal or vegetable matter or both but does not exclude unavailable forms of organic nitrogen, such as leather, hair, muck, etc.

(6) *Total available nitrogen* implies that all the nitrogen in the fertilizer is available as plant-food, but gives no information as to whether it is quickly or slowly so.

Terms used in stating guarantee of phosphorus as phosphoric acid.—While all the needed information in regard to the amount of phosphorus in a fertilizer can be stated in two or three forms, twice as many are often used. Among the various forms of statement that have been used are the following: (1) Phosphoric acid, (2) soluble phosphoric acid, (3) citrate-soluble phosphoric acid, (4) "reverted" phosphoric acid, (5) "precipitated" phosphoric acid, (6) available phosphoric acid, (7) soluble and available phosphoric acid, (8) insoluble phosphoric acid, (9) total phosphoric acid, (10) phosphoric acid equal (or equivalent) to bone phosphate of lime.

Phosphorus as phosphoric acid or phosphate compounds is usually present in fertilizers in the form of three compounds, (*a*) water-soluble acid calcium or mono-calcium phosphate (p. 33), (*b*) di-calcium or "reverted" phosphate (p. 33), insoluble in water but soluble in dilute ammonium citrate solution and (*c*) tri-calicum or insoluble phosphate (p. 34), not soluble in either water or ammonium citrate solution. The *total* phosphoric acid includes all the phosphorus in the fertilizer whether present in one, two or, as is usually the case, all three of these forms. *Available* phosphoric acid includes the *sum of the water-soluble and "reverted" or citrate-soluble,* and is equal to the *total* phosphoric acid minus the *insoluble.* Therefore, all the needed information can be furnished by stating simply the amount of phosphoric acid in these three forms—soluble, citrate-soluble and insoluble, and can usually be further simplified by specifying only "available."

(1) *Phosphoric acid,* as used to state the amount of phosphorus in the form of phosphate compounds, is a compound which consists of phosphorus and oxygen (P_2O_5) never found by itself in fertilizers but only in combination with calcium (lime), ammonium or potassium. As already explained (p. 32), it is simply an arbitrary term long used by chemists to state the percentage of phosphorus. Following the usage of substituting nitrogen for ammonia, it would be really simpler and more consistent to state the amount as actual phosphorus than phosphorus in combination with oxygen, as we now do.

(2) *Soluble phosphoric acid* means the phosphorus existing as acid calcium phosphate ($CaH_4(PO_4)_2$, p. 33), (which is soluble in water) or as ammonium phosphate or potassium phosphate.

(3) *Citrate-soluble, "reverted"* and *"precipitated,"* are terms used to indicate phosphorus when present as di-calcium phosphate ($Ca_2H_2(PO_4)_2$, p. 33).

(4) *Available phosphoric acid,* or *soluble and available,* includes the sum of water-soluble and citrate-soluble (the mono- and di-calcium phosphates as well as ammonium and potassium phosphates).

(5) *Insoluble phosphoric acid* applies to that present as tri-calcium phosphate ($Ca_3(PO_4)_2$, p. 33).

(6) *Total phosphoric acid* means the sum of the soluble, reverted and insoluble, or all phosphorus present whatever the form.

(7) *Phosphoric acid, equal to bone phosphate of lime (B. P. L.),* is an expression which usually means simply nothing more than

insoluble. The expression is misleading, since it is often taken by farmers to mean that the phosphorus is derived from bone. The most it means is that if the given amount of phosphorus were present as insoluble or tri-calcium phosphate, there would be the percentage of bone phosphate of lime indicated. It is misleading in another way, in that it gives a percentage more than twice as large as that given for phosphoric acid and gives many farmers the impression that they are getting much more phosphorus than they are. The practice is objectionable on every ground from the farmer's standpoint, and should be abandoned by all manufacturers, as it has been already by the more progressive.

Terms used in stating guarantee of potash.—It would be simpler, and more consistently accurate if we were to abandon the term potash and use simply potassium in place of it (p. 38). The compound designated as potash, potassium oxide (K_2O), is never found in fertilizers and is merely an arbitrary method of expressing results of chemical analysis. Fertilizers contain potassium mainly in the form of chloride (muriate) and sulfate, also occasionally as carbonate and phosphorus. While many terms were formerly used in stating the amount of potash in a fertilizer, the present usage is, for the most part, confined to the word potash, followed in parenthesis by sulfate, muriate, chloride, or carbonate to indicate the specific compound. When more than one of these compounds is present, this is indicated, as, for example, potash (muriate and sulfate).

The potash guaranteed is for potassium compounds dissolving in water, according to legal requirements. If this requirement were not made, it would be possible to use ground minerals containing insoluble, potassium compounds unavailable and therefore worthless as fertilizing material.

Suggestions to farmers regarding guaranteed analysis.—Whenever a farmer purchases either a mixed fertilizer or separate fertilizing materials, *he should purchase only under a guarantee of specified percentage of each plant-food constituent.* Moreover, he should insist that the guarantee be in the following terms:

(1) *Nitrogen,* and this alone, should be given for all forms of nitrogen. An added statement as to whether the nitrogen is in the form of nitrate, ammonia, cyanamid, urea, dried blood, cottonseed-meal, ground leather, etc., is desirable for full information.

(2) *Available phosphoric acid (or phosphorus),* or preferably, water-soluble and reverted separately, should be given for superphosphate and other soluble phosphate materials, while for bone, untreated rock-phosphate, etc., only total phosphoric acid (or phosphorus) is really necessary.

(3) *Potash or potassium oxide (or potassium)* should always

be given, though it is desirable to know, in addition, whether the potassium is present as sulfate, free from chlorine, or as chloride only.

(4) *In examining the guaranteed analysis of a fertilizer, give attention primarily to the nitrogen, available phosphoric acid (or phosphorus) and potash (or potassium) and require the guarantee to be in these terms before purchasing.*

How to find equivalent values of a plant-food constituent in various compounds.—If a guarantee-statement of analysis is given only in the form of ammonia, it is always desirable to know just how much nitrogen the given amount of ammonia is equal to; similarly, if only bone phosphate of lime ($Ca_3(PO_4)_2$) is guaranteed, it is important to know how much phosphorus (P) or phosphoric acid (P_2O_5) it is equal to; and if potassium sulfate (K_2SO_4) only is guaranteed, we should know how much potassium (K) or potash (K_2O), it is equal to. The conversion factors below enable one easily and quickly to make the arithmetical calculation necessary in the cases that more commonly occur. A larger number of factors is found on pages 197, 198.

To illustrate the use of the factors, suppose, for example, we wish to change 10 pounds of ammonia into an equivalent amount of nitrogen; we multiply the amount of ammonia (10 pounds) by the number found in the last column, 0.82, opposite (1) in the first column. The result is 8.2 pounds of nitrogen.

Or, suppose we wish to know how much phosphoric acid (P_2O_5) is in 10 pounds of tri-calcium phosphate (bone phosphates of lime, B.P.L.); we multiply 10 by factor (4), namely, 0.46. The result is 4.6 pounds.

CONVERSION FACTORS FOR SHOWING EQUIVALENT AMOUNTS OF A PLANT-FOOD CONSTITUENT IN DIFFERENT FORMS

	To find from a given amount of		*Multiply by*
(1)	Ammonia the equivalent amount of nitrogen,		0.82
(2)	Nitrogen " " " " ammonia,		1.21
(3)	Tri-calcium phosphate ("B.P.L.") the equivalent amount of	phosphorus,	0.20
(4)	" " " " " " " "	phosphoric acid,	0.46
(5)	Phosphoric acid (P_2O_5) " " " "	tri-calcium phosphate,	0.22
(6)	" " " " " " "	phosphorus,	0.44
(7)	Phosphorus (P) " " " "	phosphoric acid,	2.3
(8)	" " " " " "	tri-calcium phosphate,	5.0
(9)	Potash (K_2O) " " " "	potassium,	0.83
(10)	" " " " " "	potassium chloride,	1.60
(11)	" " " " " "	potassium sulfate,	1.85
(12)	Potassium chloride " " " "	potash,	0.63
(13)	" sulfate " " " "	potash,	0.54

BRAND NAMES OF FERTILIZERS

Commercial fertilizers are commonly sold under special names or brands. Some names indicate that the fertilizer is intended for some special crop, such, for example, as "Complete manure for potatoes and vegetables," "Canner's special pea and bean fertilizer," "Universal grain grower," "Grass and lawn top-dressing," etc.; others are simply fanciful names without any special significance, designed only to catch the imagination of buyers, as, for example, "New rival," "Climax phosphate," "Farmer's friend," "Electric phosphate," "Excelsior guano," "Golden harvest," "Hustler," "Reliance," etc., etc. Some years ago, manufacturers of fertilizers offered for sale in New York State nearly 2,300 brands, but, as soon as it was required by law to pay a license fee on each brand, the number at once dropped below 500.

Misleading names.—Attention has already been called to the misuse of names in connection with superphosphate, which was formerly often sold under some name stating or implying it to be dissolved bone. To illustrate further, a so-called "natural fertilizer" once on the market contained, all told, about 0.75 per cent. of plant-food and none in easily available form; another, "Special for Lawns and Fruits," contained 0.52 per cent. as its total. At one time ground phosphate rock ("floats") was sold extensively at $20 to $28 a ton under the name of "Natural Plant-Food."

Misuse of names for special-crop fertilizers.—When a brand-name indicates that a fertilizer is especially adapted for a particular crop, the question can properly be asked: To what extent does a "potato" fertilizer, for example, meet all the great variety of fertility conditions in the many localities where it is sold? To take a single illustration, which fairly represents conditions that quite commonly prevail in the fertilizer trade, there were sold in one state to farmers during one season 10 brands that were, according to name, designed for the exact needs of potato crops. The composition varied widely,—the nitrogen from 1.65 to 5.75 per cent., the phosphoric acid from 6 to 10 per cent., and the potash from 4 to 10 per cent. A similar condition prevails with reference to other special-crop fertilizers.

Proper method of using brand-names.—Fertilizer manufacturers, in increasing numbers, are adopting a method of using for a brand-name a combination of figures representing the composition, as, for example: "Five-Eight-Eight," "Four-Eight-Seven," "3-8-4," "2-9-5," etc. The first number stands for nitrogen, the second for available phosphoric acid and the third for potash. From every point of view this is a desirable system. The emphasis is placed on the composition, as it should be, and not on a name which may mean nothing.

MEANING AND USE OF COMMERCIAL VALUATION OF FERTILIZERS

The commercial valuation of a fertilizer or of any fertilizing material consists in calculating the approximate value or money-cost of the chief plant-food constituents (nitrogen, phosphorus and potassium) in one ton under market conditions.

Retail cost and commercial valuation.—The chief factors that enter into the retail cost of fertilizers, or the price the farmer pays, are: (1) Retail cash cost of unmixed materials, as, for example, sodium nitrate, dried blood, superphosphate, potassium chloride, sulfate, etc.; (2) cost of manufacture (mixing and bagging); (3) freight; (4) storage, commissions to agents and dealers, selling on long credit, losses by bad debts, etc. While these factors enter into the total or retail cost, a commercial valuation includes only the first factor, viz., the retail cash cost of unmixed materials in the market. A commercial valuation, therefore, usually gives a figure considerably lower than the selling price of a fertilizer.

How a commercial valuation is made.—In order to calculate the cost of the plant-food constituents in one ton of fertilizer, it is necessary to know two things, first, the number of pounds of each plant-food constituent and, second, the price of one pound.

(1) *To find number of pounds of a plant-food constituent in 100 pounds* one must have the statement of analysis, either the figures given in the guarantee-analysis, or, preferably, the figures given in an official state analysis. The figures giving the per cent. are, of course, the pounds per 100 of nitrogen, phosphoric acid and potash.

(2) *Price of each constituent* is obtained by consulting a price list, which is issued annually by the experiment station or some state department in charge of fertilizer inspection; this list is commonly known as a "schedule of trade-values adopted by experiment stations" or "commercial state values of fertilizers." Table 39 gives such a schedule published by the Massachusetts Experiment Station for 1931. The figures in this schedule represent the average prices at which, in the six months preceding March, the respective ingredients, in the form of unmixed materials, could be bought at retail for cash in Boston, New York, Philadelphia, etc. The figures adopted by experiment stations usually present wholesale prices plus about 20 per cent., except for available phosphoric acid. These prices represent approximately what the farmer would have to pay for the plant-food constituents if he were to buy them in unmixed

forms in Boston, New York, or some other trade center. It must, of course, be kept in mind that these trade values, especially for nitrogen, change from time to time, so in making commercial valuations, one should obtain the latest schedules from the local experiment station.

TABLE 39—SCHEDULE OF TRADE-VALUES FOR PLANT-FOOD CONSTITUENTS IN FERTILIZEPS FOR 1931

Form of Plant-Food	*Price per pound*	*Price per unit (20 pounds)*
NITROGEN	*cents*	
In ammonia compounds	11.25	$2.25
In nitrates	15	3.00
Organic nitrogen in urea and calurea	14	2.80
" " " fish	38	7.60
" " " blood, meat and hoof-meal	24.5	4.90
" " " fine[1] bone and tankage	26.5	5.30
" " " coarse bone and tankage	19	3.80
" " " pulverized manures	19	3.80
" " " mixed fertilizers	27	5.40
" " " cottonseed meal, castor pomace, etc.	27	5.40
PHOSPHORIC ACID		
Soluble in water	5.5	1.10
Soluble in neutral ammonium citrate (reverted)	4.5	0.90
In fine[1] bone, tankage and fish	5.5	1.10
" coarse[1] bone and tankage	4.5	0.90
" pulverized manures, seed residues and ashes	4.0	0.80
Insoluble, in mixed fertilizers	2.0	0.40
POTASH		
As sulfate	5.9	1.18
" muriate	4.4	0.88
" carbonate	10.0	2.00
" nitrate	4.0	0.80
Total in pulverized manures, seed residues, etc.	4.0	0.80

[1] Fine bone and tankage refer to particles which, as sampled, will pass through a sieve with circular openings one-fiftieth of an inch in diameter. Coarse bone and tankage refer to that portion which will not pass through such a sieve.

Knowing (1) the number of pounds of nitrogen, phosphoric acid and potash in a fertilizer and (2) the price of each of these constituents, one can make a commercial valuation by using either of two methods, by pounds or by units.

1. *Method by pounds.* (*a*) Multiply the given per cent. (pounds per 100) of each constituent (nitrogen, phosphoric acid, potash) by the respective schedule price; (*b*) add the results and (*c*) multiply the sum by 20.

Example: What is the commercial valuation of one ton of a commercial fertilizer, having the guaranteed analysis indicated below?

	Per cent. or pounds per 100		*Trade-schedule price Cents per lb.*		*Valuation for 100 pounds Cents*
Nitrogen	4	×	15	=	60
Available phosphoric acid	8	×	5.5	=	44
Potash	10	×	4.4	=	44
Total valuation for 100 pounds					148

Total valuation for one ton or 2000 pounds = 148 cents × 20 = $29.60

2. *Method by units.* The popular term "unit," as applied to fertilizers, means 1 per cent. on the basis of 2,000 pounds, that is, 20 pounds. One unit of nitrogen, for example, is 20 pounds or 1 per cent. of a ton. To use the unit method of calculating the valuation of a fertilizer, multiply the percentage of each plant-food constituent by the cost of one unit and add the results. When nitrogen costs 15 cents a pound, a unit is worth $3.; when available phosphoric acid costs 5.5 cents a pound, a unit costs $1.10; when potash costs 4.4 cents a pound, a unit is worth 0.88. Applying these values to the preceding examples, the commercial valuation is thus indicated:

	Per cent.		*Cost of unit*		
Nitrogen	4	×	$3.00	=	$12.00
Phosphoric acid	8	×	1.10	=	8.80
Potash	10	×	0.88	=	8.80
Valuation of one ton				=	$29.60

This method can be simplified by using the factors in Table 40. 1. Multiply the figure that represents the per cent. of each plant-food constituent by the factor found opposite the price of one pound in Table 40.

2. To make use of Table 40 consult the schedule of prices (Table 39), select from it the figure showing the given price and multiply this by the number giving the percentage of that constituent in the fertilizer. For example, when nitrogen in nitrate sells for 16 cents a pound, multiply the percentage of nitrogen by 3.2, which gives the commercial valuation of the nitrogen in one ton of the fertilizer. When nitrogen in any form costs 20 cents a pound, multiply the percentage by 4; the percentage of any constituent costing 10 cents a pound would be multiplied by 2; 5 cents a pound, by 1; 4 cents a pound, by 0.8, etc.

Valuations based on guarantees made in other than usual terms.—In considering methods of making valuations, we have assumed that guarantees are given in terms of nitrogen, phosphoric acid and potash. If, however, other terms are used, such as ammonia, bone phosphate of lime, sulfate of potash, etc., it will be necessary to calculate the figures given in the guarantee into equivalent amounts of nitrogen, phosphoric acid or potash before going on with making the valuation. The method of calculating equivalent amounts is given in Table 40 below or by Table 45 (p. 352). The same tables can be used for calculating the amounts of phosphorus equivalent to phosphoric acid, potash to potassium, etc.

TABLE 40—FACTORS TO USE IN MAKING VALUATIONS FOR VARYING SCHEDULE PRICES OF PLANT-FOODS

Price per pound of plant-food constituent Cents	*Factor by which to multiply percentage*	*Price per pound of plant-food constituent* Cents	*Factor by which to multiply percentage*	*Price per pound of plant-food constituent* Cents	*Factor by which to multiply percentage*
2	0.4	6	1.2	15	3.0
2½	0.5	6½	1.3	16	3.2
3	0.6	7	1.4	17	3.4
3½	0.7	7½	1.5	18	3.6
4	0.8	8	1.6	19	3.8
4¼	0.85	9	1.8	20	4.0
4½	0.9	10	2.0	21	4.2
4¾	0.95	11	2.2	22	4.4
5	1.00	12	2.4	23	4.6
5¼	1.05	13	2.6	24	4.8
5½	1.10	14	2.8	25	5.0

Commercial value in relation to agricultural value of fertilizers.—The *agricultural value* of a fertilizer is measured from the standpoint of the farmer, by its *crop-producing power*. The commercial value does not necessarily have any relation to crop-producing power. This is well illustrated in the case of nitrogen in dried blood, ground fish-scrap and sodium nitrate. According to commercial value at the present writing, a pound of nitrogen in fish costs 38 cents, in dried blood, 24.5 cents, and in nitrate, only 15 cents, whereas, in most cases, the nitrate nitrogen is of superior agricultural value. Again, a fertilizer that contains large amounts of available phosphorus and a small amount of nitrogen, may have less money value than a fertilizer rich in nitrogen, and yet the phosphorus and not the nitrogen might be the constituent that would be most profitable for a given farmer to purchase.

Use of commercial valuations.—When used with a proper understanding of its limitations, a commercial valuation is helpful as a means of comparing fertilizers. It enables one to make a comparison on the basis of the selling price of a fertilizer and the commercial value of its unmixed constituents. This is illustrated in the case of four fertilizers representing different grades of analysis.

TABLE 41—COMPARISON OF SELLING PRICE AND COMMERCIAL VALUATION

In 100 pounds of fertilizer				*Selling Price per ton*	*Commercial Valuation*	*Difference between selling price and commercial valuation*
Nitrogen	*Phosphoric acid*	*Potash*	*Total plant-food*			
Lbs.	*Lbs.*	*Lbs.*	*Lbs.*			
1.20	8.20	2.60	12.00	$23.00	$15.22	$7.78
1.70	8.50	3.50	13.70	24.85	17.95	6.90
2.50	8.80	6.00	17.30	28.30	23.18	5.12
4.00	9.00	7.20	20.20	32.80	29.10	3.70

In the low-analysis fertilizer, the selling price is $7.78 above the cost of the ingredients; the difference decreases as the fertilizer grows richer in plant-food constituents, until in the high-analysis fertilizer, the selling price is only $3.70 above the cost of the constituents. Expressed another way, the purchaser is getting a great deal more plant-food for his money in the high-analysis fertilizer, each pound of plant-food costing less. *Generally speaking, those fertilizers cost least to the farmer in which there is least difference between selling price and commercial valuation.*

While the comparative values are wholly commercial, they furnish valuable information, which is in most cases a safe guide. To be of greatest use, commercial valuations must be used and interpreted in connection with the composition, as given by the analysis. A choice between different fertilizers must not rest wholly on a comparison of valuation and selling price; the composition of the fertilizer in relation to the farmers' needs must be considered. For example, a fertilizer with a high percentage of nitrogen and a low percentage of phosphoric acid would show a higher valuation than one low in nitrogen and high in phosphoric acid, and yet the phosphoric acid might be just the plant-food that would give best yields on a given field and crop.

One disadvantage connected with the use of commercial valuations is the fact that ordinary chemical analysis does not show the forms or sources of the materials; this really applies only to nitrogen and more particularly to the organic forms of nitrogen. Thus, when an analysis gives no information as to whether the nitrogen is dried blood or ground leather or muck, a commercial valuation may be misleading to this extent. But in actual use it has been found that in a large majority of cases a commercial valuation is a safe guide in respect to the agricultural value when applied to brands with the same general composition.

It is worth while for farmers to learn how to make commercial valuations and apply them before purchasing fertilizers; it will have a broadening and educating effect and will increase one's interest in, and intelligent appreciation of, the resemblances and differences between commercial fertilizers.

FERTILIZER LAWS

States in which commercial fertilizers are used extensively have passed laws regulating their sale. This has been necessary as a matter of protection to farmers, for few commodities offer so tempting a field for fraud. It was the discovery of numerous cases of gross imposition that led to the passage of laws. While these laws vary somewhat in details, they are alike in essentials. At one time fertilizers were sold without any statement regarding their composition; so a farmer was wholly unable by himself to know whether he was getting plant-food or useless dirt, and it was too costly to pay for a chemical analysis.

Provisions of fertilizer laws.—Among the various items required by fertilizer laws, the chief point of interest is a guaranteed statement of analysis. Manufacturers are required to give a statement of chemical composition, (p. 302) guaranteeing the percentage of nitrogen (or ammonia), available phosphoric acid (total phosphoric acid in bone, tankage, etc.), and water-soluble potash, contained in each brand offered for sale. In some states calcium (lime) materials are included in official inspection, the minimum percentage being expressed as calcium oxide and magnesium oxide. This statement of composition is usually made on the bag or parcel that contains the fertilizer or on an official tag attached to the bag. The enforcement of the law is usually in the hands of experiment stations, but in some states it is managed by a department of agriculture. Samples of the fertilizers sold in the state are collected annually, generally from goods in the hands of dealers and consumers. These samples are analyzed and the results of chemical analysis published for free distribution.

Manufacturers are required to pay a tax which may be in the form of a license-fee for each brand offered for sale or in the form of a tax on each ton sold. The proceeds are used to pay the cost of carrying out the provisions of the law.

In some cases manufacturers are required to state the source and character of the materials contained in each fertilizer, a highly desirable provision.

In some states legal prosecutions are instituted in those cases where any constituent is below the guarantee more than a specified amount. In some states, the manufacturer is required to refund to the purchaser an amount sufficient to cover any deficiency found in the guaranteed amount of plant-food. In most states, reliance for the protection of farmers is placed upon the effect of publicity in widely distributing the results of chemical analysis instead of upon legal prosecutions.

Advantages of fertilizer laws.—In general, fertilizer laws afford a large measure of protection to farmers, as well as to honest manufacturers. Fraudulent fertilizers are exposed and driven from the market. Careless or dishonest manufacturers suffer either temporary or complete loss of business. Larger amounts of high-grade plant-food materials are now used than formerly, as the result of legal inspection. There is greater uniformity in the composition of each brand of commercial fertilizers, owing to more effective means and care used in mixing. The number of brands of fertilizers offered for sale, as well as the number of manufacturers in proportion to business, has been diminished. Of especial importance has been the value of the educational effect upon the many farmers who insist on seeing and studying bulletins that give official analyses of fertilizers before they make purchases. The chemical analysis of a fertilizer, which was formerly meaningless to most farmers, is now appreciated to the extent, at least, of distinguishing nitrogen, available phosphoric acid and potash. The interest awakened by consulting bulletins of fertilizer analysis has led many farmers to study the various forms and uses of plant-food materials and this has served as an introduction leading to a desire for widened information relating to other details of crop-growing. Thus, the fertilizer bulletin has often proved a source of genuine intellectual stimulation, a fact which should not be underestimated.

ADVANTAGES AND DISADVANTAGES OF COMMERCIAL MIXED FERTILIZERS

In closing discussion of this subject, some of the more important facts will be briefly summarized relating to the advantages and disadvantages in the use of commercial fertilizers.

Advantages of using commercial mixed fertilizers are: (1) Convenience, (2) opportunity for choice, (3) uniformity, (4) completeness of mechanical mixture.

(1) *Convenience.*—Commercial fertilizers are widely distrib-

uted, are easily obtained, and can be purchased in any desired amount and at any time.

(2) *Opportunity for choice.*—Fertilizers of different brands represent a great variety of proportions of plant-food constituents, among which one can generally find what he needs for his soil and crops, provided he has ascertained by experience what combination gives him best crop returns.

(3) *Uniformity.*—There is an advantage in being able to obtain the same mixture from year to year, when it best suits one's needs, and to feel assured of having the same kinds of plant-food materials in the same proportions. Generally speaking, the same brand of fertilizer, especially in case of the large manufacturers, has been running very uniform from year to year.

(4) *Completeness of mechanical mixture.*—It is possible for manufacturers to prepare fertilizers evenly and thoroughly mixed, finely ground, dry, and in condition for convenient use. Occasionally, however, a small manufacturer through carelessness or inadequate machinery, may fail to make good mechanical mixtures, so the plant-food constituents are not uniformly distributed, nor the materials sufficiently fine and dry. Fertilizers sent out when incompletely dry soon harden in the package and are extremely difficult to handle.

Disadvantages of using commercial mixed fertilizers.—Definite disadvantages that accompany the exclusive use of commercial fertilizers include (1) lack of knowledge of materials, (2) lack of economy, (3) lack of educational incentive.

(1) *Lack of knowledge of materials.*—Farmers are unable without help to know whether they are purchasing in a mixture the materials claimed to be present. So far as the guarantee-analysis gives information regarding the source of nitrogen, the farmer in most cases is purchasing blindly. This objection is overcome in those cases, in which the purchaser has a chance to buy mixed fertilizer, accompanied by the statement of the various ingredients used in making the fertilizer and the number of pounds of each ingredient in one ton of the mixed product.

(2) *Lack of economy.*—In general, the most expensive way to purchase plant-food is in the form of complete commercial fertilizers. Results of an investigation made by the writer showed that when nitrate nitrogen could be purchased by farmers at 14 cents a pound, the nitrogen in mixed fertilizers cost on an average, over 20 cents a pound, and, in the grades of fertilizers most extensively sold, the cost varied from 18 to 36 cents. Similar disproportions were found in case of phosphorus and potassium.

(3) *Lack of educational incentive.*—The most serious disad-

vantage in the use of commercial fertilizers, as actually used in most cases, is that farmers are not stimulated to acquire needed information in regard to plant-foods and their proper use.

General reliability of commercial fertilizers.—In closing this chapter, it is only fair to say that it is the writer's belief, based on observation and experience derived from connection with fertilizer inspection for forty years, that the manufacture of commercial fertilizers today is more carefully managed, and the products more reliable in every way, than at any previous time. This relatively satisfactory condition is in part due to the National Fertilizer Association, which has been effective in co-operating with other state and national official agencies to require accurate statement of the composition of commercial fertilizers.

HIGH-GRADE AND LOW-GRADE FERTILIZERS

Commercial fertilizers vary greatly in respect to the percentages of nitrogen, phosphoric acid and potash they contain. Mixed fertilizers or unmixed fertilizer materials containing large amounts of these constituents are known as *high-grade* or *high-analysis;* those that contain small amounts, as *low-grade* or *low-analysis.* Between these are all possible gradations of composition. As an approximate generalization, complete mixed fertilizers that contain in 100 pounds less than 15 pounds of nitrogen, available phosphoric acid and water-soluble potash are now regarded as *low-grade;* those that contain 15 to 20 pounds, as *medium-grade;* those that contain 20 to 30 pounds, as *high-grade;* those that contain over 30 pounds, as *concentrated* or "multiple-strength."

Since 1920 there has been a marked increase in the number of mixed fertilizers that contain more than 15 pounds of plant-food in 100 pounds, and since 1925 those classed as high-grade have rapidly gained in number. Since 1927 mixed fertilizer of the concentrated grade have appeared in increasing numbers, some containing as high as 60 pounds of plant-food in 100 pounds, 1,200 pounds in one ton.

Cost of plant-food lowest in high-analysis fertilizers.—It is generally more economical to buy commercial fertilizers that contain the larger amounts of plant-food. As a rule, the higher the analysis, the lower is the cost of a pound of plant-food to the purchaser. In illustration of this statement, four fertilizers are taken for comparison from a price list and data given in Table 42.

The obvious conclusion based on the figures in Table 42, is that the plant-food in the highest-grade fertilizers costs least to the purchaser. The lower cost of high-analysis fertilizers results from decreased cost for bags and freight.

In some states laws have been passed requiring that mixed complete fertilizers shall contain not less than a definite total percentage of plant-food constituents (nitrogen, phosphoric acid and potash). In one state, for example, this minimum requirement is 16 per cent. This means that no complete fertilizer that contains less than a total of 16 per cent. of plant-food constituents can be sold.

TABLE 42—COMPOSITION OF FERTILIZERS AND COST OF PLANT-FOOD

In 100 pounds of fertilizer					*Cost of 1 pound of plant-food*		
Nitrogen	*Phosphoric acid*	*Potash*	*Total plant-food*	*Retail price of 1 ton*	*Nitrogen*	*Phosphoric acid*	*Potash*
Lbs.	*Lbs.*	*Lbs.*	*Lbs.*	*Dollars*			
1	8	4	13	19.50	19.1	7.0	5.6
2.50	8	5	15.50	24.90	18.0	6.6	5.3
3.30	10	6	19.30	30.00	17.2	6.3	5.0
4.12	15	5	24.12	35.55	16.0	5.9	4.7

XXV

HOME-MIXED FERTILIZERS

COMMERCIAL FERTILIZERS are made of well-known materials, most of which can be purchased by farmers. The question was raised some years ago whether farmers could purchase unmixed plant-food materials, such as sodium nitrate, ammonium sulfate, dried blood, ground fish, superphosphate, ground bone, potassium chloride and sulfate, etc., and mix these materials for their own use. Several experiment stations have given careful attention to the study of home-made mixtures, pronounced their use to be satisfactory, and progessive farmers have followed their guidance. The number of farmers who mix their own fertilizers has increased from year to year. The amount of money possible to save by home-mixing depends upon several factors, such as the amount of fertilizer needed, the cost of labor, efficiency of mixing, formulas and materials used.

We will discuss the subject under the following divisions:

1. Utensils needed to mix materials.
2. Method of mixing.
3. Use of fillers and driers.
4. Purchasing unmixed materials.
5. Materials that should not be mixed together.
6. Advantages and disadvantages of home-mixing.
7. Co-operative purchase of unmixed and mixed fertilizers.
8. How to calculate amounts of materials for home-made mixtures.

UTENSILS NEEDED IN MIXING MATERIALS

The conveniences needed to mix fertilizers on the farm are: (1) A tight, hard, dry floor under shelter; (2) platform scales; (3) shovel; (4) iron handrake or hoe; (5) tamper to break hard lumps; (6) sand or coal-screen; (7) grinding-machine; and (8) mixing-machine. A few of these conveniences call for special attention.

(1) *The tamper* can be made by nailing two upright handles to a piece of wood 6 inches thick and 15 to 18 inches long. If one has

a good grinding-machine, breaking of coarse lumps by pounding may not be necessary.

(2) *The screen* can be 5 to 6 feet long, 1½ to 2½ feet wide, with 3 to 6 meshes to the inch.

(3) *Grinding-machine.*—For making coarse material finer, a good feed-grinding mill will answer. Where one has power on the farm and mixes large amounts of fertilizing materials, it will pay

Implements for home mixing.

to purchase a special grinding-machine. Some machines will grind limestone fine. The principal use of a grinding-machine is to pulverize lumpy or caked materials. The materials used in fertilizers have all been ground once, but may become lumpy on standing. They can usually be broken by being passed through a feed-mill regulated so as to permit wheat to pass. For lumps of superphosphate such a mill should be set so as to crack corn but not wheat. If one tries to grind superphosphate, ground tankage or fish-scrap, the material is likely to become sticky and unmanageable. This trouble can be avoided by setting the feed-mill sufficiently open to disintegrate lumps without grinding the material.

(4) *Mixing-machine.*—While mixing can be done by shoveling the materials over and over, it is easier to use a rotating box. A large, revolving churn with slats across the inside can be used, or a box or a drum with small projecting inside shelves which carry the materials around to the top and then drop them from the upper half to the bottom repeatedly. When large amounts of materials are to be mixed, a rotary mixer, such as used for mixing concrete, might be found economical to purchase.

Method of mixing materials.—The various operations in mixing fertilizer materials are: (1) Weighing materials, (2) screening out lumps, (3) crushing lumps, (4) mixing, (5) storing.

(1) *Weighing materials* is the first operation, the desired amount of each material being weighed, separately. Calculating amounts is considered on pp. 328-335.

(2) *Removing lumps.*—If any material is lumpy, it is sifted through the sand-screen and the coarse portions separated. The driest material is screened last. Screens should be kept clean.

(3) *Crushing lumps.*—The lumpy portion is made fine with a tamper or a shovel, or preferably by passing through a grinding-machine. The powdered lumps are added to the screened portion. The mill should always be cleaned at once after using. Some lumpy materials like sodium nitrate easily disintegrate when sprinkled lightly with water, and allowed to stand some hours and then raked over.

(4) *Mixing.*—When the materials are properly fine they are ready for mixing. When a mixing-machine is not used, the most bulky material, usually superphosphate, is spread out upon the floor in a uniform layer about 6 inches deep; upon this is spread evenly the material next in bulk, usually that containing nitrogen; and then the least bulky material, generally potassium salts. Care is taken to make the material in each case cover the one beneath evenly over the whole surface. Beginning at one side and working across, one should shovel the whole pile over, pushing the shovel into the bottom of the pile every time. When the whole pile has been carefully shoveled over once in this way, the mixed pile is leveled, the scattered material about the edges swept to the pile and the whole again shoveled over. This process should be repeated three or four times or until uniformity of color and absence of streaks indicates that the mixing has been thorough. The mixture can finally be passed through the screen once, if desired.

When a mixing-machine is used, the various materials in proper amounts, when ready for mixing, are put into the machine, the door closed and the machine rotated for some minutes. The amount mixed in a machine depends chiefly on the size of the mixer. Care should be taken never to fill such mixers more than half full; otherwise the mixing may not be thorough.

(5) *Storing.*—For storage and general convenience in handling, it is well to weigh the mixed fertilizer into sacks that hold 100 to 150 pounds. The filled bags should be stored in a dry place.

USE OF FILLERS AND DRIERS

There is a popular belief that all fertilizers contain some inert, valueless material added solely for the purpose of making weight. While this may be true in some cases, it is not necessarily true in most fertilizers. This belief in the universal use of fillers in fertilizers is based upon the methods of stating the results of chemical analysis. For example, a high-grade fertilizer that shows an analysis of 5 per cent. of nitrogen, 10 of phosphoric acid and 10 of potash contains, all told, only 25 pounds of plant-food constituents in 100 pounds of material, or to use the figures for phosphorus and potassium, only 17.7 pounds of plant-food constituents. The question is: What are the other 75 pounds? This seeming difficulty is easily explained when we consider that the method of stating the results of chemical analysis gives only a part of what is in the fertilizer. For example, nitrogen is present in fertilizers only in combination with other elements; in sodium nitrate, for each pound of nitrogen used, we add to a fertilizer about 1.5 pounds of sodium and 3.5 pounds of oxygen; or, expressed in another way, sodium nitrate consists of only one-sixth nitrogen so, in order to get one pound of nitrogen, it is necessary to use 6 pounds of sodium nitrate. Similarly, it requires about 10 pounds of dried blood to furnish one pound of nitrogen (p. 290). Each material used is a compound, or diluted form, of nitrogen, phosphorus or potassium. For a variety of reasons it is impossible to use either pure nitrogen, phosphorus or potassium in fertilizers, so we must use those compounds or diluted forms best adapted to feeding plants and at the same time be sufficiently plentiful and cheap to make their use practicable.

While worthless materials may be added as fillers to make low-grade fertilizers, there is also a legitimate use of so-called fillers, namely, as driers or conditioners for the special purpose of insuring a suitable mechanical condition in order to make the fertilizer distribute easily when applied. Materials like the nitrate of ammonium, calcium and sodium, ammonium chloride, urea, etc., absorb moisture so the mixtures in which they are present are likely to harden on standing. This can be prevented by using in the mixture some dry, inert material that will absorb moisture and keep the mass from caking. For this purpose nothing is better than *fine, dry muck* or *peat*. One can also use *finely-ground rock-phosphate, garbage tankage, cocoa shells, sewage sludge, castor pomace, fine sand, gypsum, coal ashes, fine cinders, sawdust* and even *fine, dry earth*. It will not do, however, to use quicklime (calcium oxide), slaked lime (calcium hydroxide) or wood-ashes in certain fertilizers (p. 323). When a mixture contains sufficient amounts of ammoniated superphosphate, bone-meal, fine tankage, cottonseed meal or calcium cyanamid (about 2 per cent.), no filler or drier will be needed, since these materials generally insure a good mechanical condition. When a home-mixed fertilizer is prepared some weeks or months before it is to be used, especially when it contains considerable amounts of nitrates or urea or ammonium chloride, it will usually be well to put into the mixture some material to act as a drier in order to prevent caking.

PURCHASING UNMIXED MATERIALS

In buying unmixed materials, attention should be given to certain points often the subject of inquiry among farmers. Some of the more important are: (1) Where to purchase, (2) how, (3) when, (4) what grade, (5) co-operative buying, (6) use of "unit" system, (7) fineness and dryness.

Where to purchase unmixed materials.—Agricultural colleges and experiment stations and the United States Department of Agriculture, as well as county farm-bureau agents, usually furnish on request a list of the addresses of reliable dealers in plant-food materials. Many dealers in fertilizer materials located in commercial centers throughout the United States make a specialty of handling unmixed materials. Owing to the increasing demand for these materials, local dealers have increased in number in localities where fertilizers are extensively used.

How to purchase unmixed materials.—Before purchasing, farmers are advised to obtain for themselves prices at which they can actually buy specific plant-food materials from several different sellers, local and state. This precaution will prevent the possibility of paying more than the prevailing price at the time.

When to purchase materials.—By watching market variations reported in agricultural papers, it is possible to save something. It often happens that the lowest prices prevail in early winter. On all accounts, this will be found the best time to buy and mix materials, since home-mixing should be completed before the beginning of spring work. It is well to obtain information about the purchase of materials in November or December. In those sections where garden crops are grown mostly during the winter months, action will, of course, be taken earlier.

What grade of materials to buy.—It will invariably be found more economical to purchase high-analysis plant-food materials. While bulk is often desirable in applying fertilizers, the object in purchasing should be to obtain the most available nitrogen, phosphorus and potassium for one dollar.

Co-operative buying.—In purchasing unmixed plant-food materials, considerable economy can be effected by several farmers in a community ordering supplies together. This method has been advantageously used by many granges and farmers' clubs. The unmixed materials are bought in quantity and distributed to individuals, each of whom does his own mixing. This method effects saving of freight and may obtain the benefit of lower prices on car lots. Such methods always depend upon cash payment and usually result in the most economical purchase of plant-food possible. Co-operative organizations are increasing in number and efficiency for furnishing fertilizer materials, mixed fertilizers and other supplies to farmers at advantageous prices.

Use of "unit" system in purchasing materials.—Fertilizer materials should always be purchased on the basis of a guaranteed statement giving the amount of nitrogen, phosphoric acid and potash. Two methods are used in buying plant-food materials. One is based directly on the percentage composition, the other on the "unit" system (p. 310). In quoting prices, a dealer offers potassium sulfate, for example, at $1 per unit of potash, 20 pounds of potash, which is equal to 5 cents a pound. We get at the same result on a ton basis and a specific percentage guarantee. For example, potassium sulfate that contains 50 per cent. of potash is quoted at $50 a ton; 50 per cent. of a ton is 1,000 pounds, so a pound costs 5 cents. The unit system is somewhat simpler, because, to find out the price of one pound, it is necessary only to divide the price of one unit by 20. The unit basis is the one commonly used in trade quotations and in commercial transactions.

Fineness and dryness important.—Materials to be used in mixing should be in finely-powdered form and dry so that they may be easily and completely mixed and that the mixture be in good mechanical condition for applying. Usually, unmixed materials that have stood long cake more or less and will need screening and grinding before mixing. The precaution should be taken to store unmixed and mixed materials in a dry place.

MATERIALS THAT SHOULD NOT BE MIXED TOGETHER

Plant-food materials cannot be mixed indiscriminately without, in some cases, incurring risk of loss or unfavorably affecting the mechanical condition; so it is important to know what mixtures should be avoided for the following reasons: (1) Loss of ammonia nitrogen, (2) loss of nitrate nitrogen, (3) reversion of soluble phosphoric acid, (4) unfavorable mechanical effect.

Mixtures that cause loss of ammonia nitrogen.—Alkaline compounds easily set free ammonia from ammonium compounds, such as ammonium sulfate. If in sufficiently concentrated form they may also decompose animal matter, setting free ammonia, especially when stored in a damp, warm place. When mixtures of such materials are immediately placed in the soil, the ammonia is held (p. 000), but when they stand in the air, the ammonia is lost. The following list indicates materials that should not be mixed for this reason:

Calcium oxide (quicklime)........	Should not be mixed with	Ammonium sulfate, nitrate, chloride, phosphate
Calcium hydroxide (slaked lime)...		Animal materials (bone, tankage, blood, fish scrap, farm manure, etc.)
Potassium carbonate (in wood-ashes)		Nitrogenous guanos
Basic-slag phosphate		Urea, calurea
Calcium cyanamid		Leunasalpeter, leunaphos, etc.
Basic calcium nitrate		

Loss of nitrate nitrogen.—Under one condition, now rare, loss of nitrate nitrogen may occur. When moist superphosphate, so carelessly made as to contain considerable amounts of free sulfuric acid, is mixed with sodium, ammonium, or calcium nitrate and kept in a warm place, some free nitric acid may be formed and escape into the air. Any such possibility of loss can be obviated by adding to the materials before mixing a sprinkling of finely powdered limestone, which will neutralize any sulfuric acid present.

Reversion of soluble phosphoric acid.—When certain calcium compounds are mixed with a soluble phosphate, as contained in any superphosphate, and in ammonium or potassium phosphates, the soluble phosphate is changed to di-calcium phosphate (reverted) or to tri-calcium phosphate or both (p. 33). The following form of statement indicates what mixtures are to be avoided on this account:

Calcium oxide	Should not be mixed with	Soluble phosphates
Calcium hydroxide ...		
Calcium carbonate ...		
Wood-ashes		
Basic calcium nitrate..		
Calcium cyanamid ...		
Basic slag phosphate..		

For the same reason, similar magnesium, iron and aluminum compounds should not be mixed with soluble phosphates.

Unfavorable mechanical effect.—Several of the materials used in making fertilizers absorb moisture. When materials that contain quicklime or slaked-lime, are mixed with such moisture-absorbing materials, they tend, on standing, to harden and make a solid mass, difficult to handle. If, however, the mixture is distributed soon after being made, this trouble is avoided. Another method to prevent hardening in such cases is to add one-fourth or more of its weight of fine, dry muck, or 100 to 240 pounds of some organic material, such as cottonseed meal, ground fish or animal tankage, to the mixture. The materials which for this reason should not be mixed, or mixed only just before using on the soil, are:

Calcium oxide (quicklime)..... Calcium hydroxide (slaked lime) Basic calcium nitrate........... Basic slag phosphate	Should not be mixed with, except just before application to soil	Sodium nitrate Calcium nitrate Potassium salts Kainite Potash manure salts Urea

ADVANTAGES AND DISADVANTAGES OF HOME-MIXING

Among specific advantages that may result from purchasing unmixed materials and mixing them on the farm, are: (1) Economy, (2) definite knowledge of materials used, (3) variations of mixtures to suit specific conditions, (4) educational value.

(1) *Economy in home-mixed fertilizers.*—The average retail price of one ton of a mixed commercial fertilizer is, considerably higher than the retail cost of the unmixed materials, (p. 000). This difference, which represents the cost of mixing, freight, profit, commissions, etc., varies according to the grade of fertilizer; in the case of high-grade goods, the difference averages less than $5, while in low-grade fertilizers it averages $7 or $8. It is obvious that at least a considerable portion of this difference can be saved by purchasing unmixed ingredients.

To the cost of materials, farmers can properly add the cost of their labor in mixing. In reality this is a small matter, since this work can be done when there is least demand on the farmer's time. For the farmer who purchases low-grade mixed commercial fertilizers, there is a possible saving of $10 a ton, and it is safe to count on $5 at least. For those who purchase high-grade fertilizers, the saving will usually be $5 or more a ton, and not less than $3. For those who use 10 tons or more of fertilizer, the saving effected by home-mixing can be seen to be considerable in the aggregate.

(2) *Definite knowledge of materials used.*—When separate materials are purchased, the farmer can more easily ascertain whether

they are what they are purchased for. Ground leather cannot be passed off for sodium nitrate or ammonium sulfate. In mixed goods, all mysteries of inferior composition are beyond detection by the eye. Unmixed materials are usually more uniform in composition than the mixed. The farmer may get better forms of plant-food in unmixed than in mixed products.

(3) *Variation of conditions to suit specific needs.*—By careful observation and some experimenting, each farmer can make mixtures with those proportions of plant-foods which best meet the conditions of his soil and crops. In purchasing unmixed materials, one can apply a single plant-food constituent at a time, a method found useful under some conditions. This practice is becoming more common, for example, in the top-dressing of some crops with a nitrate.

(4) *Educational value.*—There is little of educational value in using an unknown mixture. To purchase intelligently unmixed fertilizing materials ultimately leads in most cases to a well-grounded knowledge of agricultural science. One will seek to know what the various forms of plant-food materials are, what they do, from what sources they are derived, how they can be obtained and how they can be used to best advantage. One becomes a student and an investigator and, of necessity, takes a deeper interest in his work. His entire system of farming is lifted to a higher plane and his intelligently applied labor yields more profitable results.

Disadvantages of home-mixing.—The disadvantages suggested in connection with home-mixing are: (1) Difficulty of purchasing small amounts of unmixed materials, (2) injury to crops from incomplete mixing, (3) imperfect mechanical condition, (4) injury to crops by application of concentrated materials.

(1) *Difficulty of purchasing small amounts of unmixed materials* exists no longer; for dealers are ready to sell materials in small or large amounts. The advantage of purchasing large amounts can readily be realized by co-operation of farmers in purchasing their supplies together.

(2) *Injury to crops from incomplete mixing.*—Farmers are told by sellers of mixed fertilizers about the difficulty of mixing fertilizing materials, with the consequence of uneven application to the soil, some portions getting little or no plant-food and others too much, the results of which make uneven crops and reduced yields. The fact that a farmer who once intelligently undertakes mixing his own fertilizers seldom goes back to commercial mixed fertilizers is sufficient evidence of accomplished success.

(3) *Imperfect mechanical condition.*—While it is true in the best-made commercial fertilizers that the degree of fineness may be better than in home-mixed materials, home-made mixtures with the

Wheat in a five year rotation on limed land. During thirty-five years the left plot averaged 11.7 bushels; the right plot, 26.5 bushels. The former was unfertilized; the latter received nitrogen, phosphorus and potash at a cost of $6.20. This fertilizer also increased the average total hay yield by one ton and a third.

precautions given can be made sufficiently fine for all practical purposes and fully as satisfactory as those found in many mixed fertilizers.

(4) *Injury to crops by application of concentrated materials.*— When materials like sodium nitrate, superphosphate and potassium chloride or sulfate are put together, we have a mixture of concentrated materials, which may do harm to seeds when the mixture is applied. This difficulty is overcome by placing the fertilizer where it does not come into direct contact with seed and also by using fertilizer distributors which can place small amounts.

CO-OPERATIVE PURCHASE OF FERTILIZERS

In some towns, farmers combine to buy unmixed plant-food materials, each one doing his own mixing. In other cases, a farmers' organization decides upon one or more definite formulas, found by experience to suit the local needs, and sends manufacturers specifications for furnishing the same ready mixed and sacked, letting the contract to the lowest responsible bidder. The mixtures are checked by official analysis.

Successful co-operative organizations furnish their members with unmixed fertilizer materials and also mixed fertilizers at a marked reduction of cost.

DEFINITION OF ANALYSIS AND FORMULA

Formerly the terms, analysis and formula, as applied to fertilizers, were used with the same meaning. A distinction is now made.

Analysis means the statement that gives the percentage of nitrogen, phosphoric acid, and potash, in their various forms, in a fertilizer material or mixture.

Formula is applied to a statement that gives the amount and composition or grade of the various constituent materials contained in a fertilizer mixture. The term, *open formula,* is used of a commercial fertilizer mixture, the ingredients of which are stated. For example, an open-formula commercial fertilizer is stated to contain 200 pounds of sodium nitrate, 150 pounds ammonium sulfate, 600 pounds of 16 per cent. superphosphate, 400 pounds of 2-20 bone meal and 650 pounds of kainite.

HOW TO CALCULATE FORMULAS FOR HOME-MADE MIXTURES

It is necessary to decide first what the percentages of nitrogen, available phosphoric acid and potash shall be in the desired fertilizer mixture and then what materials shall be used in making the mixture. The usual percentages in fertilizer materials are given in Table 43, p. 332.

For example, a farmer wishes to make a fertilizer that contains:

4.0 per cent. of nitrogen
8.0 " " " available phosphoric acid
8 " " " potash

and for this purpose has the following materials:

Sodium nitrate containing 15 per cent. of nitrogen
Superphosphate containing 16 per cent. of available phosphoric acid
Potassium chloride, containing 50 per cent. of potash

How much of each of these materials is it necessary to take to make one ton of a mixture with the above analysis?

The process necessary to solve the problem may be conveniently divided into four steps or operations, as follows:

(1) The first step consists in finding the number of pounds of nitrogen, phosphoric acid and potash in one ton of a fertilizer having the analysis given above (4—8—8). This is done by multiplying the number representing the per cent. by 20, as follows:

	Per cent. (pounds per 100)				*Pounds per ton*
Nitrogen	4	×	20	=	80
Available phosphoric acid.....	8.0	×	20	=	160
Potash	8	×	20	=	160

(2) The second step consists in finding (*a*) how many pounds of sodium nitrate containing 15 per cent. of nitrogen will be required to furnish 80 pounds of nitrogen; (*b*) how many pounds of acid phosphate containing 16 per cent. of available phosphoric will be required to furnish 160 pounds of available phosphoric acid; and (*c*) how many pounds of potassium chloride containing 50 per cent. of potash will be required to furnish 160 pounds of potash.

(*a*) Since 100 pounds of sodium nitrate contains 15 pounds of nitrogen, it will require, to furnish 80 pounds of nitrogen, as many times 100 as 15 is contained in 80, or 5.33. This calculation may be represented as follows:

$$80 \div 15 \times 100 \text{ or } \frac{80 \times 100}{15} = 533$$

pounds of sodium nitrate.

(*b*) Since 100 pounds of superphosphate contains 16 pounds of available phosphoric acid, it will require as many times 100 as 16 is contained in 160, to furnish 160 pounds of available phosphoric acid:

$$160 \div 16 \times 100, \text{ or } \frac{160 \times 100}{16} = 1000$$

pounds of superphosphate.

(c) Since 100 pounds of potassium chloride contains 50 per cent. of potash, it will require, to furnish 160 pounds of potash, as many times 100 as 50 is contained in 160;

$$160 \div 50 \times 100, \text{ or } \frac{160 \times 100}{50} = 320$$

pounds of potassium chloride.

(3) The third step consists in adding the amounts of plant-food materials (533 + 1000 + 320), which equal 1,853 pounds. The amount can be made up to 2,000 pounds by adding 147 pounds of filler or drier (p. 321).

The method of operation can now be stated in the form of rules.

Rule: To find amounts of materials to use to make fertilizers.—(1) *Find the number of pounds of nitrogen, available phosphoric acid and potash contained in one ton of the mixture by multiplying in each case, respectively, the number giving the per cent. by 20.*

(a)	Per cent. nitrogen	× 20 =	pounds of nitrogen		in one ton of fertilizer
(b)	" " phosphoric acid	× 20 =	" "	phosphoric acid	" " " " "
(c)	" " potash	× 20 =	" "	potash	" " " " "

(2) *Find the number of pounds of given material required to furnish the number of pounds of each plant-food constituent in one ton of mixture, as follows: Multiply by 100 the number of pounds of each plant-food constituent contained in a ton and divide the result by the number giving the percentage of the respective plant-food constituent in the material used.*

(a) Lbs. nitrogen in 1 ton × 100 ÷ % nitrogen in material used = Lbs. nitrogen material required

(b) " phos. acid in 1 ton × 100 ÷ % phos. acid in material used = Lbs. phos. acid material required

(c) " potash in 1 ton × 100 ÷ % potash in material used = Lbs. potash material required

(3) *Find the total amount of plant-food materials by adding together the results obtained in the preceding step. If the sum is less than 2,000, add enough fine, dry muck or other inert material to make 2,000 pounds.*

Calculation of materials based on acre application.—Suppose a farmer wishes to apply per acre 30 pounds of nitrogen, 80 pounds of available phosphoric acid and 50 pounds of potash, in the respective forms of sodium nitrate, superphosphate and potassium chloride, containing the usual percentages (15 of nitrogen, 16 of available phosphoric acid and 50 of potash), how can he determine the amount of each material necessary to use? This practically involves only the second step in the preceding process. The operation is indicated briefly, as follows:

(*a*) 30 (lbs. nitrogen) × 100 ÷ 15 (% nitrogen in nitrate) = 200 lbs. of sodium nitrate
(*b*) 80 (lbs. phos. acid) × 100 ÷ 16 (% phos. acid in superphosphate) = 500 lbs. of superphosphate
(*c*) 50 (lbs. potash) × 100 ÷ 50 (% potash in chloride) = 100 lbs. of potassium chloride

The results added together make an application of 800 pounds of the mixture per acre. In such cases as this, if one desires to make larger amounts, each material is increased two, three or as many times as suits one's purpose or convenience.

Short method for calculating amounts of materials.—By using the figures in Table 43, considerable labor can be saved, especially when one knows the percentage of plant-food constituents in the materials. The tabular figures can best be understood by giving an example. Suppose one wishes to make a mixture like the one already given (p. 329) as an illustration (4—8—8), using the same materials (nitrate with 15 per cent. of nitrogen, superphosphate with 16 per cent. of phosphoric acid, and potassium chloride with 50 per cent. of potash). To find how much sodium nitrate to use, look down the percentage column in Table 43 until 15 is reached and find opposite it the number 133.3; we then multiply this by 4, the per cent. of nitrogen that the mixture is to contain. For the 16 per cent. phosphoric acid in acid phosphate, we find 125, and multiply this by 8, the per cent. of phosphoric acid in the mixture; and for 50 per cent. of potash we find 40, which we multiply by 8, the per cent. of potash in the mixed fertilizer. The operations are indicated as follows:

Plant-food constituent	*Per cent. in mixed fertilizer*		*Factors found in table*		*Amount of materials to use*
Nitrogen	4	×	133.3	=	533 pounds of sodium nitrate
Phosphoric acid	8.0	×	125	=	1000 " " acid phosphate
Potash	8	×	40	=	320 " " potassium chloride
Total material					1853

The data in Table 43 are based on the "unit" system, in which each per cent. represents 20 pounds of the plant-food constituent in 2,000 pounds of fertilizer material (p. 310). Each factor in the table shows the number of pounds of material that must be put into a ton of fertilizer, in order to make the fertilizer contain one per cent. of the plant-food constituent (nitrogen, phosphoric acid, or potash). One per cent. is the same as 20 pounds per ton.

Some additional illustrations of use of the table are given: (1) How much dried blood that contains 12 per cent. of nitrogen should be used to furnish nitrogen for one ton of a fertilizer that contains 3 per cent. of nitrogen? Opposite 12 (per cent. column) we find the factor 166.7, which, multiplied by 3, gives 500 pounds as the amount of dried blood to use. (2) How much superphosphate, that contains

TABLE 43—FACTORS FOR QUICK CALCULATION OF AMOUNTS OF MATERIALS TO USE IN MIXING FERTILIZERS

Per cent. of plant-food constituent in material used	*Factor to use in multiplying*	*Per cent. of plant-food constituent in material used*	*Factor to use in multiplying*	*Per cent. of plant-food constituent in material used*	*Factor to use in multiplying*	*Per cent. of plant-food constituent in material used*	*Factor to use in multiplying*
1	2000	9	222.2	16.5	121.2	26	77
1.5	1333.3	9.5	210.5	17	117.7	28	71.5
2	1000	10	200			30	66.7
2.5	800						
3	666.7	10.5	190.5	18	111	32	62.5
3.5	571.5	11	182	18.5	108	34	58.8
4	500	11.5	174	19	105.3	35	57.2
4.5	444.5	12	166.7	19.5	102	36	55.6
5	400	12.5	160	20	100	38	52.6
5.5	363.6	13	154	20.5	97.6	40	50
6	333.3	13.5	148.2	21	95.3	42	47.6
6.5	307.7	14	143	21.5	93	44	45.5
7	285.7	14.5	138	22	91	45	44.4
7.5	266.7	15	133.3	23	87	46	43.5
8	250	15.5	129	24	83.3	48	41.7
8.5	235.3	16	125	25	80	50	40

46 per cent. of available phosphoric acid, must be used to furnish phosphoric acid for one ton of a fertilizer that contains 16 per cent. of phosphoric acid? Opposite the figure 46 we find 43.5 which, multiplied by 16, gives 696 pounds of superphosphate. (3) How much kainite that contains 12½ per cent. of potash, should be used to furnish potassium in a fertilizer that contains 5 per cent. of potash? Opposite 12½ we find the factor 160, which, multiplied by 5, equals 800 pounds of kainite. (4) How much ammonium nitrate that contains 32 per cent. of nitrogen, should be used to furnish nitrogen in one ton of fertilizer that contains 8 per cent. of nitrogen? The factor opposite 32 per cent. is 62.5, which multiplied by 8 gives 500 pounds.

Calculations to be made when composition is stated in other than usual terms.—In our previous discussion we have assumed that the composition of materials is stated in terms of nitrogen, available phosphoric acid and potash. If, however, the guarantee of nitrogen is given as ammonia, the amount of nitrogen can be easily calculated by consulting Table 46 (p. 356). Similarly, if other changes are to be made to get the amounts of plant-food in terms of nitrogen, phosphoric acid, phosphorus, potash or potassium, the data for making the needed calculations can be found in the same table. In most cases, however, guarantees supplied by manufacturers are in usual form.

Added illustrations of home-made mixtures.—The procedure is more complicated when we use materials that contain more than one plant-food constituent, as bone-meal, fish scrap, potassium nitrate, ammo-phos, etc.; or when we use more than one material to supply one plant-food constituent, as nitrogen in part from nitrate, in part from ammonia and in part from urea.

Example 1. To make one ton of a mixture that contains 3 per cent. of nitrogen, 6 of phosphoric acid and 3 of potash (3—6—3), in which one-half (1.5) of the nitrogen is in the form of sodium nitrate and one-half, ammonium sulfate, the phosphoric acid being supplied by 20 per cent. superphosphate and the potash by high-grade potassium chloride, how many pounds of each material must be used?

First, consult Tables 35, 36, 38 to ascertain the percentage composition of materials used (sodium nitrate, 15 per cent. of nitrogen; ammonium sulfate 20.5 per cent of nitrogen; potassium chloride, 50 per cent. of potash). Then make use of Table 43 as already directed. The process is indicated as follows:

(1)	1.5	×	133.3	=	200	pounds of	sodium nitrate
(2)	1.5	×	97.6	=	146	" "	ammonium sulfate
(3)	6	×	100	=	600	" "	superphosphate
(4)	3	×	40	=	120	" "	potassium chloride
	Total			=	1066	"	
	Filler				934	"	
					2000	"	

Example 2. Make one ton of a mixture that contains 4 per cent. of nitrogen, 8 of phosphoric acid and 4 of potash (4—8—4), in which one-fourth of the nitrogen is in the form of sodium nitrate, one-fourth, ammonium sulfate, one-half, fish scrap, the phosphoric acid being supplied by 40 per cent. triple superphosphate and the potash by kainite. The operations are indicated as follows:

(1)	1 ×	133.3	=	133.0	pounds of	sodium nitrate
(2)	1 ×	97.6	=	98.0	" "	ammonium sulfate
(3)	2 ×	250	=	500.0	" "	fish scrap
(4)	8 ×	50	=	400.0	" "	triple superphosphate
(5)	4 ×	166.7	=	667.0	" "	kainite
	Total		=	1798.0	"	
	Filler			202	"	

Amount of material that contains one pound of plant-food.—It is often a matter of convenience to know without making detailed calculations how many pounds of fertilizing material are required to furnish one pound or more of plant-food (nitrogen, phosphorus or phosphoric acid, and potassium or potash). In Table 44 is given in one column the percentage of plant-food in the fertilizing material, and in the next the number of pounds of such material required to furnish one pound of plant-food.

TABLE 44—AMOUNT OF FERTILIZING MATERIAL THAT CONTAINS ONE POUND OF PLANT-FOOD

Per cent. of plant-food constituent in material	*Pounds of material required to furnish one pound of plant-food*	*Per cent. of plant-food constituent in material*	*Pounds of material required to furnish one pound of plant-food*	*Per cent. of plant-food constituent in material*	*Pounds of material required to furnish one pound of plant-food*	*Per cent. of plant-food constituent in material*	*Pounds of material required to furnish one pound of plant-food*
1	100	11	9.1	21	4.76	34	2.94
2	50	12	8.33	22	4.55	35	2.86
3	33.33	13	7.7	23	4.35	36	2.78
4	25	14	7.15	24	4.17	40	2.5
5	20	15	6.67	25	4	42	2.38
6	16.67	16	6.25	26	3.85	44	2.27
7	14.3	17	5.9	27	3.7	45	2.22
8	12.5	18	5.55	28	3.57	46	2.17
9	11.1	19	5.27	30	3.33	48	2.08
10	10	20	5	32	3.13	50	2

To show how Table 44 is used, the following illustrations are given: (1) How many pounds of ammonium sulfate that contains 20 per cent. of nitrogen are required to furnish (*a*) 1 pound of nitrogen, (*b*) 10 pounds, (*c*) 100 pounds? Ans. (*a*) 5 pounds, (*b*) 50 pounds, (*c*) 500 pounds. (2) How many pounds of superphosphate that contains 16 per cent. of available phosphoric acid furnish (*a*) 1 pound of phosphoric acid, (*b*) 20 pounds, (*c*) 100 pounds?

Ans. (*a*) 6.25 pounds, (*b*) 125 pounds, (*c*) 625 pounds. (3) How many pounds of kainite that contains 12 per cent. of potash furnish (*a*) 1 pound of potash, (*b*) 30 pounds, (*c*) 100 pounds? Ans. (*a*) 8.33 pounds, (*b*) 250 pounds, (*c*) 833 pounds.

In order to find how many pounds of fertilizing material are required to furnish any number of pounds of plant-food, multiply the figure given in Table 44 by the number of pounds of plant-food desired.

XXVI

ROTATION OF CROPS

CROP-ROTATION is a systematic arrangement for growing various crops successively on the same soil. The general object is to utilize the plant-food in the soil most economically and, at the same time, maintain the soil in such condition, physical, chemical and biological, as will promote the growth of maximum crops. The particular system of rotation to be used must depend upon kind of soil and character of farming. A fruit-grower, a dairyman, a wheat-grower, a cotton-grower, etc., would each follow a different system of rotation.

The rotation of crops as a system is in contrast with the one-crop system by which the same kind of crop is grown continuously year after year on the same soil. Under the one-crop system, soils sooner or later show decrease in crop production, usually due, in part, to a diminished supply of available plant-food and, in part, to deterioration of the soil in such a way as to render it unfavorable in one or more of its relations, physical, chemical, and biological.

We shall consider the subject under the following divisions: (1) Reasons underlying crop-rotation, (2) effect on crop production, (3) general principles of crop-rotation.

REASONS THAT UNDERLIE CROP-ROTATION

The practice of growing different crops in systematic succession on the same soil is the outgrowth of observation and experience. The beneficial effects in the way of increased crop production were observed long before the reasons for the effects were understood. Some of the advantages of crop-rotation are probably not clearly understood yet. We can better comprehend the reasons if we know what such a system can accomplish when effectively controlled; among the principal things it may do are: (1) Changes location of feeding-range of plants; (2) changes demand for individual plant-food constituents; (3) makes most advantageous use of remains of preceding crops; (4) provides economical supplies of nitrogen; (5) maintains supply of organic matter in soil; (6) keeps soil in good physical condition; (7) provides advantageous means of utilizing both farm manure and commercial plant-foods; (8) keeps soil advantageously occupied with crops most of the time; (9) prevents or reduces injury caused by toxic substances in soils; (10) aids in controlling injuries done by insects, weeds and fungi; (11) often saves labor; (12) systematizes farming.

Changes in location of feeding-range.—A properly planned rotation provides for alternation of crops whose root systems occupy different portions of the soil—deep and shallow. The plant-food is thus taken from different layers, also more or less from subsoil. By this treatment the available plant-food in any one layer is less quickly reduced and, therefore, applications of fertilizers are called for in smaller amounts or less frequently. In some soils potassium and calcium are more largely in the subsoil, while nitrogen and phosphorus are mainly in the upper layer. To illustrate the two types, wheat crops obtain their supply of food from surface soil, clovers go deeply for more or less of their food.

Variation in demand for special forms of plant-food.—Some crops use more nitrogen than others, some more potassium, etc. The continuous growing of one crop which consumes large amounts of potassium would sooner or later tend to reduce the available supply in relation to phosphorus and nitrogen; if such a crop were followed by one that requires less potassium, the latter might grow satisfactorily when the former did not. This was formerly given as the chief reason for crop-rotation. While it may be valid in some cases, it is not now regarded as of great fundamental importance.

Advantageous use of preceding crop residues.—After a crop of clover, it is better to grow wheat and then oats rather than the reverse, because the ability of oats to acquire plant-food is greater than that of wheat. If oats were grown first, wheat might be at a disadvantage unless fresh supplies of plant-food were furnished. Crop residues are therefore best utilized by a judicious arrangement of crops.

Economical supply of nitrogen.—In a four- or a five-year rotation system with one application of good farm manure, a leguminous crop will maintain the soil nitrogen practically undiminished under ordinary cropping conditions. Such a system reduces to a minimum the purchase of commercial forms of nitrogen.

Maintenance of organic matter in soil.—When hoed or grain crops are continuously grown, the organic matter of the soil gradually becomes less, because the crop residues are insufficient to take the place of the organic matter that has undergone complete decomposition. Loss of organic matter, if not replaced, sooner or later unfavorably affects the physical, chemical and biological conditions of the soil, with consequent reduction of crop yield. The use of green-crop manures and of farm manure in a rotation provides means for maintaining an abundant supply of organic matter.

Good physical condition of soil.—Each class of crops exercises a specific action upon the mechanical condition of soils. Grass crops tend to make a soil compact; grain and hoed crops tend to keep the soil structure in open condition. Proper rotation gradually alternates these conditions. The frequency and extent of compacting and loosening can be made to vary with the character of the soil. Sandy soils need crops that compact; clay soils need more those crops that lighten.

Advantageous use of farm and commercial plant-foods.—Proper systems of crop-rotation furnish economical results when farm manure is applied to the soil once at least in a rotation of four or five years. As previously suggested, most of the additional nitrogen needed can be supplied by leguminous crops. The balance of crop-feeding being mainly with potassium and phosphorus compounds, little commercial nitrogen will be required. When commercial plant-foods are used, they are usually applied to best advantage on the shallow-rooted crops, the so-called weak feeders.

Occupation of soil by crops.—Crop-rotation makes it possible to keep the land occupied with a crop of some kind most of the time. This is an advantage in preventing the loss of plant-food by leaching. Thus, when grass seed is put in with a grain crop, the grass crop occupies the land when the grain crop is removed. When corn follows grass, the land is kept covered by the grass until plowed for corn. Where long seasons permit, it is possible to grow in rotation more than one crop in a year.

Antidote to soil toxins.—Soils on which crops are grown may contain substances toxic to plants (p. 201), products of decomposition of soil organic matter. Change of crops in some way modifies conditions so that these toxins do not injure the succeeding crop when of a different kind.

Control of injuries from insects, weeds and fungi.—Each crop appears to be associated with certain weeds that grow with it; continuous growing of the same crop favors increase of the weeds characteristic of that crop. By changing from one crop to another different in character, as for example, from small grains to a cultivated crop, the weeds that flourished with the previous crop are often killed easily by cultivation. Similarly, rotation frequently is a great aid to farmers in preventing or lessening destruction caused by insects and plant diseases. Diseases and insects that infest one crop cannot flourish when a crop of another kind takes the place of the first. Where one crop is grown continuously, the propagation of insects and disease germs carries them through from one year

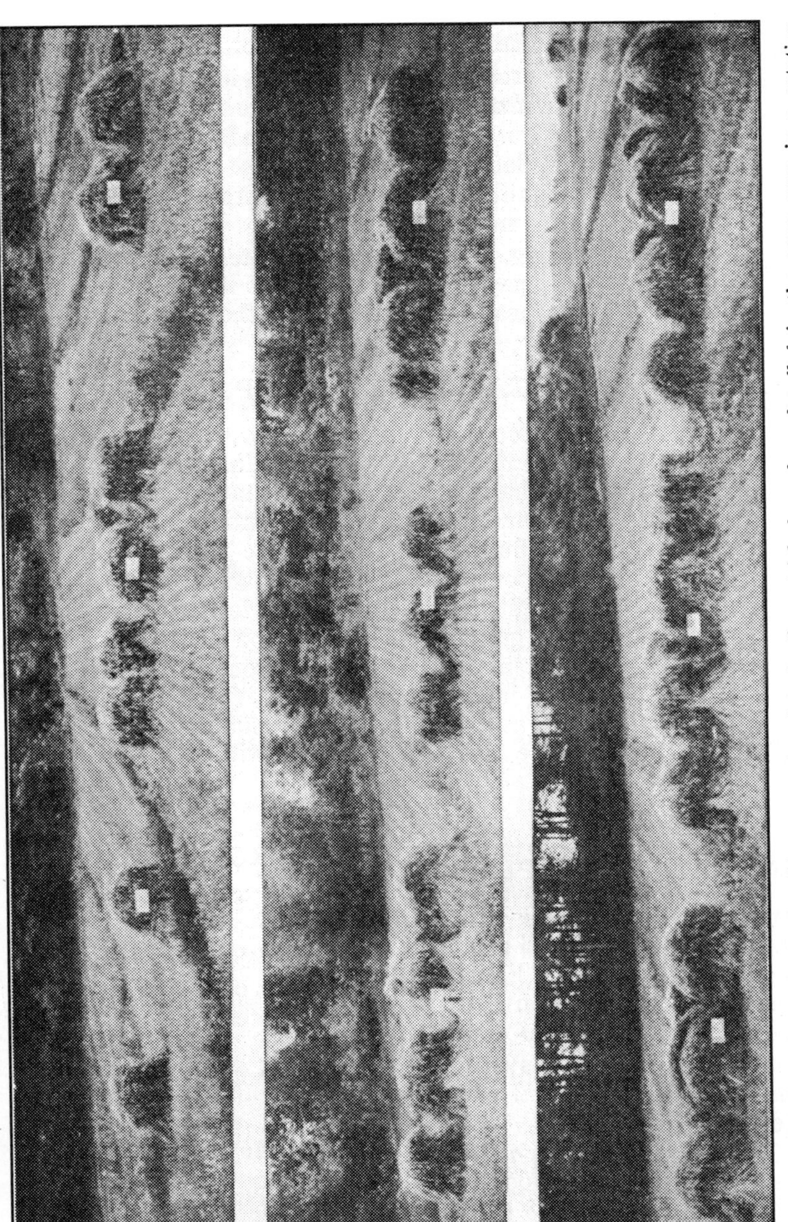

Effect of fertilizer on wheat. For fifty years the plots in these trials have been handled in the same way in a rotation system that has not varied. Top (left) potassium only; (middle) phosphorous and nitrogen; (right) potassium and nitrogen. Center (left) phosphorus and potassium; (middle) nothing; (right) phosphorus, potassium and nitrogen. Bottom (left) complete fertilizer containing forty-eight pounds of nitrogen; (middle) manure, ten tons; (right) complete fertilizer containing seventy-two pounds of nitrogen.

to another. If crops of other kinds occupy the soil three or four years, these destructive enemies disappear more or less completely.

Saving of labor by crop-rotation is shown in various ways. For example, when a grain crop is seeded to grass, the two crops are served by one operation of preparing the land. Crop-rotation means variety of crops in place of one special crop. With a variety of crops, labor is more uniformly distributed through the season instead of coming all at the same time.

Systematizing farming.—When once a system of crop-rotation is established, the management of a farm is much simplified. It means planning years ahead for each field of the farm.

EFFECT OF ROTATION ON CROP YIELD

Judicious use of crop-rotation invariably proves more profitable than the one-crop system. One of the best illustrations known is the experience of the Rothamstead experiment station in England, where experiment on the same field have been continuously carried on for over 60 years. While it has been possible with the use of fertilizers to grow good root crops without rotation, grains like wheat and barley are benefited by rotation but leguminous crops run out completely when grown continuously on the same soil. Wheat on the one-crop system gave a yield of about 12 bushels an acre, while in comparison in a 4-year rotation (turnips, barley, clover, wheat) the yield was over 28 bushels; similar differences were obtained in the case of barley grown with and without rotation.

GENERAL PRINCIPLES OF CROP-ROTATION

From statements made regarding the beneficial effects to be realized by well-planned rotations, we can derive a few general principles, as follows: (1) Essentials of a rotation system, (2) conditions that affect rotation systems, (3) order of crops in rotation, and (4) examples of successful rotation systems.

Essentials of rotation system.—In planning rotations under usual conditions, four general classes of crops should be provided for, as follows:

(1) *One crop to be sold, or money crop,* usually removing considerable amounts of plant-food.

(2) *One leguminous crop* to add nitrogen, collect subsoil phosphorus and potassium and furnish organic matter.

(3) *One crop to feed farm animals,* the plant-food of which is largely restored to the soil in manure.

(4) *One cultivated or hoed crop* to destroy weeds and improve physical condition.

In illustration of these statements, we can use the familiar three-year rotation of potatoes, wheat and mixed clover and timothy. Potatoes and wheat furnish two money crops; potatoes, the cultivated crop; wheat-straw and hay, two feeding-crops; clover, the crop for supplying nitrogen, etc.

Conditions that affect rotation systems.—In addition to other requirements a system of rotation must obviously be adapted to special conditions among which are:

(1) *Climate* definitely prescribes the particular kinds of crops that can be selected for a given locality.

(2) *Soil.*—Attention has already been called to the fact that certain crops give better results on particular types of soil.

(3) *Kind of farming.*—The crops in rotation systems will vary according to whether the farming is general or special and according to each specialty. Different details are demanded, for example, by dairying, stock-raising, grain-raising, cotton-growing, fruit-growing, etc.

(4) *Commercial conditions.*—One is limited also, in addition, to crops that can be sold readily and at profitable prices. Accessibility to market, involving transportation facilities, must also be taken into consideration.

(5) *Condition of soil* in relation to weeds, injurious insects and plant diseases must be considered.

Order of crops in rotation system.—No strict arrangement of specific crops in rotation can be prescribed, but the facts presented already furnish some helpful suggestions; for instance:

(1) Shallow-rooted and deep-rooted crops should be alternated.

(2) Crops which furnish organic matter should be alternated with those which favor its rapid decomposition.

(3) There should be in every rotation system not less than one leguminous crop in order to increase the supply of plant-food in the soil.

(4) Crops in rotation should vary as much as possible in such respects as the amounts and kinds of plant-food requirements, character of root growth, and the time of year during which they cover the soil.

(5) In connection with the preceding statements, we may add that whenever commercial plant-foods are used, they should be applied to the special crop that will be most benefited by them, usually the shallow-rooted, weak-feeding crop, such as wheat.

EXAMPLES OF SUCCESSFUL ROTATION SYSTEMS

Well-known systems in extensive use are the following crops in rotations:

(1) *Three-year rotation in potato-growing:* (1) Potatoes, (2) wheat or oats, (3) clover and timothy.

(2) *Four-year rotation in corn-growing:* (1) Corn, (2) corn, (3) oats or wheat, (4) clover and timothy.

(3) *Five-year rotation for general farming:* (1) Corn, (2) oats, (3) wheat, (4) clover and timothy, (5) clover and timothy.

(4) *Seven or eight-year rotation where alfalfa is grown:* (1) Alfalfa 4 years, (2) corn 2 years, (3) wheat or oats 1 or 2 years.

(5) *Three-year rotation in cotton-growing:* (1) Corn accompanied by cowpeas planted between rows or sown broadcast when last cultivated, (2) oats followed by cowpeas the same season, (3) cotton.

(6) *Five-year rotation in dairy-farming:* (1) Corn, (2) corn, (3) oats seeded with clover and timothy, (4) hay, (5) pasture.

PART V

PRACTICAL USE OF FERTILIZERS IN GROWING INDIVIDUAL CROPS

XXVII

PRELIMINARY CONSIDERATIONS ABOUT THE PRACTICAL USE OF FERTILIZERS

Some of the fundamental conditions that underlie successful crop growing are suggested by the following topics:

(1) Factors that affect the efficiency of fertilizers.

(2) Small and large applications of plant-food.

(3) Relative amounts of plant-food constituents in fertilizers.

(4) Limitations of usefulness of fertilizer formulas.

(5) Plant-food formulas for crops.

(6) Methods and times of applying fertilizers.

FACTORS AFFECTING THE EFFICIENCY OF FERTILIZERS

In order that plants may make efficient use of plant-food and grow satisfactorily, the following conditions are essential:

(1) Good seed.

(2) Sunshine (light and warmth).

(3) Adaptation of crops to special conditions of soil and climate.

(4) Sufficient rainfall, properly distributed.

(5) Suitable physical structure of soil in respect to (*a*) mellowness and firmness, (*b*) receiving, distributing, holding and giving up water, (*c*) circulation of air, (*d*) absorbing and holding heat.

(6) Beneficial micro-organisms in the soil and conditions favorable to their growth.

(7) Absence of substances injurious to plants, especially (*a*) acid compounds for most cultivated crops and (*b*) alkalis in soils where blueberries and other acid-requiring plants are to be grown.

(8) Abundance of available plant-food within reach of growing roots.

Most of these conditions are under control. The most effective and economical controls are:

(*a*) Seed selection with reference to variety, vitality, purity, and freedom from disease.

(*b*) Adaptation of crops to soil and climate.

(*c*) Where irrigation is practicable, moisture supply is under control.

(*d*) The physical structure of the soil is controlled, according to special conditions, by (1) drainage, (p. 108), (2) thorough tillage of proper character and at suitable times (p. 106), (3) proper rotation of crops (pp. 336-342), (4) periodical incorporation of organic matter (pp. 256-265), (5) maintaining supply of calcium carbonate in soil (p. 195).

(*e*) The presence in soils of beneficial micro-organisms (nitrifying, ammonifying, nitrogen-gathering) is secured when necessary by inoculation through treatment with farm manure, use of soil, or special preparations, according to circumstances (pp. 150, 151). Conditions favorable to maintain desirable organisms in soils (pp. 89-100) are provided by keeping the physical structure in good condition.

Assuming that the preceding conditions are satisfactory, crops are able to make efficient use of plant-food within reach. Unfavorable conditions in any one respect may prevent a crop from utilizing the plant-food ready for use. When climate and the physical conditions of soil are suitable for crop growth, the primary problem of crop production is crop-feeding. Sufficient supply of available plant-food to meet crop demands is maintained most effectively: (1) By the careful saving, management and use of farm animal excrements, liquid and solid, and organic farm refuse (pp. 215-254); (2) desirable rotation of crops (pp. 336-342); (3) use of leguminous crops or crop residues as green-manure to add nitrogen and organic matter to the soil as well as change insoluble plant-food materials into available form (pp. 256-265); (4) maintaining in the soil a sufficient supply of calcium carbonate (pp. 205-210); (5) the use of plant-food materials bought in such amounts and forms as to supplement most economically local needs.

SMALL AND LARGE APPLICATIONS OF PLANT-FOODS

In practice farmers use fertilizers in amounts varying from 100 to 3,000 pounds an acre and in some cases even more. While the usual purpose in using fertilizers is to supplement the soil's supply of plant-food, there are two points of view (1) helping crops to start, and (2) helping them during growth.

Fertilizers to start crops.—It is well-known that amounts of plant-food materials as small as 100 pounds an acre, properly placed, often have a marked effect in starting crops. The explanation commonly accepted is that in temperate climates. the available plant-food, especially of nitrate nitrogen, is at its lowest on cropped soils early in spring, and that it may be insufficient to give the new crop what it needs for a quick start; therefore, to help the young crop until the soil can furnish new supplies, a small application of quickly available plant-food is advisable. It must also be kept in mind that the roots of young plants have a limited reach and that they best utilize the most soluble forms of plant-food. Therefore, the application of comparatively small amounts of easily soluble plant-food materials at this time, just where the young rootlets can quickly reach it, furnishes nutrition when most needed.

The oat crop illustrates the foregoing explanation. Soon after germination, there is usually a week or ten days when growth is at an apparent standstill, known to farmers as the "pouting" time. Generous application of nitrate nitrogen and soluble phosphates usually shortens or prevents this behavior, which suggests the absence of soluble plant-food within reach of the rootlets. After the feeding-roots have gone deeper and the various agencies have begun actively to furnish available plant-food, the soil's supply is expected to meet the crop demands during the remainder of the season.

Fertilizers to help crops continuously.—In some cases, farmers practically ignoring the supply contained in the soil, apply as much available plant-foods as a large crop will use, or even in considerable excess of this, in order to make sure of enough. This is a still more popular practice in the case of market-garden and greenhouse crops to insure not only large crops, but often to hasten maturity in order to get the advantage of high prices early in the season. This system, however, is not profitable in the case of ordinary farm crops.

RELATIVE AMOUNTS OF PLANT-FOOD CONSTITUENTS IN FERTILIZERS

In using fertilizer for crop production, compounds that contain available nitrogen, phosphoric acid and potash, must be in sufficient supply and, further, these constituents must be present in relative proportions that are not too wide. In discussing the functions of plant-food constituents (p. 84) attention was called to the fact that, while each one exerts on the growth of plants special effects peculiar to itself, these constituents are mutually interdependent, balancing, supplementing and limiting one another. Excess or deficiency of any one may interfere with the ability of the plant to make normal use of one or more of the others. In the preparation and use of fertilizers, the relative proportions or balance of nitrogen, phosphoric acid and potash are, therefore, important, as they affect the results and the consequent economy of their use.

It is of interest to know some of the methods used to determine the proportions of these plant-foods in making fertilizers for various crops and to discuss the significance of such variations. While chemical analysis furnishes helpful suggestions, it is not a satisfactory or practical basis when used without reference to various other conditions. Another principle, still influential, was to make the application of fertilizers correspond to the chemical composition of the plant (p. 281). Much emphasis has been placed by some investigators upon the two phrases, "feeding the soil" and "feeding the crop," meaning the use of fertilizers on the basis of soil analysis and of plant analysis. Some fertilizer manufacturers claim that their analyses are based upon feeding the plant and not the soil. A knowledge of the amounts of plant food material used by various crops is desirable, useful and necessary in studying the needs of crops, but, for reasons already pointed out, this knowledge is insufficient to serve by itself as a basis to enable us to determine the relative amounts of plant-food constituents to use.

The present use of plant-food materials, in relation to the amounts applied, is based, (1) in part, upon the amounts of plant-food removed by crops, (2) in part, upon the character of soil, and, (3) in larger measure, upon laboratory investigations and actual field tests first carried out by agricultural experiment stations and then applied in successful farm practice by progressive farmers.

If we compare the proportions of nitrogen, phosphorus and potassium in most fertilizers with the proportions of these constituents in crops to which applied, we notice that the nitrogen is usually much less, the phosphorus much more, than the proportions called for by the composition of the crops. For example, phosphorus is used by crops in smaller amounts than either nitrogen or potassium and yet, in nearly all fertilizer analyses, we find that it is generally in considerable excess of the other two constituents. In departing from the proportions of constituents found in plants, certain facts are taken into consideration, among them the following:

(1) A smaller amount of nitrogen is applied in fertilizers than the crop uses, on the assumption that the soil will furnish considerable, especially as the growing season advances and conditions become favorable to decomposition and nitrification processes. This assumption, to be safe, should be based upon actual knowledge of previous soil behavior and treatment. The use of farm manure and leguminous green-manure, for example, justifies the assumption. Generally speaking, clay soils contain more nitrogen than sandy soils, while muck or peaty soils, though rich in nitrogen, do not furnish any appreciable amount of quickly available nitrogen

until they have been tilled and, in many cases, limed. Soils that lack organic matter usually need nitrogen.

(2) The amount of phosphorus in a fertilizer is usually increased much above the proportion called for by plants, especially in the case of cereals, because most soils contain phosphorus in relatively small amounts and much of it generally in forms not quickly available except as the result of treatment with organic matter, lime, etc. The fact that a soluble phosphate reverts in most soils before it is used by crops probably has had something to do with this use of relatively large amounts of phosphorus.

(3) The proportions of potassium are made to vary with the character of the crop as well as with the character of the soil. On most clay soils, when kept well supplied with calcium carbonate and organic matter, or when treated generously with sodium or calcium nitrate, little potassium need be used in fertilizers. Larger amounts are called for by leafy crops, root crops and some others than by cereals. Light, sandy soils are often deficient in available potassium, while muck soils are generally found to respond favorably to generous applications.

From the preceding discussion, it can be readily appreciated that making a properly balanced fertilizer, best adapted to soil and crop is a matter of complexity. The same fertilizer may not be equally useful on different farms even for the same crop.

LIMITATIONS OF USEFULNESS OF FERTILIZER MIXTURES

When we come to designating a definite analysis or a specific mixture of plant-food materials for a particular crop, we want to make it clearly understood now, if we have not succeeded in doing so before (p. 275), that no particular analysis or mixture of plant-foods can be prescribed for any one crop under all conditions or for every crop on a given soil. In the very nature of the case the question of quantities and proportions of plant-foods to be used must always remain more or less a matter of individual experience and observation. Moreover, it must be remembered that one's practice cannot remain continuously the same. The soil of each field is undergoing constant change in the extent and character of its needs. For example, the fact that the application of superphosphate gives large crop yields for one year does not mean that this treatment will continue to give the same results indefinitely. Sooner or later crop yields will drop, unless attention is given to the supply of nitrogen or potassium or calcium, or organic matter, or all of these, or, perhaps, some other constituent like sulfur or manganese.

The chief use of a fertilizer with a specific analysis is as a suggestion or starting-point. Beginning with a suggested analysis, the progressive farmer will constantly study results and let the experience of one year suggest to him his plan for the year following. It is unwise to be permanently attached to any one combination of plant-foods, unless one has proved by long-continued experiment that it gives the best possible results under conditions that prevail on the individual farm or field for particular crops. In making variations in the amounts and forms of plant-foods to be used, home-

On acid land limestone makes a difference between a good crop of sweet clover and no crop. When this forty acre field was limed a strip was left unlimed as a demonstration.

mixing becomes important. The disadvantage of being wedded to a single mixture of plant-foods is well illustrated in the case of a certain market-garden crop raising community. The farmers had all come to use one specific plant-food mixture, until carefully conducted experiments proved that they were using twice as much potassium as they needed and in many cases were applying the mixture in amounts twice as large as was consistent with the best profit.

STANDARDIZATION OF ANALYSES

The number of fertilizer mixtures that contain nitrogen, phosphorus and potassium compounds formerly aggregated several thousand. In the interest of both consumers and manufacturers it was found that a large proportion of these mixtures could be grouped under a comparatively small number of analyses, systematized and standardized. As the result of several regional conferences begun in 1922 and continued at intervals since, agronomists, fertilizer-control chemists, representatives of the fertilizer industry and others interested agreed that specific analyses, limited in number, be selected as standard. These standard analyses are simply advisory and have varied greatly from time to time. Farmers have been urged to use them and manufacturers have been requested to co-operate as far as practicable in furnishing them. In several individual states or groups of states standard analyses have been adopted for such recommendation, suitable for the crops and conditions in each. The movement is still in a transitional stage but it is probable that, ultimately, in the states using commercial fertilizers, individual states will settle upon a limited number of analyses, found by experience as best adapted to the crops and conditions of each.

PLANT-FOOD RATIOS IN FERTILIZERS

In connection with the adoption of definite fertilizer analyses, it was found desirable, for the sake of uniformity, to suggest a series of plant-food ratios, on which could be based the various analyses adopted. For example, a ratio indicated as "1—2—1" means 1 part of nitrogen, two parts of available phosphoric acid (P_2O_5) and 1 part of water-soluble potash (K_2O). On the basis of this ratio, we could have such fertilizer analyses as 3—6—3, 4—8—4, 5—10—5, 6—12—6, etc. On the basis of other ratios, such as 1—1—2, 1—2—3, 1—3—2, 2—3—1, etc., a countless number of ratios and analyses is possible. Therefore, in order to have some definite system by which to work out ratios and analyses and avoid having an excessive number, the "triangle" system has been suggested as furnishing a simple and practical basis. Without going into details, it is sufficient to state that 63 different ratios are possible, as a basis for preparing fertilizer analyses, as shown in Table 45.

In Table 45 each figure in all the ratios is a whole number and the sum of the constituent figures of each ratio equals 10.

It is not proposed that all of these ratios be used but that from them can be selected all that are desired as a basis for fertilizer mixtures needed in actual farm practice. The analyses of very few commercial fertilizer mixtures in use correspond to any of these ratios except in the case of some high-analysis mixtures, which have recently come into market, but some modifications can bring many of the present mixtures under the triangle system.

Regional conferences of Central Western States and of New England have selected fertilizer ratios based on the triangle method, which include the following: 1—3—6, 1—5—4, 1—6—3, 1—7—2, 2—4—4, 2—3—5, 2—5—3, 2—6—2, 3—4—3, 4—2—4, 4—3—3, 5—3—2, 2—8—0, 0—2—8, 0—5—5, 0—6—4, 0—7—3. The analyses based on these ratios contain 20 per cent. or more of nitrogen, phosphoric acid and potash, taken together. For example, the lowest analysis based on the 1—3—6 ratio is 2—6—12; on the 2—4—4 ratio, 4—8—8; on the 4—2—4 analysis, 8—4—8, etc. In further illustration, we could have the following analyses based on the 2—6—2 ratio: 3—9—3; 4—12—4, 5—15—5, 6—18—6, etc.

TABLE 45—POSSIBLE FERTILIZER RATIOS BY TRIANGLE SYSTEM

$N—P_2O_5$	$N—K_2O$	$P_2O_5—K_2O$	$N—P_2O_5—K_2O$			
1-9-0	1-0-9	0-1-9	1-1-8	2-2-6	3-4-3	5-2-3
2-8-0	2-0-8	0-2-8	1-2-7	2-3-5	3-5-2	5-3-2
3-7-0	3-0-7	0-3-7	1-3-6	2-4-4	3-6-1	5-4-1
4-6-0	4-0-6	0-4-6	1-4-5	2-5-3	4-1-5	6-1-3
5-5-0	5-0-5	0-5-5	1-5-4	2-6-2	4-2-4	6-2-2
6-4-0	6-0-4	0-6-4	1-6-3	2-7-1	4-3-3	6-3-1
7-3-0	7-0-3	0-7-3	1-7-2	3-1-6	4-4-2	7-1-2
8-2-0	8-0-2	0-8-2	1-8-1	3-2-5	4-5-1	7-2-1
9-1-0	9-0-1	0-9-1	2-1-7	3-3-4	5-1-4	8-1-1

METHODS AND TIMES OF APPLYING FERTILIZERS

While this subject will be considered in more detail in connection with individual crops in succeeding chapters, it is desirable to present here some of the main points of general importance.

The efficiency of the plant-food in a fertilizer is lost for the season's growing crop in so far as it does not come within the feeding range of the plant roots. The soluble constituents of fertilizers diffuse through the soil mostly up and down, slightly in a lateral direction, so the method of application is best which distributes the plant-food materials most thoroughly and uniformly throughout the portion of soil in which the feeding-roots forage. The various phases of the subject are receiving much experimental attention from agronomists, manufacturers of fertilizer-distributing machinery and producers of fertilizers. The points to be noticed here are: (1) Special machinery, (2) placement of fertilizers, (3) precautions in applying concentrated fertilizers, (4) application of soluble materials, (5) distribution of insoluble fertilizers, (6) time of application.

Special machinery for distributing fertilizers.—The importance of distributing fertilizers in the most efficient manner has been fully appreciated. Extended studies have been made, covering the various conditions that affect distribution by machinery. The drillability of fertilizers is greatly affected by such conditions as (1) the degree of moisture of the air, varying with the amount of material in the fertilizer that absorbs moisture easily; (2) the size, shape and uniformity of the particles of the fertilizer material. The evenness of distribution varies with the kind of distributor and the drillability of the fertilizer. Implement manufacturers have devised desirable forms of drills and planters provided with special attachments for efficient distribution, and several kinds of machines have been designed to distribute fertilizers either in the rows, beside the rows, or broadcast uniformly over the entire surface, while some forms are adjustable to make use of all three methods. Wagons are made for the distribution of farm manure (p. 253) some of them provided also with arrangements for distributing fertilizers and lime.

Placement of fertilizers.—In the placement or application of fertilizers, four methods are employed, depending on the kind of crop and fertilizer: (1) Broadcast, during the preparation of the soil before planting; (2) in the seed-row, before or at the time of planting; (3) top-dressing or side-dressing, at some time during the growing season; (4) in solution. For crops whose root systems extend sideways in every direction and occupy a horizontal layer of soil, broadcasting after plowing followed by harrowing before planting, is desirable. This applies, for example, to fertilizers used for cereal crops, orchards, green-manure crops, forage crops, meadows and pastures.

In the case of crops whose root systems extend downward rather than horizontally and which are grown in hills or rows, distribution in or near the plants is usually preferred, as for potatoes, roots, crops, many garden crops, etc. In applying fertilizers in the seed-row, the fertilizer should be spread in a band six inches wide or more, not in a narrow line; and, under no circumstances, should there be direct contact of the fertilizer with the seed, especially in case of concentrated fertilizers. Direct contact of any soluble fertilizer constituent with seed is sure to prevent germination or burn the sensitive rootlets in the first stage of growth. To prevent such injury and yet to place the fertilizer where it can be reached by the rootlets, it can be placed a little below the level, on one or both sides, of the seed, one to three inches from the seed.

Top-dressings or side-dressings with fertilizers are applied to advantage at stages of growth which differ with various crops.

Application of fertilizers in solution is used in some cases. In the intensive growing of garden crops, overhead irrigation is employed in some places and the fertilizer is applied in the irrigation water. The soluble fertilizer, such as nitrate, ammonium salts, urea, soluble phosphates, potash salts, etc., is placed in a bag, suspended in the tank of irrigation water. As much as 200 pounds of fertilizer per acre can be effectively applied by this method. After the application, pure water is run on the leaves of the growing plants to wash off the solution and prevent possible injury from drying of the solution on the leaves.

In transplanting such plants as cabbage, celery, strawberry, sweet potato, tomato, tobacco, etc., soluble fertilizer can be applied in the water used on the plants at the time of resetting. The use of a nitrate (sodium or ammonium), or urea has been found helpful to stimulate growth—one or two ounces of the plant-food material dissolved in four to eight gallons of water.

Precautions in applying concentrated fertilizers.—Special attention should be called to certain dangers that may result from careless use of concentrated fertilizer materials and mixed fertilizers. This has increased with the concentration of plant-food constituents. Formerly, the plant-food materials in a mixed fertilizer rarely exceeded 18 to 20 per cent. At present the total often exceeds 20 per cent. but in some cases as high as 60 per cent.

When a concentrated or high-grade fertilizer or any plant-food material likely to burn vegetable tissue is applied, it is essential to keep from coming into direct contact in the soil with the seed. If, for example, a small lump of such a fertilizer should be left near seed in the soil, a strong solution of the fertilizer will soon form and reaching the seed, be likely to prevent germination. In a similar way, if a young rootlet pushes into such a concentrated solution, injury follows. It is obvious that the proper way to avoid such trouble is to see that the fertilizer is in a drillable condition, that it is uniformly distributed through the soil and that it is not applied in so large amounts as to make meeting these conditions difficult.

If one has not the machinery for distributing concentrated fertilizers uniformly, a good way of handling them before application is to *dilute them by mixing thoroughly equal parts of the fertilizer and some inert substance* like fine, dry muck, peat, sand, sifted coal ashes, etc. Such a mixture can be distributed more uniformly and with much less risk of injuring seeds or roots than in concentrated form. This method of dilution is especially desirable when one wishes to make only small applications. For example, it is difficult to distribute uniformly 75 pounds of sodium nitrate in dry form over an acre. Dilute it by thorough mixing with 200 or

300 pounds of fine, dry inert material, and its uniform distribution ceases to be difficult.

In top-dressing partly grown crops with unmixed plant-food materials or with concentrated mixed fertilizers, it is highly desirable that the materials be diluted with inert matter. It is also essential that these surface applications be made not too near the growing plant and under no circumstances touching any part of the plant. Those who have not had experience in handling these active compounds have often scattered the material over the ground near the plant, after which a heavy shower has come, dissolved it suddenly, spattered the concentrated solution over the plant, caused more or less serious burning wherever plant tissue is touched and usually caused the death of the plant. It is always desirable to work a top-dressing of such strong materials lightly into the soil; this will prevent the possibility of the destructive effects caused by heavy rain. Generally speaking, it is safer to make two or three light top-dressings than one heavy application. It is generally not desirable to apply sodium nitrate and similar materials alone as a top-dressing in amounts exceeding 100 pounds an acre at one time on most crops.

Distribution of soluble fertilizers.—Readily soluble materials can be distributed over the surface and lightly worked in. When the first rainfall comes, they are distributed throughout the soil completely and uniformly.

Distribution of insoluble fertilizers.—Materials largely insoluble are preferably well mixed through the soil. Thus, dried blood, bone-meal, tankage, fish-scrap and other organic materials of similar character are best placed at a sufficient depth in the soil to promote decomposition and nitrification. If too deep decomposition may be slow or the nitrate nitrogen, when formed, may be leached and carried beyond the reach of the roots.

Time of application.—Readily soluble plant-food materials which are not reasonably well retained by the soil, are best applied only when the crop is ready to utilize them. If put on too long before crop growth is well started, there is danger, especially on light, open soils, of their being carried more or less beyond the range of the crop roots. Nitrate nitrogen, in case of light soils, is to be treated with special care in this respect. Ammonia is liable to a similar loss, as soon as converted into nitrate. Nitrates are especially useful for early spring applications in case of crops that need an early, prompt start. Care should be taken not to apply them too late in the season in case of most crops, since the maturing of the crop will be retarded and there will be an excessive growth of stems and leaves.

Fertilizers which do not dissolve readily are preferably applied early in the season and worked into the soil so they can undergo decomposition. It is important that nitrogen-containing materials be controlled so they shall exercise their maximum effect before the maturing season of the crop arrives; otherwise, the season of growth may be prolonged so late in the fall as to result in injury to the crop either in yield or quality or both. In case of trees, liability to injury by severe cold is increased.

TABLE 46—FACTORS FOR CALCULATING AMOUNTS OF A PLANT-FOOD CONSTITUENT IN DIFFERENT FORMS

1. In compounds containing NITROGEN

To find from a given amount of Ammonia (NH_3) the equivalent amount of:	*Multiply by*
Nitrogen	0.824
Ammonium nitrate	2.35
Ammonium sulfate	3.88
Calcium cyanamid	2.35
Calcium nitrate	4.82
Potassium nitrate	5.94
Sodium nitrate	5.00
Urea	1.76
From Ammonium nitrate *($NH\ NO_3$)*	
Ammonia	0.425
Ammonium sulfate	1.65
Calcium cyanamid	1.00
Calcium nitrate	2.05
Nitrogen	0.175
Potassium nitrate	1.26
Sodium nitrate	1.06
Urea	0.75
From Ammonium sulfate *($(NH_4)_2SO_4$)*	
Ammonia	0.257
Ammonium nitrate	0.61
Calcium cyanamid	0.61
Calcium nitrate	1.24
Nitrogen	0.21
Potassium nitrate	0.76
Sodium nitrate	0.64
Urea	0.45
From Calcium cyanamid *($CaCN_2$)*	
Ammonia	0.425
Ammonium nitrate	1.00
Ammonium sulfate	1.65
Calcium nitrate	2.05
Nitrogen	0.35
Potassium nitrate	1.26
Sodium nitrate	1.06
Urea	0.75

To find from a given amount of Calcium nitrate ($Ca(NO_3)_2$) the equivalent amount of:	*Multiply by*
Ammonia	0.21
Ammonium nitrate	0.50
Ammonium sulfate	0.80
Calcium cyanamid	0.50
Nitrogen	0.17
Potassium nitrate	0.62
Sodium nitrate	0.52
Urea	0.37
From Nitrogen *(N)*	
Ammonia	1.214
Ammonium nitrate	2.86
Ammonium sulfate	4.71
Calcium cyanamid	2.86
Calcium nitrate	5.86
Potassium nitrate	7.21
Sodium nitrate	6.07
Urea	2.14
From Sodium nitrate *($NaNO_3$)*	
Ammonia	0.20
Ammonium nitrate	0.94
Ammonium sulfate	1.55
Calcium cyanamid	0.94
Calcium nitrate	1.93
Nitrogen	0.165
Potassium nitrate	1.19
Urea	0.70
From Urea *(CON_2H_4)*	
Ammonia	0.567
Ammonium nitrate	1.33
Ammonium sulfate	2.20
Calcium cyanamid	1.33
Calcium nitrate	2.73
Nitrogen	0.467
Potassium nitrate	1.68
Sodium nitrate	1.42

TABLE 46—*Continued*

2. In compounds containing PHOSPHORUS (phosphoric acid)

To find from a given amount of Phosphorus (P) the equivalent amount of:	*Multiply by*
Phosphoric acid	2.29
Mono-calcium phosphate	3.77
Tri-calcium phosphate, bone phosphate of lime (B.P.L.)	5.00

From Phosphoric acid (P_2O_5)	
Phosphorus	0.437
Mono-calcium phosphate	1.65
Tri-calcium phosphate	2.18

To find from a given amount of Mono-calcium phosphate ($CaH_4(PO_4)_2$) the equivalent amount of:	*Multiply by*
Phosphorus	0.265
Phosphoric acid	0.61
Tri-calcium phosphate (B.P.L.)	1.32

From Tri-calcium phosphate ($Ca_3(PO_4)_2$) (Bone phosphate of lime, B.P.L.)	
Phosphorus	0.20
Phosphoric acid	0.46
Mono-calcium phosphate	0.76

3. In compounds containing POTASSIUM (potash)

To find from a given amount of Potassium (K) the equivalent amount of:	*Multiply by*
Potash	1.205
Potassium carbonate	1.77
Potassium chloride	1.90
Potassium nitrate	2.59
Potassium sulfate	2.23

From Potash (K_2O)	
Potassium	0.83
Potassium carbonate	1.47
Potassium chloride	1.59
Potassium nitrate	2.15
Potassium sulfate	1.85

From Potassium chloride (KCl) (Muriate)	
Potash	0.63
Potassium	0.524
Potassium carbonate	0.926
Potassium nitrate	1.355
Potassium sulfate	1.17

To find from a given amount of Potassium nitrate (KNO_3) the equivalent amount of:	*Multiply by*
Potash	0.465
Potassium	0.386
Potassium carbonate	0.683
Potassium chloride	0.738
Potassium sulfate	0.86

From Potassium sulfate (K_2SO_4)	
Potash	0.54
Potassium	0.448
Potassium carbonate	0.793
Potassium chloride	0.855
Potassium nitrate	1.16

Every farmer should regard every fertilizer as an experiment or, perhaps, rather as the beginning of a series of experiments, which will sooner or later lead him to make the most economical use of plant-food materials that his conditions will permit.

All things considered, the ideal system is to apply plant-foods not as mixtures but as separate materials. In this system the farmer applies nitrogen, phosphorus and potassium separately, usually in different seasons, in accordance with the results of observations and experience. The successful application of such a system demands a close study, not only of crops, but of the science of plant-feeding. It has the advantage of enabling the farmer to make the most economical use of all plant-food materials in crop-growing.

XXVIII

LEGUMINOUS CROPS

PLANTS botanically classed as the *Leguminosæ,* commonly called the bean or pulse family, are numerous in kind and widely distributed. Those of agricultural interest are clovers, beans, peas, lentils and vetches. For several reasons, they are of the highest value and possess characteristics which distinguish them from all other crops.

GENERAL CHARACTERISTICS

The more important points of general interest relating to leguminous crops are considered under the following divisions: (1) Relation to nitrogen, (2) relation to subsoil, (3) uses, (4) adaptation, (5) agricultural varieties.

Relation to nitrogen.—These plants possess the peculiar power or ability to make use of atmospheric nitrogen and thus to add to the soil's supply of this plant-food (pp. 146-152). By the adoption of some leguminous crop as part of a rotation, it is easily possible to reduce appreciably the amount of costly plant-food nitrogen that one needs to purchase. However, it is essential that the soil be provided with the proper nodule-forming organisms (pp. 146-152).

Relation to subsoil.—Many legumes are deep-rooted so serve the important purpose of collecting phosphorus and potassium compounds, which are carried into the upper portions of the plants and given to the upper layers of the soil by decomposition.

Various uses.—Leguminous crops, more than any others, are devoted to a great variety of efficient uses in agriculture: (1) They are the most valuable green-crop manures known; (2) they are among the most useful forage crops; (3) many of them furnish hay of the highest nutrition; and (4) some of them are useful as pasturage and in lawns.

Adaptation.—Some species can be grown under a wide range of soil and climatic conditions; others are much more limited. In the Northern United States, the species most commonly grown are clovers, field peas, alfalfa and common vetch; in regions having a mild to warm climate, cowpeas and crimson clover are extensively grown; in many parts of the West, those most grown are alfalfa, clovers, cowpeas and soy-beans.

Agricultural varieties.—The number of leguminous plant varieties grown is large, though in many cases only they are used in a limited way; the varieties best known and most extensively used are the following: Clovers (red, mammoth red, alsike, crimson, white,), alfalfa, field-peas, cowpeas, soy-beans, velvet-beans, peanuts and vetch.

SOIL CONDITIONS

Under this head, we will consider: (1) Calcium compounds, (2) organic matter, (3) kinds of soils, (4) preparation.

Calcium compounds.—Leguminous crops do not thrive on soils appreciably acid (p. 149). Therefore, to grow them, the soil should be well supplied with calcium (lime) carbonate (pp. 205-211), applied with the crop or preferably the season previous. They are also relatively large consumers of calcium as plant-food.

Organic matter.—A generous supply of decomposing organic matter is also a requisite for the most abundant yield.

Kinds of soil.—The kinds of soil on which legumes grow best vary widely. Some thrive only on one kind, others grow well on various types. In general, they require well-drained soils, but mammoth-red, alsike and white clovers grow well in lands too moist for other clovers. Deep-rooted crops, like red and mammoth red clovers, alfalfa, sweet clover, cowpeas, soy-beans, do best only when the subsoil is free from hardpan and fairly open.

Preparation of soil.—In general, much care is required in preparing the soil. For the small-seeded crops, the soil should be in fine tilth; so it is a good plan, when practicable, to let them follow a hoed crop. This also has the advantage of lessening the growth of weeds, which often give serious trouble in such newly-seeded crops as alfalfa and soy-beans.

TABULATED DATA ABOUT LEGUMINOUS CROPS

In Table 47 are data as a convenience for quick reference. More detailed specifications are given when needed in connection with a discussion of each special crop.

While some species are perennial, they run out sooner or later when left undisturbed as in meadow or pasture, the time depending upon character of the soil, frequency of cutting, amount and kind of fertilizers applied, etc.

The amount of seed used varies greatly; less is used when sown in rows than when broadcasted or when sown with some other crop. Generally, larger amounts of seed are used for green-crop manures

TABLE 47—DATA RELATING TO LEGUMINOUS CROPS

Variety	Soil	Root system	Duration	Amount of seed per acre: Green manure	Amount of seed per acre: Forage, etc.	Usual season of planting	Depth of planting
			Years				Inches
Clover, red	Well-drained clay to light	Deep	2-3	12-15 lbs.	8-12 lbs.	Early spring and late summer	½-1
" mammoth red	Moist clays and loams	Deep	3-4	15-20 "	12-15 "	Early spring and late summer	1-2
" alsike	Moist clays and loams	Not deep	3-5	12-15 "	8-15 "	Early spring	½-1
" crimson	Light	Not deep	1	15-20 "	12-15 "	June-October	1-2
" white	Moist clays and loams	Not deep	4-5	12-15 "	10-12 "	Early spring	½-1
" sweet	Clay to light	Deep	2	2-4 pks.	10-12 "	March-September	½-1
Alfalfa	Well-drained clay to light	Very deep	4-10	25-35 lbs.	20-25 "	Early spring or late summer	1-1½
Field-pea	Moist clays and loams	Not deep	1	2-3 bu.	2-3 bu.	Early spring	2-4
Cowpea	Generally on soils not too wet	Deep	1	1½-2 "	1-1½ "	May-June	1-2
Soy-bean	Clays, loams, etc.	Deep	1	1-1½ "	2-3 pks.	May-June	1-2
Velvet-bean	Light, sandy	Deep	1	1-4 pks.	1-4 "	Spring	1-2
Vetch	Clays, loams, etc.	Not deep	1	1-1½ bu.	1-1½ bu.	May-August	1-2

than for forage or hay and less when the crop is grown for seed. More seed is needed on soils in poor tilth or lacking in available plant-food, than on fertile soils.

In respect to depth of planting, it should be remembered, in general, that small seeds need only slight covering; larger seeds usually deeper.

The time of planting is often governed by the special purpose in raising the crop, whether as green manure, green forage, seed, hay, etc.

SUGGESTIONS ON USE OF FERTILIZERS

Leguminous crops require abundant supplies of phosphorus, potassium and calcium compounds in order to utilize atmospheric nitrogen. It has been stated that no plant-food nitrogen should be

Alfalfa field prepared for seeding by corrugated roller.

furnished, because, when available nitrogen is at hand in the soil, these plants will utilize this in preference to atmospheric nitrogen. While this is true with reference to the main period of growth, it is usually desirable in case of most soils, and always in case of poor soils, to furnish some available nitrogen, such as nitrate, in order to give the young plants a prompt and vigorous start. Application of nitrogen is essential on soils not provided with the proper nodule-forming organisms, but every effort should be made to insure the presence of these bacteria (p. 147) to raise a leguminous crop to best advantage. Whatever the special purpose for which the crop is grown, some application will generally be found useful at the start and should be applied, preferably, broadcast and harrowed in before sowing seed. On clay soils, well provided with calcium carbonate and organic matter, the application of 300 to 500 pounds of superphosphate (16 per cent.) at seeding time, is often found satisfactory, with minimum amounts of potassium; on sandy soils both potassium and phosphorus can commonly be added to advantage.

On soils deficient in organic matter, farm manure at the rate of 8 to 12 tons an acre may be used, preferably with the crop preceding in order to avoid weeds in the leguminous crop; for direct application, this trouble can be avoided by using farm manure sufficiently decomposed to destroy weed seeds.

On soils well provided with organic matter, calcium carbonate and the suitable nodule-forming organisms, the following mixture is suggested:

Plant-food mixture, No. 1.—*For use in starting young crop. Analysis: 4—12—4. Formula:* Sodium nitrate 376 pounds (56.4 pounds N), potassium nitrate 182 pounds (23.6 pounds N and 80 pounds K_2O), superphosphate (18%) 1332 pounds (240 pounds P_2O_5), drier, 110 pounds (p. 321).

As limits within which to work, an application of 300 to 600 pounds an acre is suggested, more exact amounts being determined in individual cases either by previous experience or by experiment. Such an application furnishes 12 to 24 pounds of nitrogen, 36 to 72 of available phosphoric acid and 12 to 24 of potash. Subsequent treatment with fertilizers depends upon the individual crops and the purposes for which grown.

The following mixture (200 to 400 pounds an acre) can be used at the time of seeding along with a moderate application of farm manure or when farm manure has been used in generous amounts on the preceding crop; the mixture can also be used as a top-dressing on meadows and pastures, in which the leguminous plants are to be encouraged:

Plant-food mixture, No. 2.—*For use with farm manure or as later top-dressing. Analysis: 2—12—6. Formula:* Sodium nitrate 134 pounds (20 pounds N), potassium nitrate 154 pounds (20 pounds N and 68 pounds K_2O), superphosphate (16%) 1500 pounds (240 pounds P_2O_5), potassium chloride 104 pounds (52 pounds K_2O), drier (p. 321) 108 pounds.

The suggested application (200 to 400 pounds) contains 4 to 8 pounds of nitrogen, 24 to 48 of available phosphoric acid and 12 to 24 of potash.

MANAGEMENT OF LEGUMINOUS CROPS

The general methods of treatment in relation to preparation of soil, fertilization, etc., are much the same, whether leguminous crops are used as green forage or hay, or as green-crop manure (pp. 259-265). Some special considerations are involved in the growing of beans (p. 417), peas (p. 419) and peanuts (p. 471).

Red clover (medium), as a green manure, is probably the most widely grown of any leguminous crop. Even the stubble and roots alone are often effective when plowed under. About one-third of the plant-food value is contained in the roots, 35 to 40 per cent. of the nitrogen in the roots and stubble.

While adapted to a great variety of soils, red clover thrives best on well-drained, moderately heavy loams, somewhat moist but not wet, well supplied with organic matter and calcium carbonate. Even on light soils, it gives large yields when moisture and plant-food are abundant, but runs out sooner than on heavier soils.

Red clover is most commonly seeded on winter wheat or rye in spring, as soon as danger of heavy frosts is past and while the ground is moist. When put in as green manure, 12 to 15 pounds of seed can be used, the larger amount on poor soils.

Red clover is not generally grown simply as a green manure, but is allowed to grow two years, being cut for hay or seed, and the aftermath plowed under. Clover sod is recognized as an excellent preparation for corn, potatoes, oats, etc. The beneficial effects of plowed clover sod usually last for more than one season. The roots of red clover are widely distributed through the soil and undergo decomposition readily.

Red clover is a vigorous nitrogen-gatherer; an entire crop, including tops and roots, often contains per acre 100 pounds each of nitrogen and potash (83 lbs. K) and 40 to 50 pounds of phosphoric acid (18 to 22 lbs. P). Even winter-killed clover may contain 50 to 80 pounds of nitrogen and other plant-food enough to make it useful for the following crop.

When grown only as a green manure, without any other crop, the application of fertilizers should be as indicated (pp. 362-363). When grown with winter wheat or rye, these crops should be well supplied with plant-food. After they are harvested, a light top-dressing of fine stable manure may be used, or 200 to 400 pounds of **Mixture No. 2** (p. 363).

Red clover for meadows, pastures, etc. (p. 372).—When grown for seed or for green forage, hay or pasturage, the management and feeding at the start of this and other clover crops is essentially the same as when grown for green manure. In permanent pastures, red clover in common with other leguminous plants persists longer when treated with minimum amounts of nitrogen and liberal amounts of phosphorus and potassium, while the vigor of grasses is promoted by liberal feeding with nitrogen to such an extent that they run out the clovers.

Mammoth red clover, known also as perennial pea-vine and sapling clover, closely resembles red clover but is coarser in growth and requires two or three weeks longer to reach maturity. It can be grown on wet lands better than the common red. It is a vigorous grower and a good collector of plant-food in the subsoil. The yield is generally larger than with red clover and the growth does not run out as quickly. Under favorable conditions, mammoth clover, when once well established, may continue for six or seven years. The general cultural treatment is essentially the same with both crops. It is well to use 15 to 20 pounds of seed an acre.

Alsike or Swedish clover, on account of its creeping habit, is best seeded with other crops of erect growth, as for example, with red or mammoth red clover. It is well adapted to cool, moist clays and loams too wet for other clovers. It is less sensitive to lack of lime than red clover but responds quickly to applications. It may be started during the summer or early fall and turned under in spring for green manure, using 12 to 15 pounds of seed. Its root system is not deep, so to insure a good catch, the soil must be carefully prepared. An application of 200 to 400 pounds an acre of **Mixture No.** 1 (p. 363) may be used before sowing, being well worked into the upper layer of soil. When used for forage or hay, it is seldom grown alone, and often remains three to five years.

White clover is seldom used as a green manure, except incidentally. Its common use is in pasturage in connection with other clovers and with grasses (pp. 372-384). It grows well on lands of inferior character where red clover fails; it is also less sensitive to drouth and cold. It is adapted to a wide range of soils, but does best on moist land. Its treatment with fertilizers is the same as for red clover.

Sweet clover, (white melilot or melilotus, large white clover and Bokhara clover) is a legume though not a real clover. It is a familiar, hardy, roadside biennial weed, widely distributed. It grows readily on moist soils, and especially well on soils rich in calcium carbonate. Its roots are large and penetrate deeply while its tops have an abundant growth. It is one of the plants able to utilize the phosphorus of ground phosphate rock ("floats"). It generally grows without application of fertilizers. On poor soils it may be used as a preliminary crop for alfalfa or some other leguminous crop. For green manure, it is regarded as equal to red clover. Its action as a green manure is said to be most effective when plowed in during the second season just before blossoming. The high value of this plant as a green-crop manure is more fully appreciated than formerly. The fact that animals are not usually fond of it as green forage or hay probably accounts for the lack of appreciation of its value as a green-manure crop.

Crimson clover, (Italian, scarlet or German clover) is exceedingly valuable as a green-crop manure in the states along the Atlantic from New Jersey southward to Georgia. It has been grown in other states, but, owing to its sensitiveness to freezing and consequent winter-killing, it is not extensively grown north of the Middle Atlantic States. It is an annual, flourishing in the cool weather of late fall and early spring, blossoming in April or May, when it is in best condition to be plowed under. It is not suited to spring seeding, so is usually sown from July 15 to September in Eastern Central States and as late as October farther south. It has the advantage of growing on soils too light for other clovers and also of growing at the time of year when cowpeas do not thrive. It is excellent as a cover-crop to be used between regular crops without interfering with rotation. It is therefore suitable when seeded in corn, orchards, berry lots, etc., after regular cultivation is finished, and also after harvesting such crops as tomatoes, early potatoes, etc. Even when the crop is winter-killed, it will be found useful as a green manure, if it has made a six-inch growth; the organic matter in such a crop is equal to that in 4 or 5 tons of manure. The nitrogen in a full crop of crimson clover ranges from 50 to over 200 pounds an acre, while the fall growth alone may contain as much as 100 pounds. The root system is comparatively unimportant, making up only 10 per cent. of the whole crop. In improving poor lands in the Middle Atlantic section, excellent success has been attained by combining crimson clover as a winter crop with cowpeas as a summer crop.

While crimson clover is adapted to light, sandy soils, it needs to be well supplied with phosphorus and potassium compounds. On land known to be lacking in plant-food it will be well to use 200 to 400 pounds or more per acre of **Mixture No. 1** (p. 363), in order to insure the crop a good start.

Alfalfa or Lucern is grown chiefly for forage and hay and only incidentally as a green-crop manure. On account of the difficulty and expense of seeding and its slow, early growth, it is not adapted to be grown solely as a green manure. It does not reach its greatest growth until the second or third year after seeding. When well started the crop will continue growing for four to ten years or more without reseeding, the length of time depending on soil, supply of plant-food, freedom from weeds and grass, and frequency of cutting.

Alfalfa grows well on practically all good, well-drained soils; it does best on a rich, sandy loam with deep, open subsoil. It dies out in a year or two on cold, heavy clays and wet soils. It is peculiarly adapted to dry regions, since its roots go down several feet into the soil after water as do those of few other agricultural plants, cases being on record of roots 30 feet long. For this reason, it is likely to injure the roots of trees if grown permanently in orchards.

Soil intended for alfalfa must be deeply plowed, put into as fine tilth as possible and be practically free from weed-seeds and grasses. In the North and the East it is commonly sown in spring as soon as danger from heavy frost is past, best sown by itself without nurse crop. When sown broadcast, it requires 25 to 35 pounds of seed; when drilled, 20 to 25 pounds. In the South it is customary to seed alfalfa after some hoed crop in the fall in order to get the start of spring weeds. The seeds are generally harrowed in and the ground lightly rolled.

Alfalfa plants are extremely sensitive to soil acidity. Unless the soil has been recently treated, it is generally desirable to lime it at the time of, or preferably the year before, putting in alfalfa. A generous application of farm manure and ground rock-phosphate or basic-slag phosphate with the preceding crop is advantageous. Just before seeding, an application of 300 to 600 pounds of **Mixture No. 1** (p. 363) may be harrowed in lightly. After the first year 200 to 400 pounds of **Mixture No. 2** can be applied, or generous annual dressings of farm manure can be given, together with 100 to 200 pounds of superphosphate and 50 to 100 pounds of potassium chloride.

About 40 per cent. of the plant-food value of a normal crop of alfalfa is usually contained in the roots. When these decay, they leave the subsoil well fitted for drainage and circulation of air, as well as provide abundance of organic matter.

Inoculation is generally necessary on soils where neither alfalfa nor sweet clover has been grown. For this purpose soil from an alfalfa or a sweet-clover field should be used. It is often a good plan, especially on poor soils, to grow a crop of sweet clover on a soil as a preparation for alfalfa.

Field-peas.—For green manure Canada field-peas grow well even on heaviest clays but do best on clay loams. On light soils deficient in moisture, they do not thrive. The crop grows best only in moist, cool weather, such as is common in northern climates during early spring and late fall. It is, therefore, largely confined to northern states. If sown in spring, the seeding should be made as early as possible. For fall planting, the crop can be started in August or September and make a good growth before freezing weather comes; it will then be ready for use in spring plowing. It can follow such crops as are not removed until September. It has been shown that in two months it is possible to grow a crop of 13 tons an acre, containing about 2 tons of dry matter, with 120 pounds of nitrogen, 35 pounds of phosphoric acid (15 lbs. P), and 120 pounds of potash (100 lbs. K).

Field-peas make vigorous growth when put into a well-tilled seed-bed, but they will do better than most other leguminous crops on soil less well prepared. From 2 to 3 bushels of seed an acre should be used and covered 2 to 4 inches deep. They do better when drilled in than when sown broadcast. On soils not well provided with available plant-food, it is well to apply at seeding 200 pounds an acre of **Mixture No. 1.** For use as green forage and hay, with other crops, see pages 394, 396.

Cowpeas are vine-like annuals that more nearly resemble beans than peas; they grow best in warm climates, but have been adapted to cooler climates, some varieties being grown as far north as Massachusetts and Wisconsin. Varieties vary from the bush-like, rapid-growing and early-maturing kinds to those distinctly vine-like with stems several feet long.

This crop possesses some prominent characteristics which peculiarly fit it for a green manure in places where clover and alfalfa are not utilized: (1) It utilizes atmospheric nitrogen and furnishes large amounts of organic matter; (2) has unusual ability to grow on poor soils and on almost any kind of land not too wet; (3) covers the ground so thickly as to prevent growth of most weeds; (4) is a vigorous, rapid and deep-rooted grower; (5) grows best during the hot months, when it is most important to have the soil covered in order to prevent injury to nitrifying bacteria; (6) the beneficial effects of the crop, when used as green-manure, last two or three seasons if followed by such crops as corn, cotton or small grains; (7) in the South two crops a year can be grown on the same soil.

Cowpeas are sensitive to cold and should therefore not be planted until danger of freezing is past and the soil has become warm. They are often sown between rows of corn at the last cultivation or broadcasted on stubble of small grains. The soil should be put into a good tilth, the seeds sown broadcast (1½ to 2 bushels to an acre) and covered about 2 inches deep, or put in with a grain-drill. When grown for forage, hay or seed, better results are obtained by sowing in drills 20 to 30 inches apart, according to large or small size of variety used.

Soil inoculation is not usually necessary in the South, but in the North has sometimes been found needful, as shown by lack of nodules on roots. On moderately good soils fertilizers are not commonly used. However, on light soils that have been continuously cropped and are low in available plant-food, 200 or 300 pounds of **Mixture No. 1** (p. 363) may be cultivated into the soil before seeding. On clay soils, potash can usually be omitted. Unlike most leguminous plants, the cowpea tolerates a moderate degree of soil acidity.

When grown as a green manure it can be plowed under in the green state or left as a mulch on the surface until early spring and then plowed under. Generally, the most economical method is to

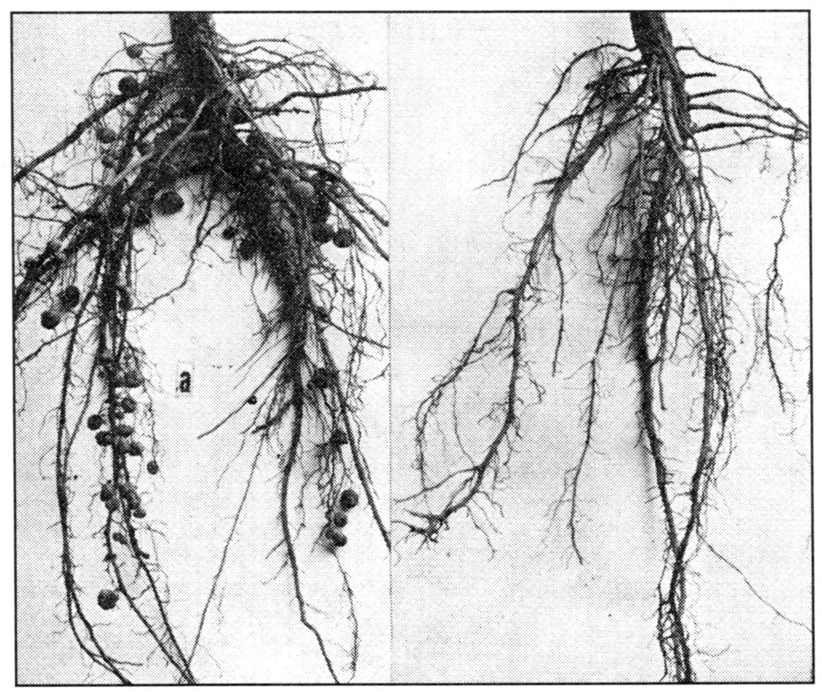

Root tubercles (left) on inoculated soy-bean plant. (Right) Uninoculated plant of same age.

use it as forage and put back the manure. The large amount of foliage makes it difficult to plow the whole crop under satisfactorily. On light soils, especially in warm climates, the whole crop should not be plowed under, since it makes the soil too open for a time. The cowpea is the most important leguminous crop used in the South as a means of soil improvement. Wherever it can be grown successfully, it is probably the most useful of all crops for this purpose, since it appears to combine nearly all the advantages possible in a green-crop manure. Cowpeas can be raised along with vetch (p. 370) as a cover-crop for orchards or in rotations on sandy soils.

A good crop of cowpeas contains about 2 tons of organic matter an acre, with 75 to over 100 pounds of nitrogen, 20 to 25 pounds of phosphoric acid (8 to 11 lbs. P) and 100 pounds or more of potash (83 lbs. K). About one-third of the nitrogen of the crop is in the roots.

Soy-beans or Japanese peas resemble cowpeas in many ways, but in some respects are more difficult to grow, and the yield is not so large. The adaptation to climates and soils is much the same as for cowpeas. The crop is grown satisfactorily as far north as Massachusetts and Wisconsin, and on soils varying from clay to sandy; but in the Northern States well-drained sandy or sandy-loam soil is desirable in order to ripen the crop earlier when grown for seed. The crop withstands drouth and is also able to grow well on rather wet soils. Owing to less rapid and vigorous early growth, soy-beans are more subject to be troubled by weeds than are cow-peas. The plant is sensitive to frost. General cultural treatment is essentially the same as for cowpeas. In starting soy-beans on new soils, it has often been found that bacterial nitrogen-gathering nodules are not found on the roots until inoculation has been done from a soy-bean field. Treatment with fertilizers is the same as for cowpeas.

A crop of soy-beans, including roots, may contain as much as 175 pounds of nitrogen, 45 pounds of phosphoric acid (20 lbs. P) and 115 pounds of potash (96 lbs. K). The results of some experiments appear to indicate that the plant is capable of utilizing atmospheric nitrogen to a greater extent than the cowpea crop does.

Velvet-beans have been found extremely useful in those parts of the South where warm, moist weather prevails, as in the Gulf States. They are adapted to light soils, but may be grown on any fairly well-drained soil. The vines, usually 10 to 30 feet in length, may grow as long as 75 feet, making the crop difficult to plow under. Yields of 10 tons and more of green matter an acre are reported, with a nitrogen content of 100 to 200 pounds. The same general cultural treatment may be followed as with cowpeas or soy-beans, with application of 200 or 300 pounds of **Mixture No. 1**, worked into soil just before seeding.

Vetch has been grown successfully as a late-fall green manure. On light, poor soils, it is usually better than cowpeas or soy-beans. Two species are used, common, Oregon or smooth vetch or spring tare (*Vicia sativa*) and hairy or sand vetch (*V. villosa*). The former is grown in orchards as a cover-crop or winter green-manure crop. In the Northern States it is likely to winter-kill. Hairy vetch closely resembles common vetch in general, but is more hardy at low temperatures than any other leguminous plant used as a green-crop manure; thus, in New York it remains green all winter and resumes growing in spring. This hardiness is a desirable quality in the North. Vetch has been found useful as a cover-crop and winter green manure in Connecticut on tobacco lands. A combination of cowpeas followed by vetch has been utilized in orchards and crop-rotations to furnish a maximum amount of organic matter for the soil.

The hairy vetch may be seeded either from the middle of August to the middle of September, or in spring from the last of April to the middle of May. In corn or cotton fields, it is sown broadcast and worked in with the last cultivation.

Inoculation is usually desirable on new soils. Soil on which garden-peas have been grown can be used for this purpose, if that is more convenient to obtain than soil from a vetch field. On poor land, vetch responds well to application of fertilizers; 200 to 400 pounds an acre of **Mixture No. 1** can be used.

TABLE 48—APPROXIMATE AMOUNTS OF PLANT-FOOD CONSTITUENTS IN ONE CROP

Crop	*Yield per acre*	*Water*	*Nitrogen*	*Phosphoric acid (P_2O_5)*	*Potash (K_2O)*
	Pounds	*Per cent.*	*Pounds*	*Pounds*	*Pounds*
Alfalfa	20,000	75	120	30 (13P)	100 (83K)
Clover, alsike	16,000	82	80	19 (8P)	48 (40K)
" crimscn	16,000	82	72	19 (8P)	64 (52K)
" mammoth	20,000	80	100	24 (11P)	80 (66K)
" red	12,000	80	66	16 (7P)	60 (50K)
" white	8,000	81	40	16 (7P)	24 (20K)
" sweet	20,000	80	110	30 (13P)	100 (83K)
Field-pea	10,000	82	50	15 (6.5P)	50 (42K)
Cowpea	12,000	84	54	14 (6P)	54 (45K)
Soy-bean	10,000	75	50	15 (6.5P)	60 (50K)
Velvet-bean	20,000	75	110	30 (13P)	110 (91K)
Vetch	10,000	84	50	10 (4.4P)	45 (37K)

P, phosphorus. K, potassium.

In Table 48 the estimates are made on the basis of what may be regarded as reasonable averages under fair conditions of growth.

XXIX

MEADOWS, PASTURES AND LAWNS

LAND on which crops, especially grasses and clovers, are grown to make hay is known as *meadow;* it is known as *pasture* when continuously used by stock for grazing. A *lawn* is a piece of land covered with grass and kept cut close, as in private grounds near dwellings or in golf-courses or polo-grounds. Permanent meadows or pastures are those used continuously for long periods; when temporary, the land is occupied with grass from one to three years, ordinarily as part of a rotation. The various kinds of seeds to be used will be determined by the object in mind, whether meadow, pasture or lawn, as well as by soil and climatic conditions. Thus, in a meadow it is important to use plants which reach their best condition for hay-making at the same time; while in pastures such species should be used as will furnish a continuous and uniform supply of feed from spring to fall. This is accomplished by selecting several different kinds of plants that reach their best condition for grazing purposes at different periods during the season and which remain in the soil continuously. For lawns, much the same general requirements exist as in the case of pastures, the selection of plants being made with reference to furnishing a dense, velvety, carpet-like appearance when kept closely cut.

The question of keeping land temporarily or permanently occupied with grass must depend upon individual conditions; for example, permanent grass-lands may advantageously be located on hillsides and uplands, heavy soils, stony land, lowland subject to overflow, swamp land, etc., while on sandy or light soils grasses run out after a short time so only temporary occupation is usually practicable.

MEADOWS

Formerly the general tendency among farmers was to regard hay as a catch-crop with small grains, little attention being given to special treatment with fertilizers. Farmers are, however, coming to regard meadows as lands that respond profitably to more particular care and fertilization, especially as to: (1) Soil conditions and preparation, (2) use of fertilizers, (3) plants used for meadows.

Soil conditions and preparation for meadows may vary widely in character, ranging from heavy clays to sandy loams. On heavier soils, meadows are more easily started and kept, while on light soils difficulty is experienced in getting plants started and they run out more quickly. One of the most important qualifications in soils for permanent meadows is the power to hold moisture.

The popular method of starting a meadow where rain is sufficiently abundant is to seed the grass along with small grains, especially wheat or rye, the seeds of the grasses being usually sown in the fall and those of clovers in the spring. The starting of grasses with an oat crop often proves unsatisfactory. The preparation of soil required for wheat or rye usually suffices for seeding of grasses along with the sowing of the grain. Best results, however, are secured in starting permanent meadows by preparing the land especially for the grass crop and then seeding without grain, for example, putting in timothy in August and clovers the following spring. The method of really thorough preparation is to plow deeply and cultivate several times thoroughly for the double purpose of destroying weeds and giving the soil a fine tilth. This treatment results in making the soil mellow for the easy penetration of the young rootlets and also tends to increase the amount of available plant-food.

Suggestions for use of fertilizers on meadows.—In supplying plant-food to meadows, some few guiding facts should be kept in mind. It is assumed that meadows are kept as free as possible from weeds. While the liberal application of fertilizers to meadows in which grasses have a good stand tends to encourage the growth of the grasses and keep down weeds, such application only promotes growth of weeds when they are already abundant.

(1) *Presence of calcium carbonate.*—If there is any doubt that enough calcium carbonate is present in the soil, an application should be made while the soil is being prepared for seeding, using 2,000 to 4,000 pounds of ground limestone or 1,000 to 2,000 pounds of burned lump-lime (pp. 205-211). The application may be repeated as a top-dressing on permanent meadows once in four or five years.

(2) *Effect of nitrogen.*—Grasses are favorably affected by nitrogen compounds, particularly nitrate. This is because of the characteristic action of nitrogen in promoting vigorous growth of stems and leaves (p. 49). Meadows benefit especially from application of nitrate early in the spring, when the available nitrogen supply is usually at its lowest and when the nitrifying organisms (p. 147) have not yet begun work in renewal. When clovers are

prominent constituents of meadows, less nitrogen need be applied than where the vegetation consists wholly of grasses. Heavy applications of nitrogen compounds tend to cause clover to run out. In case of mixed clovers and grasses, an annual spring application as top-dressing on established meadows may contain 15 to 24 pounds of nitrogen equal to about 100 to 150 pounds of sodium or calcium nitrate; on timothy and on mixed grasses, more may be used. Frequently nitrate alone gives excellent results, but it is generally better used in a mixture with phosphorus and potassium compounds. In applying nitrate alone, it should be finely ground and thoroughly mixed with three or four times its weight of inert material (p. 354), to make uniform distribution easier. When it contains small lumps or when unevenly distributed, nitrate may burn the vegetation seriously, though danger of this is at a minimum in early spring applications before the plants have started green growth.

(3) *Use of phosphorus and potassium compounds.*—To insure the continuation of clovers in meadows, there must be a generous supply of phosphorus and potassium compounds. Available phosphoric acid promotes the growth of roots and increases the number of secondary stalks, increases yield and improves quality. It may be supplied as superphosphate, ammonium or potassium phosphate, or basic-slag phosphate in amounts depending on general character of soil, previous cropping and fertilizer treatment.

(4) *Use of fertilizers in seeding meadows.*—However accomplished, it is desirable that, previous to seeding grass for permanent meadow, the ground be well provided with organic matter and be free from weeds. Both of these conditions can be realized by applying manure if fresh (10 to 12 tons) to the preceding crop, preferably some hoed crop, as corn, for example; or if sufficiently decomposed to destroy weed seeds (p. 347) directly (8 tons an acre) and worked into the soil thoroughly just before sowing. In whichever way used, 250 to 500 pounds of **Mixture No. 1** (p. 363) can be applied per acre just before seeding. Such an application will furnish approximately nitrogen, 10 to 20 pounds; phosphoric acid, 30 to 60 pounds; potash, 10 to 20 pounds.

When the soil is well supplied with organic matter one can apply also 100 pounds of ammonium sulfate, or 50 pounds of urea or 100 pounds of calcium cyanamid to insure sufficient nitrogen to start the crop and also furnish some for continued growth, especially in case of a large proportion of grasses in the mixture.

(5) *Spring application on established meadows.*—To give grass on established meadows a prompt and vigorous start early, an annual top-dressing early in spring is desirable, either of well-decomposed farm manure or commercial fertilizer. The finer the manure, the better it will serve the purpose, since it can more easily

work down into the interstices in the soil and grass; it also acts as a mulch to hold water. When practicable, it may be well to apply manure one spring and the commercial fertilizer the next.

In the case of timothy meadows, a plant-food mixture relatively rich in nitrogen is generally used. The following is suggested.

Plant-food mixture No. 3.—*For early spring use on timothy meadows: Analysis; 8—6—6; formula;* sodium nitrate, 825 pounds (124 pounds N); potassium nitrate, 275 pounds (36 pounds N and 120 pounds K_2O); superphosphate (16%), 750 pounds (120 pounds P_2O_5); drier (p. 321) 150.

If it is desired to use part of the nitrogen in the form of ammonia, the following formula could be used: 275 pounds potassium nitrate; 500 pounds sodium nitrate; 300 pounds ammo-phos (20-20); 375 pounds superphosphate (16%) and 650 pounds drier at the rate of 300 to 600 pounds an acre, furnishing 24 to 48 pounds nitrogen, 18 to 36 pounds phosphoric acid, and 18 to 36 pounds potash.

When it is desired to encourage the growth of clovers in meadows, **Mixture No. 2** (p. 363) may be used at the rate of 300 to 600 pounds an acre for top-dressing.

Plants used for meadows.—In selecting plants to use for meadows, one is governed by a variety of conditions, such as soil, climate, duration of meadow, value of grasses as food for farm animals, etc. For example, in the eastern states, the following will grow well on moist soils: Fowl meadow-grass, Italian rye-grass, meadow fescue, meadow foxtail, redtop; on heavy loams and clays, alsike clover and timothy. To these for pasturage may be added Kentucky blue-grass, white clover and meadow fescue; on average soils, timothy, redtop, Kentucky blue-grass and red clover are useful.

When land is to be used only two or three years as meadow, plants that attain full growth quickly, such as red and alsike clovers, timothy, redtop and orchard grass should be selected; for permanent meadow, the following added: Alfalfa, Kentucky blue-grass, meadow foxtail, meadow fescue, Canada blue-grass. Generally, kinds that grow strong and tall, as timothy, redtop, tall fescue, alsike and red clovers are selected. Largest yields most commonly come where only one or two kinds are grown. In Table 49 mixtures of seeds for meadows are given as suitable for those sections where the plants can be grown. The figures denote in each case the amount of seed to be used for one acre. Taking different localities and different soil conditions, an endless variety of mixtures can be made. For individual needs, one should get specific advice from local authorities.

The properties of leguminous plants used to make hay have been discussed in the previous chapter, so we will here briefly describe grasses most commonly used.

(1) *Timothy,* though primarily a hay plant, is commonly used in pastures, because its seed is inexpensive and it produces a fair crop the first year. In a few years it usually runs out in permanent pastures. It is easily adapted to good crop-rotations; grows best in good soils with plenty of rain; is responsive to generous use of fertilizers when soil conditions are good and moisture is plentiful. The best conditions are furnished by a cool, moist climate, and a cool, moist soil. While requiring less calcium than clover, timothy gives best results with fertilizers when the soil contains enough calcium.

TABLE 49—MIXTURES OF SEEDS FOR MEADOWS

Kind of grass	*Temporary meadows 1-3 years*	*Permanent meadow*	*Permanent meadow*	*For very poor land*	*For poor and wet acid soils*	*Permanent meadow*	*Permanent meadow*
	Lbs.	*Lbs.*	*Lbs.*	*Lbs.*	*Lbs.*	*Lbs.*	*Lbs.*
Timothy	15	15	15	10	—	8	—
Red clover	—	5	7½	—	—	4	4
Mammoth red clover	6	—	—	—	—	—	—
Alsike	4	3	—	4	6	2	—
Redtop	—	—	7½	5	15	2*	13
Kentucky blue-grass	—	—	—	—	—	2*	—
Tall meadow fescue.	—	—	—	—	—	—	9
Orchard grass	—	—	—	—	—	—	18

* Most prominent after a few years.

(2) *Redtop,* while not making as valuable a hay as timothy, is useful, because it grows on soils on which timothy will not grow, soils that are wet, acid or deficient in available plant-food. It does not endure severe drouth. It produces hay or pasture the year after being sown.

(3) *Kentucky blue-grass,* while of little value for hay, is of the highest value as a pasture grass on soils that are neither acid nor dry. It is not usually seeded alone, since it requires two or three years to reach full growth. It begins growing early in the season and heads early; it also grows well in autumn, but makes poor growth in hot weather.

(4) *Tall meadow fescue* does well on good land; it takes about three years to form a good sod, so is adapted only to permanent grassland.

(5) *Orchard grass* is tall, grows best on rich, deep, fairly moist

soils, but it makes fair growth on poor soils. It starts early and grows rapidly. On account of its deep root system, it withstands drouth well. It has the disadvantage of growing in bunches and making a rough sod, and also it must be cut near blossoming time to make good hay.

(6) *Perennial rye-grass* does best on stiff, wet soils, as on marshy land and is good for pastures.

(7) *Red fescue* is found in lawn and pasture grasses. It grows on dry, light soils and rather barren uplands, makes a fine, close sod and also does well in shady places.

(8) *Rhode Island bent-grass* is much like redtop. It is especially valuable in lawns and pastures. It can be grown on soils too acid for Kentucky blue-grass.

PASTURES

Probably no part of farms has been so generally neglected as the pasture-land. In most cases no attention is paid to supplying plant-food further than the incidental droppings of grazing animals, and grazing is often begun too early and allowed to become too

A pasture spotted like this will usually respond to nitrogen fertilizers.

close. This treatment results in the disappearance of perennial grasses, decreased yield and nutritive value of herbage, loss of organic soil matter, decreased power of retaining moisture, and the substitution of worthless weeds for valuable forage plants; in other words, the pastures "run out." Permanent pastures of value can be secured if proper attention is given to soil preparation, seeding, fertilizing and subsequent care, resulting in the replacement of weeds and moss by grasses and clovers of high feeding value.

Suggestions for use of fertilizers on pastures.—In the use of fertilizers for pastures, three conditions must be considered; (1)

starting permanent pastures, (2) maintaining pastures in good condition, and (3) renewing rundown pastures. In each case, it is important to have plenty of calcium carbonate in the soil, liming at the start with one to three tons of ground limestone, preferably applied late in fall, and then systematically every four or five years, where any doubt exists regarding the need of lime. When clovers do not grow well and where moss-grown areas appear in pastures, the need of lime is clearly indicated.

In starting a permanent pasture, the same kind of care should be given to conditions as in starting permanent meadows. In case of pastures badly run down, the most effective method is to plow the sod and start anew, when this is practicable.

In the maintenance of established permanent pastures, application of ground limestone should be made as indicated above and a top-dressing of 200 to 400 pounds of **Plant-food Mixture No. 3** (p. 375) an acre should be applied early each spring. In some cases excellent results follow the use of sodium or calcium nitrate, calcium cyanamid, ammo-phos, etc., but, in general, a complete fertilizer, high in nitrogen, is preferable. Proper use of fertilizers replaces weeds by desirable grasses and clovers. Light top-dressing of well-decomposed farm manure may be applied to advantage early in spring, using, in addition, 200 to 400 pounds of **Plant-food Mixture No. 2** (p. 363). Farm manure is of special value for application to poorer parts of a pasture. A liberal application of seed every few years, especially along with farm manure, mowing weeds once or twice a year, before they bloom and occasional harrowing, will aid in keeping the pasture permanently useful. If practicable, it is a good plan to go over the pasture in midsummer and fall with a harrow to distribute more evenly the animal droppings, which, if undisturbed, form the center of circles of foul, unpalatable grass and encourage growth of weeds.

The renewal of run-down pastures may be accomplished by liming if acid, harrowing, reseeding and applying farm manure and later a complete fertilizer.

In seeding permanent pastures, the plants should be selected with reference to using those which in combination will give a continuous growth of grass through the season; some should mature early to get ahead of weeds; some should contribute to make a close, dense turf; some serve as a protection to prevent injury from the treading of stock. Various mixtures of plants have been used to secure these objects, the combinations being adapted to local conditions of soil and climate. For example, such grasses as Bermuda, Dallis Carpet and Centipede find use only in the South. A numberless variety of mixtures may be used, according to adaptation to local conditions. For such combinations as experience has proved most suitable, one should obtain the advice of local authorities, such as the state agricultural college and the local county agent. For illustration, a variety of mixtures is given in Table 50, the figures indicating pounds of seed used for one acre.

TABLE 50—MIXTURES OF SEEDS FOR PERMANENT PASTURES

Kind of plant	*Hay two yrs. followed by pasture on good land*	*Permanent pasture*	*Permanent pasture*	*Temporary pasture*	*Pasture on wet land*	*Pasture on light, sandy soil*	*Pasture on poor acid soil*	*Pasture on poor land*	*Mixture for sowing on old pastures*
	Lbs.	*Lbs.*	*Lbs.*	*Lbs.*	*Lbs.*	*Lbs.*	*Lbs.*	*Lbs.*	*Lbs.*
Timothy	10	8	—	10	—	—	5	8	2-3
Red clover	4	6	—	8	—	—	—	6	2-3
Alsike	3	3	—	4	8	—	5	3	1
Redtop	—	—	8	—	14	10	5*	4	2
White clover	2*	1-2*	4	—	—	2	2*	1	—
Kentucky blue-grass	3*	4*	8	—	—	8	—	—	4
Meadow fescue ...	2	1-4	—	—	—	—	—	—	3
Orchard grass	2	1-4	—	—	—	—	—	—	—
Alfalfa	2	—	—	—	—	—	—	—	—
Perennial rye-grass.	—	—	9	—	12	—	—	—	—
Red fescue	—	—	3	—	—	20	—	—	—
Creeping bent-grass	—	—	—	—	6	—	—	—	—
Canadian blue-grass	—	—	—	—	—	—	—	4	—

* Prominent after a few years.

LAWNS

In making and maintaining lawns, haphazard and inefficient methods have too often been the rule. Even more pains should be taken, if possible, than with meadows or pastures. Private lawns are usually small, so easily permit extra attention; the main difficulty has been lack of information in regard to the proper methods of starting and maintaining them. Points to consider are: (1) Character of soil, (2) preparation and fertilization of soil, (3) starting the growth of grasses, (4) care and fertilization after seeding, (5) grasses to use for lawns.

Character of lawn soil.—The best soil conditions for lawns are essentially the same as for meadows and pastures, that is, a well-drained soil that contains considerable clay, with a tendency to be heavy and compact (not light, loose and sandy), rather retentive of moisture, keeping moderately moist without becoming too wet. These conditions are best found in a decided clay loam, or a sandy loam resting on a clay subsoil. Soils used for private lawns often consist of mixtures of building débris, surface soil and subsoil from excavations. Where filling to a considerable depth is required, no kind of coarse material, like bricks, stones, etc., should be used in the upper 4 feet. The subsoil should be made of moderately heavy clay. When the filling material varies in texture, the heaviest is placed first, evenly distributed and covered with an even layer of lighter subsoil material, care being taken to use only soil mellow enough to permit roots to penetrate deeply for water. For the upper layer of 6 to 12 inches, only surface soil of good productive capacity should be used, preferably that obtained from a cultivated field; this should be lighter than the subsoil but not radically different in general character.

Preparation and fertilization of lawn soil.—Attention must be given to tillage and treatment with lime, organic matter and special plant-food mixtures. The soil should be thoroughly plowed or spaded 8 to 10 inches deep, then made finer at the surface by raking or harrowing, then compacted by a lawn or a field-roller, after which the surface should be again loosened by a rake or a harrow, thus making a seed-bed suitable for fine seeds.

When possible, it is desirable to begin preparation and treatment the season before seeding to grass, especially in case of soils in any way abnormal. The following plan is suggested: Plow the soil, apply and harrow in 2,000 to 4,000 pounds of powdered limestone an acre (about 5 to 10 pounds for 100 square feet), or 1,000 to 2,000 pounds of quicklime slaked to fine powder; after 2 or 3 weeks harrow in well-decomposed farm manure 10 or 15 tons an acre, or about 50 to 75 pounds to 100 square feet), working it in thoroughly. In the absence of farm manure, apply per acre before or at the time of seeding 250 to 500 pounds of **Mixture No.** 1 (about ½ to 1 pound to 100 square feet). Then sow clovers, cowpeas, soybeans or some other leguminous crop suitable to local conditions. When this has made sufficient growth, plow it under. This treatment insures abundance of organic matter and lime and also aids in destroying weeds. The following season the soil is put in condition for a good seed-bed as described above, after previously working in 300 to 600 pounds of **Mixture No.** 3 (p. 375).

When the soil is not thus treated the season before seeding to grass, and when good farm manure can be obtained the following method is used: Lime as directed above, 10 to 15 tons of well-decomposed farm manure and in addition 250 to 500 pounds of **Mixture No.** 3 (p. 375). When manure can not be obtained, one can apply commercial peat (10 tons an acre) together with 500 to 1,000 pounds of **Mixture No. 3**, worked into the soil shortly before or at the time of seeding.

Starting the growth of lawn grasses.—When the seed-bed is perfected, grass-seed is sown broadcast in amounts depending on the kinds used; the seed then covered less than half an inch deep by harrowing or raking. In order to insure a uniform distribution of seed a second sowing should be made, at right angles to the first with the same amount of seed, covering as before. The ground is then compacted by a 300-pound lawn roller. Spring sowing is more commonly practised than fall sowing in the Eastern and Middle states, but in this region, late August or early September, is usually best. If seed can be sown before showers, prompt germination will be insured.

Care and fertilization of lawn after seeding.—When the grass is 3 or 4 inches high, it is clipped by a lawn-mower, and this should be repeated from time to time. It is never allowed to go to seed. Ordinarily, clipping within 2 inches of the ground is about right. It is desirable early each spring to go over the lawn with a heavy roller while the soil is still wet, but not muddy.

Trouble is often experienced with deep-rooted weeds, such as dandelion and plantain. When these persist it is necessary to remove them; an asparagus knife is a good tool for this purpose. When weeds crowd out most of the grasses, it is least trouble to plow up the ground and start anew.

Several methods of fertilizing lawns are employed. A popular one is to spread over the surface in late fall a thin coating of well-decomposed, finely divided farm manure, raking it in as close to the surface of the soil as possible. More recently stock-yard products, such as dried sheep manure, cow manure and pig manure have become popular (p. 219). In drying, these materials are heated high enough to destroy weed-seeds, which makes them more desirable than many farm manures; moreover they are finely pulverized. They can be used as a spring application at the rate of 1 to 2 ounces per square foot, or nearly 1½ to 3 tons an acre. Their high cost is a serious objection to these materials. Bone-meal or bone-flour can be applied as a top-dressing in spring, sprinkling lightly over the surface.

Another method of fertilizing established lawns is to apply some soluble compound rich in nitrogen. Sodium nitrate has been extensively used with excellent results, applied in early spring at the rate of 100 pounds an acre (about 4 ounces per 100 square feet). Two or three additional applications at intervals during the season, each at the rate of 50 pounds an acre, has been found effective in maintaining the vigor of the grasses in lawns. When used in the same amounts as sodium nitrate, calcium nitrate is equally effective. Ammonium sulfate-nitrate (Leunasalpeter) can be used at the rate of 75 pounds an acre. Good results are reported from ammo-phos at the rate of 100 pounds an acre. When the foregoing or similar materials are used, certain precautions must be observed. The grass should be dry when the application is made. The material must be distributed uniformly and in finely-powdered condition to avoid burning the grass. In order to apply evenly as small amounts as 75 or 100 pounds an acre, the material can be mixed thoroughly with two or three times its bulk of any fine, dry inert substance (p. 321). For uniform application a broadcast distributor should be used. Small ones are made for limited areas; otherwise for distribution on small areas, like ordinary home lawns, it is a good plan to apply the material just before a shower or to water the lawn thoroughly at once after application. Still better the chemicals may be dissolved in water at the rate of one ounce or a tablespoonful in two gallons, and distributed with a sprinkler over a space five feet square.

On the whole, however, more satisfactory results may be expected under most conditions from a complete fertilizer, provided sufficient care is taken to put the soil in the best possible physical and chemical condition before sowing the seed. **Mixture No. 3** is well adapted for use on lawns, applying first in early spring at the rate of 600 pounds an acre or about 2 ounces for 1 square yard. Two or three additional smaller applications can be made at intervals during the season, using 100 or 200 pounds an acre or, in place of the complete fertilizer, 75 to 100 pounds of one of the nitrogen-containing compounds mentioned above.

On golf putting greens, an application of 200 to 400 pounds an acre (½ to 1 pound for 100 square feet) may be made soon after the grass begins to grow and again three or four times during the season at the rate of 100 to 200 pounds for each application. On fairways the same application (200 to 400 pounds) may be made and the same amount again twice at intervals during the season.

Grasses for lawns must possess the following requirements: (1) Make a close turf, as in case of species that have creeping rootstocks, short joints and abundance of long, narrow leaves about the crown of the plant; (2) have a pleasing color not easily affected by changes of season; (3) resist drouth; (4) respond promptly to the marked changes from cold to warm seasons; (5) bear without injury continuous clipping by lawn-mowers. In general, best results are obtained by a mixture of various species, because it is not often that soil and climatic conditions are favorable enough to develop a single grass into a perfect lawn. The grasses best adapted to lawns in the Eastern and Northern States are Kentucky bluegrass, redtop, Rhode Island bent-grass, creeping bent-grass and white clover, all of which make a dense, deep sward. White clover is desirable upon sandy soils and in spring seeding, because it covers the ground quickly and comes back quickly after severe drouths.

Care must be taken to obtain pure grass seed of high germination test. The amount of seed to use should be sufficient to insure a thick stand and prevent the growth of weeds. When used in combination, 2 to 3 bushels an acre should be sown of such grasses as blue-grass, bent-grass and the fescues. Blue-grass alone may be sown at the rate of 2 bushels an acre; when white clover is added to the foregoing varieties, a peck per acre should be used.

Pasture rotation and intense fertilization.—Co-operative experiments have been carried on by several agricultural colleges, experiment stations and individual farmers under the supervision of the National Fertilizer Association to study improved methods of pasture management and effective fertilization. Special attention has been devoted to trials of the method developed in some parts of Europe, known as the paddock or Hohenhein (Germany) system. The details of practical application are varied to suit individual conditions. Briefly stated, this method includes (1) the permanent division of the pasture into separate fenced fields or paddocks of such size as to allow one-eight to one-fourth of an acre for each cow in the herd; (2) grazing each paddock in turn, allowing a sufficient period for new growth of grass; (3) heavy application of fertilizer, early in spring four or five weeks before grazing as, for example, 800 pounds an acre of a mixture analyzing of 8—6—6 and subsequent top-dressing after each grazing period with 100 to 150 pounds of sodium or calcium nitrate or an equivalent amount of nitrogen (15 to 20 pounds) as ammonium sulfate, calcium cyanamid, urea, calurea or cal-nitro, etc.

TABLE 51—APPROXIMATE AMOUNTS OF PLANT-FOOD CONSTITUENTS IN ONE TON OF HAY FROM VARIOUS GRASSES

Crop	*Nitrogen*	*Phosphoric acid* (P_2O_5)	*Potash* (K_2O)
	Lbs.	*Lbs.*	*Lbs.*
Kentucky blue-grass	24	8 (3.5P)	31 (26K)
Meadow fescue	19	8 (3.5P)	40 (33K)
Orchard grass	24	7 (3.1P)	35 (29K)
Perennial rye-grass	23	11 (4.8P)	30 (25K)
Redtop	24	7 (3.1P)	20 (17K)
Timothy	25	11 (4.8P)	20 (17K)
Mixed grasses	30	8 (3.5P)	27 (22K)

P, phosphorus. K, potassium.

This method is practicable only on soil adapted in every way to the satisfactory growth of grass and clover. For necessary details consult references to bulletins.

As an illustration of a mixture of seeds for lawns, we give the following:

Timothy	10 pounds
White clover	5 "
Redtop	10 "
Kentucky blue-grass	20 "
Rhode Island bent-grass	5 "
Creeping bent-grass	5 "

XXX

CEREAL AND GREEN-FORAGE CROPS

Crops most commonly grown for grain are (1) corn, pop-corn, (2) wheat, (3) rye, (4) oats, (5) barley, (6) buckwheat, (7) millet, and (8) sorghum.

Most of these crops are used also as green forage, being cut before mature and fed directly to animals. This method of feeding is known as "soiling" so the crops are often called soiling-crops. They are of special importance in connection with milk production, being used to supplement, or even to take the place of, pastures, and also to furnish succulent food during winter. Cultivable land can generally be more profitably used in this way than for ordinary pasturage.

The chief qualities desired in a green-forage crop are succulence, palatability and comparatively large yield in a moderately short growing period. Some of these crops are also used as pasturage and hay and as green-crop manure. As a matter of convenience we shall consider treatment for each of these uses (grain, green forage, pasturage, hay, and green manure). Generally speaking, the conditions of soil and tillage are essentially the same, for whatever purpose the crop is grown; some variations occur in respect to the variety and amount of seed and the treatment with fertilizers, according as the crop is grown for one purpose or another.

While the cereal crops vary considerably from one another in their feeding powers, period of growth and character of root systems, for satisfactory yields, all require good seed, and most of them good soil, good culture and a good supply of available plant-food. Abundance of organic matter and calcium carbonate is desirable.

In connection with each crop we shall consider: (1) Crop-rotations, (2) soil and preparation, (3) fertilizers, (4) seeding, (5) cultivation.

Corn is a crop of special value because, where grown under favorable conditions of soil and climate, it is capable of producing considerably more food per acre than any other grain, and is useful for both grain and green forage and as the source of valuable products and by-products.

(1) *Crop-rotations for corn.*—Among rotations in common use, are the following:

(*a*) Six years; (common in the corn-belt): Corn, 2 years; wheat or oats, 1 year; timothy and clover, 3 years.

(*b*) Four or more years; (outside the corn-belt): Corn, 1 year; oats, 1 year; wheat, seeded with timothy and clover, maintained 1 to 3 years.

(*c*) Three years; (in cotton-growing sections): Corn, 1 year; oats, in fall, 1 year; cowpeas and cotton, 1 year.

(2) *Soil and preparation.*—Though good crops are obtainable on a great variety of soils, ranging from light, sandy loams to heavy clays, for whatever purpose grown, the best soils for corn are rich, porous, well-drained, warm loams, that do not bake during drouth. Abundance of organic matter is essential. The reaction of the soil may be neutral or slightly acid. On strongly acid soil one to three tons of pulverized limestone may be used to advantage.

When corn follows grass (a common form of rotation) it is well to plow in the fall; spring plowing, when necessary, should be done as early as conditions permit. The depth of plowing should usually not exceed 6 or 7 inches. The soil should be well cultivated to insure mellowness, warmth, and circulation of air, conditions essential to promote quick and complete germination of seed and rapid, early growth of plants.

(3) *Fertilizers.*—In discussing plant-food supply, the following facts must be considered: (*a*) The crop makes most of its growth after hot weather begins, when the process of nitrification (p. 139) is most active. (*b*) The period of growth is prolonged compared with that of other common cereals. For these reasons, corn can utilize organic nitrogen as farm manure, leguminous green-crop manure, bone-meal, meat tankage, fish-scrap, cottonseed-meal, etc. (*c*) Therefore, corn well supplied with organic nitrogen does not require nitrate except in small amounts, sufficient to supply the early needs of the young plants while their root systems are limited and before nitrification actively begins. (*d*) It is, however, desirable that generous amounts of soluble phosphorus and potassium compounds be furnished. (*e*) Farm manure can be applied to excellent advantage for whatever purpose corn is grown. It is best applied, if coarse and fresh, in fall or winter on sod (8 to 10 tons an acre). When applied in spring directly to the crop, it is usually better to use partly decomposed manure especially in case of lighter soils, putting it on the plowed ground and harrowing in thoroughly before sowing. When corn follows a leguminous crop, it is well to apply, if possible, 2 or 3 tons of farm manure an acre in order to carry into the soil micro-organisms to hasten decomposition of the green manure. In the case of soils deficient in organic matter and

Effect of manure on corn. The left plot was untreated, that at the right received manure and superphosphate.

available plant-food, treatment with farm manure or leguminous crops, or a combination of the two, should be regarded as essential for a successful crop of corn.

Commercial fertilizers varying widely in amounts and proportions of plant-food constituents are used for corn, depending on the needs of the soil. When a corn crop is grown under favorable soil conditions, with an abundance of leguminous green-crop manure or good farm manure, the 4—12—4 mixture (p. 351) is suggested at the rate of 200 to 400 pounds an acre. Instead of having all the nitrogen as nitrate, half may be ammonium sulfate. When farm manure can not be obtained, 400 to 800 pounds of the mixture can be used. Where a long-growing season is assured, half the nitrogen can be in organic form, such as meat tankage, fish scrap, cottonseed meal, etc. On peat and muck soils, a fertilizer high in available phosphoric acid and potash is used, such as **Mixture No. 2** (p. 363).

In distributing fertilizer, various methods are used. It may be; (*a*) broadcasted during preparation of soil before planting; (*b*) distributed by a special machine on opposite sides of each row of corn three or four inches below the seed; or (*c*) applied in hill or drill below the seed. Whatever method is used, direct contact of seed and concentrated fertilizer must be avoided. In some cases, part of the fertilizer is broadcasted before planting and the rest applied as a side-dressing between the rows at the first cultivation, or the full application is broadcasted before planting and a side-dressing of 100 pounds of nitrate made later.

The root system of corn is such that fertilizers, when intended for use during the growing season, are most serviceable if uniformly distributed. In the course of a couple of months after planting, the roots extend not only deep into the soil but horizontally in every direction, completely underlying the surface and filling the soil below the cultivated layer.

(4) *Planting.*—The amount of seed used ranges from 6 quarts up, depending largely upon the character of the soil, more being used on good than on poor soils. When planted in hills on fertile soils, the hills may be 2½ to 3 feet apart each way and 3 or 4 kernels in a hill; on poorer soils, the hills and rows should be farther apart. Too thick planting is undesirable, because in dry spells loss of soil moisture by evaporation through the plant leaves is likely to be so great as to injure the crop. The depth of planting is usually between 1 and 3 inches.

(5) *Cultivation.*—Soon after planting in rows, the surface is cultivated 2 or 3 inches deep to keep weeds down and to form an earth mulch to retain soil moisture (p. 107). When the plants are 3 or 4 inches high, they may receive one or two ordinary cultiva-

tions of 3 or 4 inches, after which the cultivation must be shallow enough to avoid injury to the feeding-roots. Frequency of cultivation depends on weather conditions. Cultivation may be advantageously continued as long as it is convenient to go between the rows.

Corn as green-forage crop, whether for immediate use or as ensilage, must have a rapid growth to give a large yield of succulent stalks and leaves in comparison with ears. The character and conditions of soil preparation and cultivation are the same as grown for grain. For green forage when used as soilage, additional application of nitrogen as side-dressing is permissible, at the rate of 50 to 100 pounds of nitrate.

Corn is planted more thickly when grown as green forage. The thicker the planting, the smaller will usually be the size and number of ears per stalk. In order to obtain ears about one-half full size, 11 to 14 or more quarts of corn, according to size of kernels, are commonly used per acre, when the rows are about 3½ feet apart and the separate stalks in the row 6 to 8 inches apart. When planted in hills, on fertile soil, the hills may be 2½ to 3 feet apart each way, with 3 or 4 kernels in a hill.

When the corn-growing season extends from May first to October, two crops of corn can usually be grown for green forage. After the first crop is removed, the ground is harrowed deeply without plowing, as preparation for planting; this leaves it in condition to promote quick germination and rapid early growth, which is important to avoid danger from possible early frost. Treatment with fertilizer and tillage is the same as with the first planting.

When a constant supply of green-forage corn for soilage is desired, plantings are made in succession, 2 or 3 weeks apart, in amounts to meet one's needs; or one may plant early and late varieties.

In some cases, corn is sowed broadcast with the purpose of having only green forage without any ears, to feed especially in time of drouth, in August or September. In such cases, comparatively large amounts of seed are used. **Mixture No. 1** can be used at the rate of 300 to 600 pounds or more per acre.

Pop-corn is now grown as a field crop. The conditions of soil, culture, feeding, etc., are essentially the same as for sweet corn (p. 423).

Wheat is adapted to a great variety of climatic and soil conditions. Kind of seed used, methods of rotation, tillage, fertilization and seeding depend upon local conditions, especially climatic. While wheat is grown mostly for the grain, it is sometimes used as a green-forage crop (p. 392). The particular points are: (1) Rotations, (2) soil and preparation, (3) fertilizers, (4) seeding.

(1) *Crop-rotations for wheat.*—Continuous growth of wheat on the same soil has proved disastrous to soil fertility. The kind of rotation used varies with local soil and climatic conditions and with the special kind of farming practiced. The following rotations are typical where they are practicable:

(*a*) Three years: Corn, wheat, clover; or potatoes, wheat, clover.

(*b*) Four years: Corn, oats, wheat, clover.

(*c*) Five years: Corn, oats, wheat, timothy and clover, 2 years.

(*d*) Five years (used in corn-belt): Corn, 2 years; oats, wheat, clover.

A weedy fallow not only is expensive to prepare for a seed-bed but reduces the yield of wheat sown thereon.

(2) *Soil and preparation.*—Wheat does best on well-drained rich clays and heavy loams, but good yields occur on a great variety of soils, even on light ones when moisture is ample. An abundance of organic matter is essential for good crops of wheat.

It is necessary that attention be given to proper tillage of soil. General experience shows it best to plow as soon as practicable after the preceding crop has been removed and to harrow after each rainfall. This treatment compacts the soil, prevents weeds going to seed, retains moisture, and puts the soil in fine tilth. The depth of plowing varies from 4 to 8 inches according to special conditions; the aim must be to make the upper layer of 3 to 4 inches mellow, in order to insure prompt germination of seed and rapid early growth. On light, mellow soils, the ground may be only harrowed for seeding, especially when following potatoes or corn. On light soils the seed-bed may be rolled. For spring wheat, fall plowing appears to give best results.

(3) *Fertilizers.*—Among the general methods of fertilizing are: (*a*) Nitrogen supplied in part by a leguminous crop as part of the rotation, whether as green-manure, cover-crop, green forage or hay. The leguminous crop used depends on local conditions. (*b*) This is usually followed by some such crop as corn, which can utilize the plant-food contained in the green crop. A few tons of manure, when obtainable, can be applied with advantage to the leguminous crop and plowed under in late fall or early spring. (*c*) In some cases, commercial fertilizer may be applied to the preceding crop, in addition to farm manure or green manure, in amounts and forms that will leave some plant-food for the wheat following. (*d*) Generally, the best plan is to utilize manure, if fresh and coarse, or green manure for the crop or crops preceding and to apply commercial fertilizer directly to the wheat. (*e*) When manure is used directly for the wheat, it should preferably be well-decomposed, if used in considerable amounts, especially where rainfall is likely to be deficient; it is well harrowed into the soil before seeding. While the use of 20 or 30 tons of manure an acre is sometimes advised, this refers to soils deficient in organic matter and where green crops are not used. About 8 or 10 tons every 4 or 5 years will suffice where a good system of rotation is used. Farm manure is preferable on heavy soils. In some cases, when well-decomposed and in fine condition, manure is applied after wheat is sown. (*f*) In many cases, the chief reliance in feeding the crop is placed upon commercial fertilizers, usually a complete fertilizer being used, though in some cases phosphorus in the form of superphosphate is used alone for a time with good results, especially on clay soils well supplied with nitrogen by means of leguminous crops or manure.

In feeding wheat, it is found that on most soils the application of nitrogen and phosphorus is more often called for than potassium, but, in the absence of definite information, it is best to assume the need of a small amount of available potassium and to make use of a complete fertilizer. When one uses a good crop-rotation, an application of 200 to 300 pounds of **Mixture No. 2** (p. 363) is advisable.

In case wheat does not appear to have wintered well or in case of light soils, an early spring application of 75 to 150 pounds of sodium nitrate as a top-dressing about as soon as the plants begin to grow, will usually prove beneficial in giving the crop a vigorous, early start.

For wheat, fertilizers are usually applied evenly by distribution at the time of seeding.

In case liming is needed the application is better made with some crop preceding wheat.

(4) *Seeding.*—In moist soils, seed should be covered about 1 inch, and in light, dry soils 2 to 3 inches. The amount of seed used varies with soil, climate, quality and variety of seed, time of seeding, etc. Less seed is used in case of early seeding, good seed-bed, fertile soil, when drilled, where winters are mild, or when seed is small. The amount most commonly used varies from 6 to 8 pecks. The time of seeding winter wheat should be early enough to give the crop a good start before freezing weather begins. Spring wheat is put in as early as seasonal conditions permit.

Wheat as green forage.—Wheat is sometimes grown as green forage, being cut and fed at about the heading-out stage. The conditions of treatment are the same as when the crop is grown for grain except that the seeding should be a little earlier. Treatment with fertilizers should be the same, except that somewhat heavier application of nitrogen is permissible **(Mixture No. 1** p. 363), and early spring top-dressing with 100 to 150 pounds of sodium nitrate is generally desirable, the application being made when the foliage is dry and as soon as active growth appears.

Wheat, sometimes used for pasturage, gives excellent results when not cropped too closely early in the season. When cut just in full head, and carefully cured, wheat makes satisfactory hay.

Rye is useful as grain, green forage, hay, pasturage and green manure. It has a wider range of adaptability to climatic and soil conditions than any other cereal. In many general conditions that affect its growth, it closely resembles wheat.

(1) *Crop-rotations for rye* are the same as in wheat rotations (p. 390). On account of its hardiness, rye is of special value in connection with seeding of grass and clover, particularly in northern climates.

(2) *Soil and preparation.*—Rye grows on soils too poor for other cereals and it will stand more neglect than other cereals. It does best on light, rich, well-drained loams and is adapted to many sandy soils, not to wet soils or heavy clays. For best results the preparation of the soil should be the same as for wheat (p. 390).

(3) *Fertilizers,* commonly, are not used extensively with rye but their use will often be profitable. The same treatment applies as to wheat. Excessive application of nitrogen is to be avoided, because it results in early lodging, discolored straw, and shrunken grain.

(4) *Seeding.*—Winter rye is generally used; the seeding is made early enough in fall to enable the crop to get a good start before winter, especially on poor soils. Light soils may be rolled after

seeding. Drilling seed is better than broadcasting. On poor soils 3 to 4 pecks are used, on better soils twice as much.

Rye as green forage possesses several marked advantages. It is hardy, produces rapid growth earlier than other crops and is palatable. In the South, it furnishes 3 or 4 cuttings during fall and winter. The soil should be prepared as for rye to be grown for grain. Treatment with fertilizers is the same as for wheat grown as green forage (p. 392), generous nitrogenous fertilizing being desirable, especially in early spring. In seeding for forage, 2 to 3 bushels of seed should be used.

Rye may be used for light early spring pasturage, or, when started early enough, for light fall pasturage, without injuring the crop for grain. It is used sometimes as ensilage, being cut when headed out fully, and allowed to wilt slightly before putting in silo. When cut just before full heading out, it makes good hay, if cured properly.

Rye as green manure is of special value as a cover crop for fall and winter. It has the advantages (*a*) of being adapted to use late in the season following other crops, (*b*) growing rapidly, (*c*) being hardy in winter and (*d*) making rapid growth in early spring. It can be plowed under in time for the use of various crops, such as potatoes, corn, etc. In the North, this permits its use as a cover-crop in cornfields to furnish late fall and early spring pasturage, after which it can be turned under as green manure. While it is useful to supply organic matter to soils, it gives back to the soil only the plant-food constituents it has used in its growth. It is of special value to furnish organic matter on poor, light soils where leguminous crops do not grow, and is often so used as a preliminary step in preparing such a soil for legumes.

Oats, besides their value for grain, make excellent green forage and hay, grown with peas or vetch for forage. They are adapted to a cooler climate than corn, wheat or barley, but require much water and are therefore adapted to a moist climate. Growth is rapid and the amount of straw in relation to grain is large in comparison with wheat, rye, etc.

(1) *Crop-rotations for oats* include the following types:

(*a*) Four to five years (in winter-wheat sections): Corn, 1 year; oats, 1 year; wheat, 1 year; timothy and red clover, 1 to 2 years.

(*b*) Three to six years (in corn-growing sections not adapted to wheat): Corn, 1 to 2 years; oats, 1 year; timothy and red clover, 1 to 3 years.

(*c*) Three to four years: Corn with cowpeas, grown between rows and harvested for grain, 1 year; oats, followed by cowpeas used as hay, 1 year; cotton, 1 to 2 years.

(2) *Soil and preparation.*—The character of the soil may vary much more for oats than for any other cereal. In general, the soil should be fairly moist. While best crops are obtained on well-drained, rich clay loams, crops can be grown on poor clays, sandy loams and peaty soils. Abundance of organic matter is necessary to hold moisture. Light, warm soils that heat up easily and lose moisture rapidly are undesirable. When oats follow corn, the soil is often only harrowed; but plowing is necessary for compact soils. On light, dry soils, rolling, followed by light cultivation, may be needed. In order to get the crop in as early as possible in spring, fall plowing is desirable. The soil should be in good tilth, whatever method is used in preparing it.

(3) *Fertilizers.*—When fed heavily, especially with nitrogen, the growth of straw is excessive, and the yield of grain is less and of inferior quality. When oats follow corn on rich soil liberally manured, or green manured, generally no fertilizer is applied, especially if any has been used for the corn along with farm manure or green manure. Ordinarily, it may be well to apply at seeding 200 or 300 pounds of **Mixture No. 2** (p. 363) especially when timothy and clover are seeded with it.

(4) *Seeding* in the North is done as early in spring as possible. The seed (usually 2 to 3 bushels) is ordinarily covered 1 to 2 inches deep.

In the South oats are sown in the fall and extensively grown as a winter crop, for which they are especially valuable, serving as a cover crop to prevent soil washing, and furnishing grain and green forage. When thus grown, top-dressing with about 100 pounds of sodium nitrate in spring is usually advised.

Oats serve well as green forage; also for hay. For this purpose, rich loams well supplied with organic matter give best yields. When grown for green forage, the oat crop should be liberally fertilized, more nitrogen being used than for grain. An application of 250 to 500 pounds of **Mixture No. 1** (p. 363), can be used an acre. The crop is best for green forage when the grain is in milk-stage. In cool, moist seasons, the oats furnish a continuous supply of green forage when one makes a series of sowings about 2 weeks apart.

Oats and Canada field-peas grown together make a satisfactory green-forage crop, and are also useful for hay. The soil is finely prepared to a considerable depth. Plant-food may be used as manure, 8 to 10 tons an acre, or 200 to 400 pounds **Mixture No. 2** (p. 363). The soil needs an abundance of organic matter and calcium carbonate. The amount of seed to use varies from 1 to 2 bushels of each per acre, but a great variety of proportions is used, according to individual conditions and preferences. The seed should be uniformly distributed as early as possible in spring. The best time for use as forage or for hay is when the oat-grain is in the milk stage and the peas are forming pods.

Barley possesses several marked characteristics that distinguish it from other cereals. It grows under a wide range of climatic conditions; can grow with a smaller supply of moisture than any other cereal; though best adapted to a warm, dry climate, it grows well under other conditions. It is grown for grain, green forage, pasturage and nurse-crop for alfalfa, clovers and grasses. It is used as a grain for two chief purposes, as a cattle food and for malting. When used for malting, the grain should have a high percentage of starch and a low amount of nitrogen compounds; be of uniform ripeness, size and variety, conditions essential to complete even and quick germination. The grain should be harvested before ripening has gone too far. Over-ripe condition at harvesting is revealed by discolored grains. Barley is more easily injured by dew, rain or sunshine than are other cereals.

(1) *Crop-rotations.*—Barley cannot be maintained continuously on the same soil as well as other cereals, so care is required to grow it in rotation. It often replaces wheat or oats as in the following rotations:

(*a*) Three to four years: Corn, 1 year; barley, 1 year; oats, 1 year, or timothy and clover, 1 to 2 years.

(*b*) Four to five years: Corn, 1 year; barley, 1 year; wheat, 1 year; clover or clover and timothy, 1 to 2 years.

(2) *Soil and preparation.*—Barley is more sensitive to soil character than are other cereals. Sandy, well-drained soils are better than heavy clays or soils not well-drained. A fertile soil, not over-supplied with easily available nitrogen, is best, especially for malting barley. It is more important than for other cereals that the seed-bed be mellow and deep, in order to permit rapid root growth and to make available the plant-food in the upper soil.

(3) *Fertilizers.*—The general use of fertilizers, as given for wheat (p. 390), applies to barley. It is best to apply manure to corn preceding barley and, if more plant-food is needed, to apply the plant-food mixture as directed above. The short period of growth and limited range of the root system require that the plant-food be readily available. When grown for feeding, larger amounts of nitrogen can be used; but when grown for malting less nitrogen and more potassium can be supplied in order to promote the production of starch. Experiments indicate that when materials which change insoluble into soluble potassium compounds are applied to a soil rich in potassium barley is beneficially affected. Favorable results following the application of 300 to 400 pounds of salt per acre may be explained in this way. An abundance of calcium car-

bonate in the soil has a beneficial effect, in part, for the same reason.

(4) *Seeding.*—The grain should be put in generally between the time of sowing spring wheat and oats in the North. The amount of seed most commonly used varies from 2 to 2½ bushels.

Barley as a green forage.—In many Northern States, barley makes an excellent late-fall forage crop. The conditions required to grow oats as forage apply to it. For this purpose it is sown about the middle of August, 2 to 2½ bushels of seed being used. The growth may also be used as a late fall pasture. In the South, it is sown in September or October and used as fall, winter or spring pasture. Good grain crops may be obtained even after considerable pasturage.

Barley and Canada field-peas furnish excellent late forage. About 1½ bushels of each are used for seeding during the first half of August, other conditions being observed as in case of oats and peas (p. 394).

Barley as a nurse-crop has been found extremely useful in starting seedings of alfalfa, clovers and grasses, for which purpose it possesses the following advantages: (1) It seldom lodges; (2) the stems are shorter and less leafy than other cereals, thus admitting sunlight and air; (3) it uses smaller amounts of water than other cereals. When barley is so used one sows from 3 to 4 pecks of seed per acre.

Buckwheat can grow at nigher altitudes than any other cereal and more is grown in higher altitudes than others. Its season of growth, 8 to 10 weeks, is the shortest of the cereals. It does best in a cool, moist climate, its yield being seriously decreased by hot, dry weather. It is injured by frost. It is grown largely for the grain, but to some extent as a green manure.

(1) *Crop-rotations for buckwheat.*—Ordinarily no special rotation is observed in growing buckwheat, because it is commonly grown on the least valuable land, or to take the place of some other crop that has failed. Buckwheat leaves the soil mellow so is often followed by potatoes. The following is a popular rotation:

First year, clover, the first growth cut and the sod plowed for buckwheat; second year, potatoes; third year, oats or wheat seeded with clover.

(2) *Soil and preparation.*—Buckwheat does best on somewhat sandy, well-drained land, but can be grown on heavy clays and even wet soils, though not to good advantage. As the crop can mature on poor soils it is often relegated to the poorest and worst-tilled parts of a farm, but it will respond profitably tc better treatment. For good crops, the land should be plowed and treated as for other cereals; early and thorough preparation will be repaid. This is, of

Effect of lime on barley. *Left*, limed; *right*, unlimed. Both sides fertilized alike with sulphate of ammonia.

course, not practicable when the crop is put in to take the place of some other crop that has failed.

(3) *Fertilizers.*—Buckwheat lodges badly on soils richly supplied with nitrogen. It responds well to applications of phosphorus and potassium. Since its growth occurs mostly during July and August, when nitrification is most rapid, little or no nitrogen need be supplied on soils that contain fair amounts of organic matter. On light, poor soils, deficient in organic matter, 4 to 6 tons of manure can be used; but, in general, manure is best applied to some preceding crop. On ordinary soils an application of 200 to

Thick stand of buckwheat sown in July for green manure to be plowed in early September for rye to follow as winter cover crop.

300 pounds an acre of **Mixture No. 1** or **No. 2** should prove beneficial. When buckwheat follows clover, a mixture of 200 pounds of acid phosphate and 50 pounds of potassium chloride an acre is suggested.

(4) *Seeding.*—The usual time of seeding is July 15 to August 15, though it may be put in any time from May to August. The amount of seed used varies from 2 to 5 pecks; it is covered 2 to 3 inches deep.

Buckwheat as a green manure possesses specific qualifications which fit it to meet some conditions. It grows rapidly, takes up considerable amounts of plant-food materials from the soil and, when plowed under, decomposes easily. If desired, two crops can be grown in one season. It cannot make use of atmospheric nitrogen and, therefore, returns no more plant-food than it takes from the soil. Its special use as a green-crop manure is on soils too poor to raise any other crop; it is often useful to furnish organic matter to poor, sandy soils where clover fails to grow. It also serves well as a cover-crop on soils that would otherwise lie bare during the latter part of the season after July. When used as a green-crop manure, it is preferably plowed under about the time it begins to blossom.

Millet is used in this country most largely as a forage crop, but to some extent as hay and as a cover crop. It is adapted to hot weather, resisting drouth well. Under favorable conditions of moisture and warmth, the crop is ready to cut for forage in 6 to 10 weeks. It is easily killed by frost. It has been found useful as a crop to put on new lands and also as a means of overcoming weeds, especially during hot weather.

(1) *Soil and preparation.*—Millet does best on rich, mellow, well-drained sandy loams. It can grow fairly on rather poor soils. On heavy clays or wet soils, it does not grow well. The soil should be given careful preparation, much as for barley, the seed-bed being made mellow to a considerable depth in order to enable the shallow rootlets to range easily for food.

(2) *Fertilizers.*—Millet is distinctly a surface feeder, so a good crop draws large amounts of plant-food from a rather limited layer of soil and often reduces the supply of available plant-food in this layer so much as to affect the next crop unfavorably, unless heavily fed. When obtainable, 10 to 12 tons of well-decomposed manure should be worked into the soil after plowing. When commercial fertilizers are depended upon wholly, an application of 250 to 500 pounds to the acre of **Mixture No.** 1 or **No.** 2 (p. 363) can be made.

(3) *Seeding.*—Owing to the sensitiveness of millet to cold, seeding should be delayed until danger of frost is passed. This will ordinarily be after May 15 in the North; from then on until early July will be a favorable time for seeding. The amount of seed used varies commonly from 2 to 4 pecks.

Millet as a cover crop can be put in after rye or other early ripening crop, or when the hay crop fails. It can be started when too late to grow corn.

Sorghum grows wherever corn grows; but it does better on poor soils and with much less moisture than corn. The sweet varieties are used largely as forage in dry regions and are grown also to some extent to make sugar and syrup. In nearly all respects sorghum is grown for forage in the same way that corn is grown for the same purpose. It needs more attention to weeds than corn, since its early growth is slow. For this reason, it should follow a cultivated crop. For forage, the seed can be sown broadcast, 20 to 25 pounds an acre, or seeded in rows, 10 to 12 pounds an acre.

TABLE 52—APPROXIMATE AMOUNTS OF PLANT-FOOD CONSTITUENTS IN ONE CROP. GRAINS

Crop	*Portion of crop*	*Yield per acre*	*Nitrogen*	*Phosphoric acid (P_2O_5)*	*Potash (K_2O)*
			Lbs.	*Lbs.*	*Lbs.*
Corn.........	Grain	25 bu.	23.2	9.1 (4. P)	5.5 (4.6K)
	Stalks, etc.	1500 lbs.	15.0	4.5 (2. P)	21.0 (17.4K)
	Cobs	250 "	1.0	0.2 (0.1P)	1.1 (0.9K)
			39.2	13.8 (6.1P)	27.6 (22.9K)
	Green forage	20000 lbs.	60.0	26.0 (11.4P)	66.0 (55. K)
Wheat........	Grain	25 bu.	30.0	12.8 (5.6P)	6.0 (5. K)
	Straw	2500 lbs.	12.5	3.8 (1.7P)	15.0 (12.5K)
			42.5	16.6 (7.3P)	21.0 (17.5K)
Rye..........	Grain	20 bu.	19.1	9.8 (4.3P)	6.7 (5.6K)
	Straw	2000 lbs.	10.0	6.0 (2.6P)	17.0 (14.1K)
			29.1	15.8 (6.9P)	23.7 (19.7K)
	Green forage	15000 lbs.	67.5	30.0 (13.2P)	97.5 (81. K)
Oats..........	Grain	25 bu.	16.0	6.5 (2.9P)	4.8 (4. K)
	Straw	1250 lbs.	8.0	2.5 (1.1P)	15.6 (13. K)
			24.0	9.0 (4. P)	20.4 (17. K)
	Green forage	12000 lbs.	72.0	13.0 (8. P)	54.0 (44.8K)
Barley........	Grain	25 bu.	21.0	9.0 (4. P)	6.0 (5. K)
	Straw	1600 lbs.	9.6	3.2 (1.4P)	17.6 (14.6K)
			30.6	12.2 (5.4P)	23.6 (19.6K)
	Green forage	10000 lbs.	40.0	15.0 (6.6P)	50.0 (41.5K)
Buckwheat....	Grain	20 bu.	15.0	6.0 (2.6P)	3.0 (2.5K)
	Straw	5000 lbs.	62.5	7.5 (3.3P)	57.5 (47.7K)
			77.5	13.5 (5.9P)	60.5 (50.2K)
	Green crop	10000 lbs.	40.0	8. (3.5P)	35.0 (20. K)
Millet........	Green forage	20000 lbs.	60.0	20.0 (8.8P)	100.0 (83. K)
Sorghum......	Green forage	20000 lbs.	60.0	24.0 (10.6P)	60.0 (49.8K)

P, phosphorus. K, potassium.

The crop should be liberally fed, using 6 to 8 tons of well-decomposed manure an acre, well worked into the soil and 200 to 300 pounds of **Mixture No. 2** (p. 363).

When sorghum is grown for sugar or syrup, 400 to 600 pounds of **Mixture No. 2** can be applied.

XXXI

TUBER AND ROOT CROPS

TUBER and root crops include: (1) Potatoes, (2) sweet potatoes, (3) beets, (4) turnips, (5) carrots, (6) parsnips, (7) radishes, (8) salsify.

These are used largely as human food, but to some extent as food for farm animals. They possess common characteristics which distinguish them from cereals, especially in their relations to plant-food. First, the root growth is such that they are not commonly regarded as strong feeders; that is, the root system does not range widely through the soil, so the plant-food supply within reach is less than in case of plants whose roots have an extensive reach. A supply of readily available plant-food in the neighborhood of the growing roots, therefore, meets with prompt response. In most cases, it is desirable that the soil be mellow to a much greater depth than for cereals in order to permit the root to go as deeply as possible for food. Second, the food product most characteristic of these crops is carbohydrate material, especially starch and sugar. Such crops use relatively large amounts of potassium (p. 54). Many of them have a long growing season and can utilize organic nitrogen compounds. While the percentages of plant-food constituents are not large in these crops, the yields are heavy and therefore the amounts of plant-food contained in the crop grown on an acre are large in the aggregate as compared with many other crops.

Potatoes.—Though potatoes grow best in moderately moist, cool climates, they do well under a great variety of climatic conditions and are widely distributed.

(1) *Crop-rotations for potatoes.*—For several reasons, crop-rotations are essential for best results with potatoes, especially because, when grown continuously on the same soil, they are subject to scab and other diseases. They do especially well after clover and cowpeas. The following rotations are illustrations:

(*a*) Three years: Winter wheat seeded to clover in spring, 1 year; clover, hay and green manure, 1 year; potatoes, 1 year.

(*b*) Four years: Rye, seeded with clover and timothy, 1 year; hay, 1 year; potatoes, 1 year; oats or barley, 1 year.

How potash affects potato plants. In this potato-wheat-clover rotation the potato plants have on the left been fertilized with superphosphate; those on the right with muriate of potash.

(*c*) Four years: Wheat seeded with clover, 1 year; clover hay and green manure, 1 year; potatoes, 1 year; beans, 1 year.

In case of long seasons, potatoes often follow some early crop, or are put in as an early crop to be followed by something else.

(2) *Soil and preparation.*—While potatoes do well on many kinds of soil, they do best on rich, well-drained, sandy loams that contain an abundance of organic matter. When grown on heavy soils not well drained, the tendency is to produce watery tubers of undesirable quality. The soil should be mellow, deep and porous. It is desirable that heavy soils be plowed in the fall. Deep plowing is advisable, since the roots feed well down in the soil when given a chance.

(3) *Fertilizers* to use depend on (*a*) amount of organic matter in the soil, (*b*) time of season during which the crop is grown, (*c*) quality of the crop, (*d*) time and manner of application, and (*e*) amount of plant-food constituents.

(*a*) Organic matter in soil. In soils that already contain an abundant organic matter, whether furnished by farm or green-crop manure, only application of commercial plant-food constituents needs to be considered. In soils not well supplied, organic matter is furnished by manure or green manure, preferably leguminous, according to the special conditions on each farm. When manure is available and is not preferred for some other crop, it is applied at the rate of 10 to 12 tons an acre; if fresh and coarse, it is better to apply it in the fall and plow in, but, if well-decomposed, it can be applied in the spring to the plowed ground and thoroughly harrowed in. When manure is used in larger amounts, it is regarded as better practice to apply it to corn, clover or grass and then follow with the potato crop, since the direct application of manure favors the growth of scab on tubers. The soil should not be limed for a potato crop, especially with slaked-lime (calcium hydroxide) or quicklime (calcium oxide), owing to the tendency to produce scab. The application is best made for the leguminous crop in the rotation.

(*b*) Early and late potatoes require liberal amounts of plant-food and even on fertile soils respond well in yield to generous applications of available plant-food constituents. It must be kept in mind that the growing season of the early crop is desired to be short, so much of the growth comes before nitrification and other processes have reached their maximum activity and produced their greatest effects. Therefore, the forms of plant-food applied should be readily available. In the case of late potatoes, the growing period is more prolonged and comes during the part of the season when unavailable plant-food in the soil is being most rapidly made available. Therefore, those forms of plant-food constituents, especially nitrogen, can be used which become available gradually; for example, in the case of early potatoes, the nitrogen applied is preferably in the form of nitrate, while, for late potatoes, it may be distributed in the forms of nitrate, ammonia and organic nitrogen, such as sodium nitrate, ammonium sulfate, tankage, cottonseed-meal or fish-scrap, etc. Excessive application of nitrogen compound is indicated by dark-green colored potato tops, as compared with the normal color of lighter green. Deficiency of nitrogen is shown by a light yellowish color.

(*c*) Quality of the crop. Some experimental work has been done to determine the effect of potassium sulfate, as compared with chloride, on the quality of potato tubers. In respect to yield, results may not vary much, whichever compound is used; but experiments show that in some cases sulfate gives larger yields. So far as present knowledge goes, potassium chloride, especially on clay soils, appears to favor a larger percentage of water in the potatoes than when sulfate is applied. On lighter soils, there appears to be little or no appreciable difference. It is said, however, that sulfate produces tubers of cleaner appearance and more uniform size. On heavy or moist soils, it is, therefore, well to take the precaution of using the sulfate. With the decreased cost of potassium nitrate, this compound can well be used as a source of part of the nitrogen and all of the potassium.

(*d*) Time and manner of application. When large amounts of high analysis fertilizers are used for potatoes, care must be taken to keep the fertilizer from coming into direct contact with the seed pieces, in order to avoid interference with germination. The fertilizer may be all applied in the line of the rows and worked well into the soil or well covered with soil before planting. Another method is to apply one-half or two-thirds in the row before planting, the remaining portion being used as a side-dressing after the plants are 2 to 4 inches high, using special machinery to apply and cover. In some cases one-half the fertilizer is broadcast and well worked into the soil before planting and the rest put in the row at the time of planting. On light soils application in the row is favored. On heavy soils with retentive subsoil, in places subject to drouth, good results have been obtained by broadcasting and harrowing into the soil all the fertilizer before plowing, the fertilizer thus being placed well down in the soil, causing the roots to go down more deeply than when the fertilizer is applied nearer the surface. In the case of light sandy soils, as in Florida, where fertilizer is used at the rate of 2,000 to 2,500 pounds an acre, the soil is prepared in beds about 36 to 40 inches apart and the fertilizer worked in two weeks before planting.

(*e*) The amount and kind of plant-food constituents used by potato growers varies from 500 to 3,000 pounds an acre. On light sandy soils, the higher amounts are applied. On heavier soils the smaller applications, especially when supplemented by stable manure or green manure. On muck soils the amount varies from 800 to 1,500 pounds.

In composition, the potato fertilizers in use vary widely, the following analyses illustrating in part those on the market: 2—8—10, 4—6—10, 4—8—4, 4—8—7, etc. The mixture that follows comes fairly near to meeting the usage of many potato growers in respect to analysis: **Plant-food Mixture No. 4.**—*For potatoes and garden crops: Analysis:* 4—8—8; *formula;* potassium nitrate, 365 pounds (47 pounds N, 160 pounds K_2O); ammonium sulfate, 165 pounds (33 pounds N); superphosphate (16%) 1,000 pounds (160 pounds P_2O_5); drier, 470 pounds.

This fertilizer contains only quickly available constituents, so is adapted to the growth of early potatoes. For late potatoes one-half the nitrate can be replaced by organic material used, such as meat tankage, fish scrap, cottonseed-meal, etc. The formula given above could for illustration, be changed to the following, giving the same analysis approximately: **Plant-food Mixture No. 5**, ground fish, 555 pounds (40 pounds N); potassium nitrate, 165 pounds (20 pounds N and 70 pounds K_2O); ammonium sulfate, 100 pounds (20 pounds N); superphosphate, 1,000 pounds (160 pounds P_2O_5); potassium chloride, 180 pounds (90 pounds K_2O).

Sweet potatoes.—In growing this crop, certain qualities of the product must be considered. Watery sweet potatoes are undesirable; those of good edible quality should be mealy when cooked. This quality is the result of soil conditions and of plant-food supply. From the market standpoint, potatoes of medium size and well-rounded shape are preferable to large elongated ones. The crop is not grown commercially north of New Jersey in the eastern United States. The best climatic conditions are warmth and moisture during most of the growth, followed by moderately dry weather during the final maturing.

(1) *Crop-rotations for sweet potatoes.*—An excellent preparation for a sweet-potato crop is some crop like corn, tomatoes, melons, etc., with which crimson clover is seeded at the last cultivation and plowed under when 6 to 8 inches high. Crimson clover or some other legume can also follow any early crop for the same purpose.

(2) *Soil and preparation.*—Soils best adapted for sweet potatoes are well-drained, porous, warm and sandy. Good drainage is essential. Some highly crumbly and porous clays are successfully used. The land is plowed rather shallow, (5 or 6 inches), at a time in spring when it crumbles perfectly under the plow and gives a fine tilth under the harrow. Special care is required to pulverize the

soil completely in case of sandy soils that contain enough clay to form clods by baking.

(3) *Fertilizers.*—Leguminous crops to furnish organic matter, with some nitrogen and commercial fertilizers, are generally advocated as more economical than the use of manure. Large applications of nitrogen, particularly as nitrate, are to be avoided, since growth of vines is stimulated. While the yield may be thus increased, the quality of the potatoes is impaired. Large proportions of potassium and phosphorus in relation to nitrogen are generally used.

Numerous mixtures with a fairly wide variation in percentage composition are used. They contain from 2 to 4 per cent. nitrogen, 8 to 10 per cent. available phosphoric acid and 6 to 10 per cent. potash. When manure is used, it is best to apply to the crop preceding. When the crop is grown on soil well supplied with organic matter by a legume the following mixture is suggested: Plant-food Mixture No. 6.—*For use on sweet potatoes: Analysis:* 2—10—8; *formula;* potassium nitrate, 80 pounds (10 pounds N and 35 pounds K_2O); ammonium sulfate, 50 pounds (10 pounds N); fish-scrap, 250 pounds (20 pounds K); superphosphate (16%), 1,250 pounds (200 pounds P_2O_5); potassium chloride or sulfate, 250 pounds (125 pounds K_2O); drier, 120.

Cottonseed-meal, castor pomace or meat tankage could be substituted for fish-scraps to furnish organic nitrogen; the purpose in mind is to have most of the nitrogen in a form not immediately available. Such a mixture may be applied in amounts of 500 to 1,500 pounds an acre. When the smaller amounts are used, this may be applied in the furrows ten to fifteen days before the vines are planted. When amounts exceeding 1,000 pounds an acre are used, the fertilizer is drilled in after harrowing or sown broadcast and then harrowed in, or about two-thirds may be applied in this way and the remainder in the furrows a week or so before setting plants.

Beets are grown for three general purposes, as a garden crop for table use, as a field crop for cattle-food, and as a commercial source of sugar, varieties being chosen according to the particular purpose for which the crop is grown. Beets do best in cool climates with abundant sunshine, especially during the maturing period in case of beets grown for sugar.

(1) *Crop-rotations.*—Beets do well after such crops as corn, cabbage, wheat, and clover. For field-beets or sugar-beets, the following rotations are good:

(*a*) Four years: Corn, with heavy application of manure, 1 year; beets, 1 year; oats or barley seeded with clover, 1 year; clover, 1 year.

(*b*) Five years: Wheat, 1 year; beets, 1 year; clover, one crop for hay and one for green manure, 1 year; potatoes, 1 year.

(2) *Soil and preparation.*—Generally speaking, beets grow well on any well-drained, deep and mellow soil good for wheat, corn or potatoes. For best results in yield, rich clay loams are best. A fairly level soil is desirable, especially for sugar-beets. Lighter soils mature the crop earlier, which is a marked advantage in case of early beets for table use. It is generally desirable to plow in the fall, especially for early table-beets and in case of heavy soils. The plowing should be deep to permit extension of feeding roots downward. For sugar-beets and mangels, plowing should be 9 inches deep, followed by a subsoiler to loosen 6 or 7 inches more. As early in spring as conditions permit, the soil is put into fine tilth and made mellow just before planting, for whatever purpose beets are grown.

(3) *Fertilizers.*—Treatment with fertilizers varies materially according to the end in view. When manure is obtainable, it gives excellent results, if properly used and supplemented with commercial fertilizer. When heavy applications of coarse manure are made, it is better to apply to a crop preceding or to plow in during the fall preceding the beet crop. On rather heavy soils, 20 tons or more an acre of partly decomposed farm manure can be applied just before planting with satisfactory results, even on sugar-beets, but, under most conditions, such treatment cannot be regarded as economical. When manure is used, an application of not more than 10 or 12 tons is advised under ordinary circumstances.

(*a*) Fertilizers for early garden-beets. In growing early beets for market, the chief aim is to push the growth as rapidly as possible and get them into market when prices are highest. Other conditions being favorable, success in accomplishing this depends upon a liberal supply of available nitrogen and phosphorus. The early growth takes place when soil nitrates are usually at their lowest and before their formation has been actively renewed in spring. Therefore, a generous supply of nitrate is desirable. At the time of planting, a good supply of available nitrogen, phosphorus and potassium should be within reach of the young plants. The nitrate to be used, should not all be applied at planting time but some should be reserved for several top-dressings. If it were all applied at the start, there would be risk of considerable loss by leaching (p. 114) in case of heavy rainfall, which is common when the beet crop is getting started. An application of 800 to 1,200 pounds of **Mixture No. 4** (p. 406) at planting time is suggested. After the plants are well started, surface applications of 50 to 100 pounds of nitrate (p. 30) about once in ten days for three or four weeks. When earliness is not an object, the top-dressings can be smaller or omitted altogether. On lighter soils, the first application may be larger than given.

(*b*) Fertilizers for beets grown as cattle-food. Mangels or mangel wurzels (known also as mangolds, field-beets and cattle-beets) are extensively grown for cattle-food. Sugar-beets are also grown for the same purpose. In growing beets for this purpose, the chief object is to obtain a large yield, which depends on rapid and uninterrupted growth. The season of growth is prolonged and it is therefore not necessary to supply the nitrogen so largely as nitrate as in the case of early table-beets. Whichever kind of beet is used, the plant-food requirements are the same. Manure can be used to advantage, (10 to 12 tons) and an application of 500 pounds of **Mixture No. 6** (p. 407) worked into the soil just at seeding. On light soils, on which no farm manure is used, 800 to 1,000 pounds of **Mixture No. 4** (p. 406) or **No. 6** can be used.

(*c*) Fertilizers for sugar-beets. When grown for sugar, beets are valued according to richness and purity of sugar and for the yield of sugar per acre. Excessive application of nitrogen produces increased yields of large beets, but with decreased sugar; enough nitrogen, however, must be used to promote rapid growth early in the season. In the production of sugar, potassium is highly important and on light soils must be supplied liberally. Well-decomposed manure well worked into the soil before seeding has given excellent results, especially on soils inclined to be heavy. An abundant supply of quickly available phosphoric acid is to promote early, rapid growth. It is applied in considerably larger amounts than analysis of the beet indicates. The formation of sugar occurs in the latter part of the growing season, so in order that this may not be too much delayed, it is important during the early stages of growth to encourage rapid and extensive growth of leaves, since through them sugar is formed (p. 127). If slow-acting forms of nitrogen are supplied, the vegetative growth is prolonged unduly and sugar formation retarded, so the amount formed is likely to be less than where it begins earlier.

Hence plant-food should be supplied in readily available forms and in generous amounts. These requirements are met by **Mixture No. 4** (p. 406), of which an application of 600 to 1,000 pounds an acre will not usually be excessive on good soils; while on light soils the amount may be 900 to 1,200 pounds.

The plant-food in the fertilizer will be more economically used if applied in three lots, one-half being thoroughly worked into the soil immediately before seeding, one-fourth between rows or as a side-dressing after the plants are up and just before the first cultivation, and the remainder just before the second cultivation. When applied in this way, the danger of loss of nitrate by leaching in case of heavy rain is diminished, the soluble phosphate has a chance to reach the roots before becoming entirely reverted, and the direct action will likely be more manifest than when all the plant-food is applied at seeding time or before.

(4) *Seeding.*—Beet seeds are sown in rows, at varying distances apart. Garden-beet rows may be only 12 inches apart, though usually far enough to permit use of horse-cultivator; ordinary field-beets and sugar-beets, 18 to 24 inches apart; mangels, 2 to 3 feet. The seeds are covered ½ to 1½ inch. The amount of seed used varies usually from 5 to 8 pounds an acre for field-beets, but from 15 to 20 pounds for sugar-beets. It is better to use too much seed than to have the plants too thinly scattered in the row. When the plants are high enough to handle easily, they are thinned. In case of garden-beets, this is usually done when the tops are large enough to use as greens, in case of field and sugar-beets, when four leaves show. Garden-beets are thinned to 4 to 6 inches between plants; ordinary field-beets and sugar-beets are about 8 inches apart, while mangels are 8 to 10 inches.

(5) *Cultivation.*—For whatever purpose grown, beets should be as thoroughly cultivated as conditions permit, as often as a week or 10 days if possible. The early cultivation should be fairly deep to remove weeds and to enable the soil to receive and hold moisture to best advantage.

Turnips are grown for table use and to some extent for feeding sheep and cattle; they may also be utilized at the same time as a cover crop. As a garden crop, turnips are grown for early markets as well as for ordinary market purposes. They ordinarily require a damp, cool climate, but need less sunshine than beets, especially sugar-beets. Rutabagas bear about the same relation to ordinary turnips that mangels do to beets. They are grown largely for stock-feeding. Their place in crop rotations, character of soil, preparation and cultivation, is the same as beets. The general conditions of seeding are about the same as with beets, except that the amount of seed used is only 2 to 4 pounds an acre.

Fertilizers.—Turnips respond most favorably to a generous supply of soluble phosphorus compounds. When soils have been treated with 10 tons of good manure an acre, the application of phosphoric acid alone in soluble form usually is sufficient to secure good crops, especially on soils not deficient in potash. For growing early turnips for table use, the same fertilizer treatment is recommended as for early beets (p. 407). For field-turnips grown with manure on good soils, an application of 250 pounds of acid phosphate and 75 pounds of potassium chloride may be used to the acre, or 500 pounds of **Mixture No.** 1 (p. 363); on light soils, larger amounts (800 to 1,200 pounds) should be applied.

Turnips grown as cover-crops may be used after any early crop or may be seeded in corn at the time of the last cultivation. They are then available for late fall and winter feeding. Some deep-rooting variety is preferable for use as cover-crop. The seed is sown broadcast at the rate of 2 or 3 pounds an acre.

Carrots are grown for both table use and food for farm animals. The crop is hardy, being adapted to wide climatic and soil variations. Methods of treatment used in growing beets and turnips are applicable to carrots. The soil should be deep and well supplied

Typical injury to corn on newly drained alkali land.

with plant-food, a rich, mellow loam being most suitable. Since the early growth is slow, it is desirable to seed the crop on soil as free as possible from weeds. Fertilizers are used as in case of field crops of beets or turnips.

Parsnips, largely grown as a garden crop, do well on rather heavy loams and even on clay soils. Since the roots grow long, it is important that the soil be well loosened 16 or 18 inches deep. Cultural conditions that apply to beets and turnips are suited to parsnips. Excessive supply of nitrogen tends to produce coarse texture. The amount of seed used per acre varies from 4 to 6 pounds, but, if more than a year old, it is likely to fail. For table use, the roots are best left in the ground over winter, or stored in a cool place, since they acquire a sweetness and tenderness that they do not possess in the fall.

Radishes, grown both as a garden and as a greenhouse crop, are hardy and among the most rapid-growing and earliest crops. In order to be tender and crisp in quality, they must be raised under conditions which permit continuous, rapid growth. They are especially adapted to cool weather; when grown in hot weather, they tend to develop a rank growth, which results in a tough, fibrous or stringy texture, lacking in desired crispness and tenderness. The soil should be mellow, well supplied with organic matter and an abundance of easily available plant-food. For a fertilizer, **Mixture No. 4** (p. 406) is recommended, 500 pounds or more per acre worked lightly into the soil just before seeding.

Salsify or vegetable oyster, grown only for the table is a plant that resembles the parsnip somewhat in its root growth, but is not as large. When cooked, the roots have a flavor that resembles that of oysters. It is slow of growth, requiring the entire season. It is best adapted to deep, rich, moist soils. It is planted in rows 12 inches or more apart, and the plants stand 4 to 5 inches apart in the row after thinning. The amount of seed used per acre is 8 to 10 pounds. The culture and general treatment are essentially the same as for other root crops. The roots are better for cooking when left in the ground through winter. **Mixture No. 4 or 5** (p. 406) is adapted to the needs of this crop in amounts of 500 pounds or more an acre.

TABLE 53—APPROXIMATE AMOUNTS OF PLANT-FOOD CONSTITUENTS IN ONE CROP. ROOTS AND TUBERS

Crop	*Yield per acre*	*Nitrogen*	*Phosphoric acid (P_2O_5)*	*Potash (K_2O)*
		Pounds	*Pounds*	*Pounds*
Potatoes, Irish	150 bu.	31.5	13.5 (6. P)	45.0 (37. K)
" sweet	200 "	27.5	11.0 (4.8P)	55.0 (45.7K)
Beets, common	25,000 lbs.	62.5	25.0 (11. P)	125.0 (104. K)
" mangels	40,000 "	60.0	40.0 (17.6P)	140.0 (116. K)
Turnips, common	20,000 "	50.0	20.0 (8.8P)	90.0 (74.7K)
" rutabagas	20,000 "	40.0	24.0 (10.5P)	100.0 (83. K)
Carrots	10,000 "	23.0	13.0 (5.7P)	53.0 (44. K)
Parsnips	12,000 "	26.4	24.0 (10.5P)	78.0 (65. K)
Radishes	2,000 "	3.0	1.2 (0.5P)	7.0 (5.8K)
Salsify	5,000 "	25.0	8.0 (3.5P)	27.5 (22.8K)

P, phosphorus. K, potassium.

XXXII

GARDEN CROPS

The growing of garden crops on a commercial scale has become an important agricultural specialty. In market-garden regions, many crops are grown in succession on a relatively restricted area. No line of plant production depends so much on commercial fertilizers as do market-garden crops. Not only are they used to increase yield, but especially to control quality. Formerly, sole dependence was placed upon manure and composted materials to supplement the plant-food needs of soils; while this is still true in cases, market gardeners have come to depend increasingly upon commercial plant food materials, which they use in amounts that would be extravagant for ordinary crops.

GENERAL CHARACTERISTICS

While market garden crops vary greatly in individual characteristics, they have certain characteristics in common. In general, their growth is stimulated or forced so as to make quick and continuous development. So far as preparation of soil and supply of plant foods are concerned, the aim is to furnish such conditions that, with a proper amount of sunshine and rainfall, the crop will be pushed forward as speedily as possible without delay or interruption.

One edible quality which all vegetables must possess in common is that they be agreeable to the taste, appetizing, palatable; this is usually dependent on some peculiar taste or combination of taste and smell. For example, sweetness in peas, sweet corn, melons, etc., together with the additional flavor characteristic in each case, determines the palatability, while in rhubarb, tomatoes, etc., it is combination of acidity and other qualities; in the case of onions, horseradish, cabbage, turnips, etc., the characteristic flavor is due to the presence of certain sulphur compounds. Over-developed flavor, as in the case of excessive pungency found in some onions, or the strong disagreeable flavor of some celery, disagreeable bitterness in lettuce, over-acidity in tomatoes, absence of sweetness in peas, corn, etc., may, to some extent, be due to failure to regulate the supply of plant-food, although these defects are often due to variety used or to weather conditions.

Another edible quality of importance in most vegetables is that dependent on tenderness, succulence and crispness. The development of tough, fibrous texture greatly depreciates this quality, as illustrated in case of beets, celery, lettuce, radishes, etc., of poor quality. This defect is more or less due to regulation of plant-food supply, so can be largely prevented.

Generally speaking, the ideal qualities of vegetables are dependent upon conditions that make possible quick and uninterrupted growth. When growth is interrupted or retarded, some undesirable quality of flavor or texture is likely to make its appearance. When interruption of growth is due to lack of sunshine, or rainfall, one is, of course, helpless, unless irrigation can solve the latter problem. However, under favorable conditions of soil, light, warmth and moisture, the quality of vegetables can be reasonably well controlled by judicious regulation of the amounts and kinds of plant-food supplied.

The problem of successfully growing market-garden crops becomes one, therefore, of so regulating the supply of plant-food as to insure rapid and continuous growth, weather conditions being favorable.

SOIL CONDITIONS

For the successful growth of garden crops, certain essential soil conditions are common to all market-garden soils; namely, (1) Organic matter, (2) calcium carbonate, (3) influence of character of soil, (4) preparation.

Organic matter.—Garden soils should contain a generous supply of organic matter. The customary method of keeping up the supply has been to make large applications of manure yearly. With decreasing supplies, increasing dependence is placed on some quick-growing leguminous cover crop (crimson clover, cowpeas, etc.) as part of a well-planned rotation. This is easy of accomplishment where the growing seasons are long.

Calcium compounds.—The necessity of keeping a large amount of decaying organic matter in a soil involves the formation of acids (p. 94) and the removal of calcium carbonate. In order to prevent the exhaustion of this material and the development of soil acidity, it is necessary to keep market-garden soils, in case of most crops, well supplied with this compound.

Influence of character of soil.—While some vegetables grow on a great variety of soils, others are more limited. In general, the growth of individual crops is more or less localized, one or more crops being grown in one locality where the soil and climatic conditions are especially favorable. In the case of home gardens, whatever kind of soil happens to be at hand is adapted as far as possible to the growing of as great a variety of products as are used in the household. As regards the influence of the character of the soil upon the crop growth, light, porous, warm soils promote rapid growth and early ripening, while the reverse is true of heavy soils. Light soils should, therefore, be modified to hold moisture and heavy soils to become more open.

Preparation of soil.—All market-garden soils require thorough tillage. The depth required varies for the various crops, but in all cases the soil must be mellow and finely pulverized.

USE OF FERTILIZERS

To promote early, rapid and continuous growth requires nitrate and soluble phosphates in considerable amounts. Though much less phosphorus is used by garden crops than either nitrogen or potassium, its supply in most soils is much less (p. 84). Potassium is used in relatively large amounts in the leaves; therefore, when this part of the crop is removed for market, the amount of potassium taken from the soil is usually greater than when some other part of the plant is sold. Generally speaking, potassium needs usually to be applied rather generously on light, sandy soils and soils rich in muck or peat. On soils that contain fair amounts of clay, much lighter applications often suffice, especially if the soil is well supplied with calcium carbonate, and also when considerable amounts of sodium nitrate are used.

The traditional method of maintaining plant-food supplies in garden soils was to use large quantities of manure. Good manure when obtainable, can be advantageously used, but the applications should be moderate (8 to 10 tons an acre) and be supplemented by commercial plant-food, especially phosphorus.

The general rule is to apply enough of each plant-food constituent to meet or exceed the demands of large crops. The kinds and amounts vary in case of different crops. As a general fertilizer for market-garden use, **Mixture No. 4** (4—8—8, p. 406), or modifications of this have been extensively used with success.

These fertilizers can be used on practically all garden crops, adaptations being made in special cases by varying the amounts or by supplementing with additions of other materials.

We shall consider only those crops of general interest. For convenience, we will divide them, as follows, into groups, the members of which are alike in respect to their plant-food requirements: (1) Asparagus and rhubarb, (2) beans and peas, (3) cabbage, cauliflower and brussels sprouts, (4) celery, (5) cucumbers, muskmelons, watermelons, pumpkins and squashes, (6) sweet corn, (7) lettuce and spinach, (8) onions, (9) tomatoes, (10) eggplant.

Asparagus, on account of its being one of the earliest spring vegetables, is an important product for market and home gardens. It is hardy and, if once properly established and cared for, the same roots will continue to produce generous annual yields for 20 years or more. While other crops cannot be grown on the same land with asparagus, and while it is cropped only once, yet its cropping season is spread over one or two months and the yield can be made large by generous feeding and care.

(1) *Soil and preparation.*—Asparagus will grow on almost any agricultural soil, but it does best on deep, well-drained, loose, rich loams. In preparing soil for a permanent planting, the ground should be plowed 10 or 12 inches deep the fall before and well harrowed in the spring.

(2) *Planting and cultivation.*—In making a small bed for home use none but the largest one year plants should be planted; older ones suffer such losses of root as to delay their recovery one or more years. In growing roots for one's own use, seed is sown early in spring in rows 14 to 16 inches apart at the rate of one ounce for 50 feet, furnishing about 400 plants; the seed-bed should be rich, mellow and loose. The soil is kept free from weeds and well cultivated during the season. The roots can be transplanted into permanent quarters the following spring. The roots when transplanted are set deep so that the crown is 6 to 8 inches below the surface of the soil, in order to permit cultivation without injury to the roots. They are set 18 to 24 inches apart in rows 4 feet apart. The soil is regularly cultivated between rows at intervals. After the tops die in the fall and are removed, the entire surface is cultivated about 4 inches deep. Cultivation begins again in spring and is continued as before.

(3) *Cutting shoots.*—A few shoots may be cut the second year, but regular cropping should not take place before the third. Cutting should not usually continue longer than 6 to 8 weeks.

(4) *Fertilizers.*—In starting an asparagus planting, well decomposed stable manure, when obtainable at reasonable cost, should be applied to the plowed ground at the rate of 10 or 12 tons an acre and thoroughly worked into the soil before the roots are planted; or it may be applied to clover sod and turned under the fall before planting. In spring before planting 600 to 800 pounds of **Mixture No. 4** or **5,** according to soil conditions, may be applied. The purpose should be at the beginning to have an abundance of organic matter and available plant-food in a deep and mellow soil (p. 406).

In the absence of manure, use more of the fertilizer mixture and each spring apply 400 to 800 pounds of the same mixture. When well-decomposed manure is obtainable, it is a good plan to apply and harrow in early in spring and supplement with commercial fertilizer according to need. Some growers apply 100 pounds of sodium or calcium nitrate when the cutting season is about half over. Soon after the cutting season is over it is important to make another application of fertilizer, largely in readily available form—500 to 600 pounds of **Mixture No. 4** (p. 406) an acre between rows well worked in. The significance of this later application should be understood. Constant removal of shoots during the cutting period is exhaustive so the plant needs an abundance of plant-food to help it recover by making a vigorous growth during the rest of the season and storing nutritive material for the growth of shoots during the next cutting season. The application of plant-food is for two purposes: (1) To increase the yield by increasing the number and, particularly, the size of the shoots, and (2) to produce shoots of the highest possible quality in respect to flavor and tenderness. Small, insignificant-looking shoots generally indicate lack of nutrition or cultivation, or excessive cutting.

Some growers use a leguminous crop, as cowpeas, soy-beans, etc., for a late corn crop between the rows two or three weeks after the cutting season.

Fertilizers do not control the color of the asparagus shoots. The matter of white or green shoots depends upon protection from, or exposure to, sunlight. The demand of the market determines which kind is most profitable to furnish.

Rhubarb, also commonly known as pie-plant, resembles asparagus as a garden crop, because it is early and because the stems or tops are pulled for use. It is now extensively grown for canning. The methods of setting roots, character of soil and methods of cultivation are essentially the same as for asparagus. The hills are placed four feet or more apart. When once established, the plants survive much neglect, but respond profitably to good cultivation and feeding. Methods used to supply plant-food to asparagus apply equally well to rhubarb. The flower or seed-stalks are pulled out as soon as they appear.

Garden-beans.—Edible garden-beans include (1) those grown for use as string-beans, both fresh and canned, and (2) lima beans, the latter grown in both poled and dwarf forms.

(1) *Soil and preparation.*—Garden-beans respond to good soil conditions. While growing on a wide range of soil types, earlier crops are favored by warm, somewhat sandy loams; later plantings do better on heavier loams, well drained and well supplied with organic matter. Late fall or early spring plowing is desirable, the soil being carefully prepared as a seed-bed before planting.

(2) *Fertilizers.*—Garden-beans, especially the pole types, use generous applications of plant-food. Although beans are able to utilize atmospheric nitrogen, it is, nevertheless, desirable, in growing garden varieties, to supply a liberal amount of nitrogen along with phosphorus and potassium compounds, as well as calcium carbonate, in case of acid soils. Excessive application of farm manure or of the commercial nitrogen-containing compounds causes the plants to develop too much vine and leaf and decrease the yield of pods. When manure is used, it should, preferably, be well rotted and applied at the rate of 10 to 12 tons an acre to the previous crop or (especially for lima beans) in the previous fall before plowing. When manure is not obtainable, a leguminous sod or cover crop should be used. The 4—12—4 mixture (p. 406) can be employed (300 to 600 pounds an acre) on good loams; and, on poorer soils, 500 to 1,000 pounds. Applied in small amounts (300 pounds), the fertilizer is worked into the soil along lines where the rows are to run, mixing the fertilizer with the soil by harrowing or cultivating. In case of application of 600 pounds or more, the fertilizer may all be applied broadcast and thoroughly harrowed in before planting or part may be reserved and applied in the hill or the drill. Whatever method is used the fertilizer must be kept from contact with the seed.

Field-beans grown for the seeds require treatment that differs considerably from garden beans.

(1) *Rotations for field-beans.*—The following illustrate rotations employed in bean-growing sections:

(*a*) Three years: Beans, 1 year; wheat, 1 year; clover, 1 year.

(*b*) Four years: Corn or potatoes, 1 year; beans, 1 year; wheat seeded with clover, 1 year; clover cut for hay and then plowed under, 1 year.

(2) *Soil and preparation.*—Soils best adapted to field-beans are well-drained clay loams; sandy or gravelly loams give good crops when kept well supplied with organic matter and plant-food. Soils extreme in either heaviness or lightness are undesirable, also soils that contain excess of organic matter like muck or peat soils. Generally speaking, good corn or wheat land is good bean land. For best results, land should be plowed 4 to 6 weeks before planting, and be put in good mechanical condition by harrowing to prevent growth of weeds and to hold moisture.

(3) *Fertilizers.*—When beans are grown after clover or when manure is used, little nitrogen need be supplied in addition. On clayey soils, large amounts of potassium are not usually needed. As an application for field-beans, the 2—12—6 or 4—12—4 mixture (p. 351) is suggested, (200 to 400 pounds an acre) to be used

at the time of planting or just before, the application to be made in the manner indicated with garden beans.

Garden-peas.—Statements relative to garden-beans apply in most respects to garden-peas. Peas are grown as a market-garden crop and as a field crop for canning, the object being the production of a sweet, tender, well-flavored product.

Light loams favor early maturity, which is desirable for market peas while heavier soils, being cooler and holding moisture, are preferable for canning crops. Acid soils should be limed for peas. Peas are grown in a three or four year crop rotation. For fertilizer 400 to 800 pounds an acre of the 2—12—6 or 4—12—4 mixture (p. 351) can be used.

Beets. (See page 407).

Cabbage.—The growing of cabbage on a large scale has become a specialty in sections where soil and climatic conditions are favorable. Early cabbage is practically all used immediately, while the late crop is distributed, some going to the table at once, some made into sauerkraut and some stored for later use. With southern, early and late northern crops, cabbage is available at all times.

(1) *Crop-rotation for cabbage* is desirable, both for the sake of the physical and chemical factors in the soil, and for protection against insects and diseases that infest this crop. Probably nothing is better to precede cabbage than clover sod, well-decomposed; cabbage can therefore fit into many rotations. Where long seasons prevail and cabbage is grown as an early crop, it can be followed in the same season by some other crop.

(2) *Soils and preparation.*—The qualities requisite in a soil adapted for growing cabbage as a farm crop are good drainage, abundance of available plant-food and structure suited to hold moisture. A considerable variety of soils meets these conditions. In the South, sandy soils are used to grow cabbage for the early northern market because they promote rapid growth at the right season and can be cultivated at a time when heavier soils cannot be. The soil, whatever its character, should be put into mellow, well-pulverized condition before setting out the young plants.

(3) *Fertilizers.*—Large applications of manure (15 to 20 tons an acre) have been successfully used, and clover sod supplemented by 10 or 12 tons is desirable when practicable. Under such conditions, some commercial fertilizer can be used to advantage, especially one supplying phosphorus and potassium in larger proportions than nitrogen, as 400 to 600 pounds of a 4—8—8 mixture (p. 351), for farm-crop cabbages; for early market-garden crops in the North larger amounts of easily available fertilizer can be used, especially on light soils. Excessive supplies of nitrogen are to be avoided for the following reasons: (*a*) Rapid, early growth is pro-

moted to such a degree that the cabbage does not form the close, solid kind of head desired. (*b*) Even when head formation is not affected, the cabbage does not keep well because of its succulence, an objectionable property in cabbages to be shipped long distances or kept long before consumption. This quality of tenderness is, however, desirable in cabbages for immediate consumption.

In the case of market-garden crops which have to depend largely upon applied fertilizers, 1,000 pounds or more of a 4—8—8 mixture can be worked into the soil just before plant setting. On light soils, further applications of top-dressings are sometimes recommended; for example, 75 to 100 pounds of sodium nitrate and 150 to 200 pounds of phosphate when the plants begin to show good growth after transplanting; and 100 to 200 pounds more of sodium nitrate after heads begin to form.

In the case of farm-crop cabbage, an application of 600 to 1,000 pounds of a 4—8—8 mixture (p. 351) can be used without further top-dressings, except under special conditions when the soil appears to lack in plant-food.

(4) *Transplanting.*—When the young cabbage plants are being transplanted in the field, it is often useful after setting to water the soil around each plant with about one-quarter of a pint of solution made by dissolving one ounce of sodium nitrate in 3 or 4 gallons of water. This furnishes both moisture and plant-food, and makes conditions favorable for a prompt start.

(5) *Cultivation.*—Cabbage needs frequent and thorough but not deep cultivation. To destroy weeds and maintain a loose, two-inch earth-mulch which will prevent evaporation of water is the object of cultivation. As grown in the South, however, cultivation is postponed until spring, when active growth begins.

Cauliflower is managed like cabbage except that special skill, as the result of experience, is required to grow it successfully. For best results, a cool, moist season is needed.

Brussels sprouts are treated like cabbage except that the plants are set out somewhat later.

Celery, to be good, must be tender, crisp, sweet and mild in its characteristic flavor. It is of poor quality when the stalks consist largely of bundles of tough fibers, without sweetness and of so strong a flavor as to be bitter or pungent. The good quality of celery depends largely upon abundant moisture and readily available plant-food adapted to promote rapid growth.

(1) *Soil.*—Land ordinarily called "celery land" consists of reclaimed swamp. It is heavily loaded with decomposed organic matter and usually contains considerable nitrogen (largely in forms not quickly available) but is generally deficient in potassium and phosphorus. Celery grows well on non-acid uplands, the soil of

which is deep, well supplied with plant-food, mellow, loose and well-drained.

(2) *Fertilizers.*—Large amounts of plant-food materials are often supplied celery crops, 2,000 to 3,000 pounds an acre not being uncommon of such a fertilizer as a 4—8—8 mixture (p. 351) applied at setting time. A few side-dressings of sodium nitrate, 100 pounds an acre each time, at intervals of 10 to 15 days during the growing season well worked into the soil on both sides of the rows, promote the rapid growth requisite for fine quality. Side-dressing applications should be made only when the plants are dry. No fertilizer should be allowed to touch the plants. Abundance of water is just as essential as plant-food. Some growers apply the whole amount of fertilizer about a week before setting the plants; others one-half before putting in the crop, and the remainder later in two equal parts. Some growers claim that the application of 500 pounds of common salt per acre gives celery a better flavor.

Cucumbers are used for pickling, making salads, and as a relish. In some localities large quantities are grown solely for pickling. The vines are sensitive to frost. For best results, the soil should be a warm, sandy loam, though the crop does well on clay soils. Generally speaking, land suited to corn is adapted to cucumbers. The seeds are sown when danger from frost is past. The soil is thoroughly prepared as in case of corn (p. 386). When practicable, as in gardens, it is common to make a hole in the soil where the seed is to be planted and to place a shovelful or so of well-decomposed manure or compost in the hole and cover with about 2 inches of earth before planting the seed; this is especially useful in poor soils. The same result can usually be accomplished by working the manure into the soil. On soils that do not leach, the fertilizer may be worked into the soil along the line of the rows before planting. On light soils that leach, it is best to apply only a part of the fertilizer (a half or a third) before planting. The rest is worked into the soil around the hills after the plants are up and before the vines have grown enough to interfere.

Fertilizers.—Manure or leguminous green manure is necessary on poor soils, since there should be an abundance of organic matter. When cucumbers are grown for table use, the fertilizers should be such as to promote steady rather than rapid growth of vines. From 500 to 1,000 pounds an acre of the 4—8—8 mixture (p. 351) can be applied to the soil and worked in previous to planting. In case the crop is grown for pickles, the chief purpose is a large setting and quick growth of small, green cucumbers. This is promoted by applications of such a fertilizer as the 4—12—4 mixture (p. 351) at the rate of 500 to 1,000 pounds an acre.

Pumpkins and squashes are grown like cucumbers. They are now extensively grown for canning as material for pie-making. Pumpkins are also used to a considerable extent as cattle-food. They need plenty of calcium carbonate (lime) in the soil. **Mixture No. 4** or **5** can be applied at the rate of 500 to 1,000 pounds an acre, being well worked into the soil in line of rows before planting. Summer squashes are rapid growers, for which **Mixture No. 4** is adapted, while winter varieties have a much longer growing period; the latter can, therefore be supplied in part with organic nitrogen as in **Mixture No. 5** (p. 407).

Distributing mixed fertilizer by hand around cucumber hills.

Muskmelon or cantaloupe growing has become a truck-growing specialty in some localities where soil and climate are specially favorable. The muskmelon does best in warm, sandy loams. It is a crop that requires a long season for maturity, so plant-food nitrogen can be supplied largely in organic forms. Muskmelons require abundant organic matter in the soil. Manure can be placed in a hole and covered with soil 3 or 4 inches deep, the seeds planted on this and covered with 2 or 3 inches of additional soil. A more convenient method is to apply the manure liberally to the parts of plowed ground where the rows are to be planted and work it into the soil thoroughly. In addition, 600 to 1,000 pounds of **Mixture No. 4** or **5** (p. 407) can be used. In the absence of manure, some leguminous green manure should be used to furnish organic matter and well-decomposed before planting. In this case, an application of 600 to 800 pounds of **Mixture No. 4** or **5** can be broadcasted and harrowed in before seeding and an additional handful worked into the soil where the seed is to be planted. This can be followed by a side-dressing of **Mixture No. 4** worked into the soil around each hill about 4 weeks after planting.

Watermelons thrive under the same treatment as muskmelons, but are less sensitive to soil acidity.

Sweet corn, primarily grown for table use when fresh, is extensively grown for canning. Most of it thus used is grown as a regular farm crop. It is sometimes grown for fresh green forage and for silage, but, on account of its smaller yield than field corn it is not so well adapted for green forage. When raised for canning, the stalks and leaves can be fed after the ears have been removed. Sweet corn of fine flavor and satisfactory sweetness can be raised only in a northern climate.

In respect to crop-rotation, soil preparation and cultivation, the same conditions apply as to field corn (p. 386). A succession of planting of an early variety or of different varieties planted simultaneously will secure a continuous supply. When grown as a market-garden crop, it is important to push it to as early maturity as possible in order to obtain best prices.

Fertilizers.—There is a marked difference in the objects in view as between field corn and sweet corn. Field corn is grown for a ripened crop. Much of the growth is made during July and August, and the growing season is relatively long. It can therefore, utilize forms of plant-food that become available gradually during the growing season. Sweet corn is harvested before it reaches maturity, at a time when field corn has made comparatively small development. Like most other garden crops, it should make early rapid growth. Part of its nitrogen, therefore, should be nitrate, in order to supply this element before nitrification becomes active. Phosphorus in soluble compounds must be supplied in considerable amounts to promote early and continuous growth. Potassium is essential to produce starch and sugar. An application of the 4—12—4 mixture (p. 351) is recommended at the rate of 500 to 800 pounds an acre, harrowed into the soil just before planting, when manure is not used. With manure, smaller amounts (200 to 400 pounds) may suffice. The same methods of distributing fertilizer are applicable as in case of field corn (p. 386).

Lettuce of good quality is tender and crisp. These qualities can be secured only by an abundant supply of plant-foods which promote rapid and continuous growth. Interruption of steady growth results in fibrous, tough leaf and bitter taste. Clay as well as sandy loams with abundance of organic matter and non-acid muck soils are adapted to lettuce growing. Sandy loams are better adapted for early crops. Some growers use 1,500 to 3,000 pounds or even more an acre. Ordinarily, on good soils, of 800 to 1,200 pounds an acre of the 4—8—8 mixture (p. 351) will be sufficient, especially if supplemented by two or three applications of sodium nitrate (100 pounds to the acre) after the crop has well started, at intervals of 10 to 15 days; it can be distributed broadcast between the rows and well worked in. Such applications should be made, in case of head lettuce, before the heads begin to form. In using large amounts some growers put complete fertilizer all on the soil in one application about one week before putting in the crop.

Lettuce in sand with nutrient solutions testing respectively pH 3.0, 4.0 and 6.0.

Others use the same amount in three applications, putting on one-half a week before putting in the crop and the remaining half in two subsequent equal applications.

Spinach.—The conditions of soil, growth, cultivation and treatment with fertilizers are essentially the same as for lettuce.

Onions are now grown as a specialty by market-gardeners and where soil and climatic conditions are especially favorable to some extent by farmers engaged in raising general crops.

(1) *Crop-rotation for onions.*—Generally speaking, the rotation for onions should be such that the soil will be as free as possible from weeds and well provided with organic matter. A second crop of clover, well covered with manure in the fall and plowed, followed by potatoes the next season, makes a good preparation for onions the year following.

(2) *Soil and preparation.*—Rich, well-drained loams, with abundance of organic matter, are best adapted to onions. Some

soils suited to celery are also used, especially deep, compact, non-acid, rich muck soils under crop for several years, and in which the water-level is about 2½ feet below the surface of the soil. The soil should be made mellow and well pulverized.

(3) *Fertilizers.*—Manure, when used, is best applied to a preceding crop. With clay and sandy loam that contains an abundance of organic matter, 800 to 1,200 pounds and acre of **Mixture No. 4** (p. 406) can be used, being applied broadcast and well worked in before planting. In the case of early onions, where rapid, early growth is important, **Mixture No. 4** can be used, supplemented with 50 to 100 pounds of sodium nitrate an acre after the crop is well started, and another similar top-dressing 2 or 3 weeks later. In the case of reclaimed muck soils already rich in nitrogen but deficient in phosphoric acid and especially in potash, mixtures in variety, may be used, some containing only phosphoric acid and potash, such as 0—10—10, 0—12—18, 0—16—24, etc., used at the rate of 800 to 1,200 pounds an acre.

(4) *Seeding.*—The seed should be put in the ground as early as the land can be properly prepared. When done in the ordinary way, the rows are 12 to 14 inches apart for hand cultivation or 30 inches for horse cultivation. About 18 to 30 seeds are sown to the foot and covered 1 inch. From 3 to 6 pounds an acre are used.

Another method is to start the onions with seed sown in a greenhouse or in hotbeds about 6 weeks ahead of usual time for seeding and then transplant in the field as soon as soil conditions are suitable.

Early small table onions served raw are generally grown from sets, or little bulbs, produced by drilling 30 to 40 pounds of seed an acre; the thick seeding results in the production of small bulbs, which are harvested in August or September. The thick seeding is to prevent the sets becoming too large under generous application of plant-foods. Another method of seeding for sets is to sow the seed thickly in rows about 3 inches wide, about 8 inches between rows. When these little onions, called sets, are put in the field 2 or 3 inches apart and about 3 inches deep, a new bulb is formed which is ready for the table early in summer. In the South, sets may be planted in the fall. As stated above, rapid, early growth is desirable and should be promoted by liberal applications of readily available plant-food **Mixture No. 4** (p. 406), followed by a side-dressing of 100 pounds of sodium nitrate an acre, when growth is well started.

Tomatoes are grown widely in the United States both in home gardens and as a field crop. They are used as market-garden crops and for canning.

(1) *Crop-rotations for tomatoes.*—Generally speaking, tomatoes can replace potatoes in rotation with other crops (p. 402).

(2) *Soil and preparation.*—Tomatoes grow on a great variety of soils from heavy clays to light, sandy soils. For field crops to be used in canning, good clay loams furnish large yields. For early tomatoes, warm, light, well-drained, rich sandy loams are requisite. Ordinarily, any soil that raises good crops of potatoes is adapted to growing tomatoes. Methods of preparing the soil are essentially the same as for potatoes.

(3) *Fertilizers.*—For early tomatoes, on soils which from previous treatment are known to be well supplied with phosphorus and potassium, an application of 400 to 600 pounds of **Mixture No. 3** (p. 375) may be used in the row at the time of transplanting; or one-half of this amount may be used then and the other half 3 or 4 weeks later. When soils are in good mechanical condition but deficient in available plant-food apply 500 to 1,000 pounds of **Mixture No. 4** (p. 406) an acre previous to setting the plants and work it well into the soil; then, after the plants are set, apply around the plants 100 pounds of sodium nitrate an acre. In 3 or 4 weeks, apply 100 to 150 pounds more of nitrate an acre around the plants, taking the usual precautions. This method insures quick growth from the start without risk of excessive growth of vine that comes from exclusive or excessive use of sodium nitrate.

When farm manure is used, it should be distributed uniformly over the entire surface and thoroughly worked in before plants are set, using not more than 10 or 12 tons an acre. A 4—12—4 mixture can be used at the rate of 400 to 600 pounds an acre. Excessive use of fresh manure must be avoided for early tomateos, since the urinary nitrogen tends to promote excessive growth of leaves and stems at the expense of fruit and also delays the formation and ripening of fruit. Generous applications of phosphorus and potassium compounds should be made to prevent the effects of too much nitrogen (p. 406).

For late tomatoes, on good soils, 500 pounds of **Mixture No. 4** (p. 406) can be used; on poorer soils, 1,000 pounds of **Mixture No. 5** (p. 406).

Eggplant.—In general treatment and culture, eggplant resembles the tomato, with the exception that its growing season is longer. On this account **Mixture No. 5** is used.

TABLE 54—APPROXIMATE AMOUNTS OF PLANT-FOOL CONSTITUENTS IN ONE CROP. VEGETABLES

Crop	*Portion*	*Yield per acre*	*Nitrogen*	*Phosphoric acid* (P_2O_5)	*Potash* (K_2O)
			Lbs.	*Lbs.*	*Lbs.*
Asparagus	Tops	2000 lbs.	7.0	2.0 (0.9P)	5.0 (4. K)
Beans, string ..	Green pods	4000 "	10.0	3.2 (1.4P)	12.0 (10. K)
Beans, field	Seeds	25 bu.	60.0	18.0 (8. P)	19.5 (16.2K)
" "	Straw	2000 lbs.	28.0	6.0 (2.6P)	38.0 (31.5K)
Total			88.0	24.0 (10.6P)	57.5 (47.7K)
Beets, garden ..	Tops & young beets	5000 lbs.	17.5	5.0 (2.2P)	32.5 (27. K)
Cabbage	Heads	20000 "	60.0	20.0 (8.8P)	80.0 (66. K)
Cauliflower ...	"	15000 "	42.0	15.0 (6.6P)	50.0 (42. K)
Celery	Tops	10000 "	25.0	20.0 (8.8P)	75.0 (62. K)
Corn, sweet ...	Ears	4000 "	18.0	8.0 (3.5P)	12.0 (10. K)
Cucumbers	Edible part	100 bu.	5.5	3.3 (1.4P)	11.0 (9.1K)
Eggplant	" "	4000 lbs.	8.0	2.0 (0.9P)	12.0 (10. K)
Lettuce	Leaves	8000 "	20.0	6.4 (2.8P)	36.0 (30. K)
Onions	Bulbs	300 bu.	39.3	15.4 (6.8P)	37.6 (31.2K)
Muskmelons ...	Fruit	10000 "	22.0	8.0 (3.5P)	40.0 (33. K)
Peas, garden ...	Green seeds	1000 "	11.5	3.0 (1.3P)	4.5 (3.7K)
" "	" pods	4000 "	10.0	2.0 (0.9P)	8.0 (6.6K)
Total		5000 "	21.0	5.0 (2.2P)	12.5 (10.3K)
Pumpkins	Fruit	20000 "	32.0	14.0 (6.2P)	52.0 (43. K)
Rhubarb	Stems	20000 "	20.0	8.0 (3.5P)	70.0 (58. K)
Spinach	Stems & leaves	10000 "	50.0	15.0 (6.6P)	25.0 (43.5K)
Tomatoes	Fruit	250 bu.	30.0	10.5 (6. P)	52.5 (21. K)
Watermelons ..	Fruit	20000 lbs.	34.0	12.0 (5.3P)	60.0 (50. K)

XXXIII

GREENHOUSE CROPS

CROPS grown under glass for commercial purposes include vegetables, flowers and non-flowering ornamental plants. The vegetables most commonly raised as forcing-house crops are lettuce, tomatoes, cucumbers and radishes. With the rapid development of truck-gardening in the South, the number of greenhouse vegetables that can be grown to commercial advantage in the North has been necessarily limited. The flowers most extensively grown are roses, carnations, violets and chrysanthemums, also a wide assortment of potted plants for household decorations.

Economy of time is an important factor in relation to forcing-house management, the main object being to push a crop to its marketable condition as rapidly as consistent with quality of product. The general purpose is somewhat the same as for early market-garden crops, but in the greenhouse every effort is made to have all conditions under strict control so as to promote rapid growth. The amount of soil used is small, especially when we consider the vegetable growth often taken from it in comparison with the same area of field soil. Such rapid, large growth of plant material in the quickest time possible means generous amounts of quickly available plant-food supplies. While nitrogen, phosphorus and potassium form only a small fraction of the actual substance of the plant itself, we must remember that the processes of assimilation, upon which plant growth depends, are largely controlled by the amounts of these plant-food constituents ready at just the time needed and in sufficient amounts. In order to grow quickly and satisfactorily, greenhouse crops must be generously and suitably fed.

Soils.—For growing plants under glass, soils should be rich in well-decomposed organic matter or humus (p. 100). This is essential to a mellow and porous soil, properties which forcing-house soils must possess. Organic matter may be supplied by means of grass sod, horse or cow manure, commercial dried animal manures (p. 219), leaf-mold, commercial peat, muck, etc. Whatever its source, it is essential that it be completely decomposed. Too much clay produces too great compactness in the soil, interferes with the extension of growing rootlets, and reduces circulation of air, holds water too tenaciously, hardens on the surface, etc. Undecomposed organic matter and manure are undesirable, since they cause too much looseness, which results in excessive ventilation and too rapid drying. Greenhouse soil is, however, essentially artificial and is, therefore, under much better control than the soil of fields.

Good soil is ordinarily made by combining rich garden loam or decomposed sod, sharp sand, and well-decomposed organic matter that contains considerable nitrogen, as in case of manure (the material commonly used). Dried commercial animal manures are costly for such use. There may be added also some fine bone-meal or bone-flour to supply additional phosphorus and nitrogen. Finely-ground charcoal is used in some cases to increase porosity as well as water-holding power. Other materials are used for special purposes in case of individual crops. The proportions of materials are made to vary to suit the special requirements of each crop.

As a type of preparation commonly used for general greenhouse purposes, the following mixture will serve: One third, each, (1) mellow garden soil, or well-decomposed grass sod, (2) sharp sand, and (3) well-decomposed cow, horse, or mixed farm manure. In preparing the decomposed sod and manure, each material may be put in a separate pile to decompose by itself (p. 240). These are forked over occasionally and finally mixed when ready for use.

Another, usually preferable, method is to compost the sod and manure together, forming layers of sod 6 inches to a foot thick and alternating with four inches of manure, building to a height of five or six feet. When partially decomposed the pile is sliced vertically downward and a new pile made with the outside of the original pile thrown inside of the second. When fully decayed, the soil is put into beds or benches to the depth of 6 to 8 inches; the same kind of soil is used for potted plants. On account of the large amount of hand labor required and resulting costliness, another method is employed when large amounts of soil are used. This consists in applying a coating of farm manure to sod land, or to a green-manure crop, in the field during the summer and plowing it under in the fall, then mixing thoroughly by harrowing. The soil is later plowed again before freezing and left until spring when it is disked several times to make the mixing complete before transference to the greenhouse benches. A layer of 4 to 6 inches of this soil is taken for use.

It is common to clean out the benches and change such soil every year, because, as a result of cropping, it is likely to be impaired in physical properties and chemical composition and, in part, because it is usually infested with plant diseases and insects or their eggs. Remaking is usually easier and safer than trying to restore used soil. A prevalent practice is to restore used greenhouse soils by sterilization with steam or by some chemical compound, both before its first use and also for renewal. This has the advantage of killing insect pests, undesirable micro-organisms and weed seeds; but desirable forms of life are also destroyed, such as decomposing, ammonifying and nitrifying organisms (pp. 138-139). These can, however, be easily introduced again by treatment with good stable manure.

When the use of steam is not practicable for sterilizing soil, one pint of formalin (commercial 40 per cent. formaldehyde solution) in 50 gallons of water can be employed, six quarts to one square foot. After application, the soil is covered with paper or canvas for 24 hours to keep the formaldehyde gas from escaping too quickly. When the soil is uncovered, it is thoroughly stirred to cause escape of the gas. After this treatment the soil should not be used for seeds or plants for 10 to 14 days.

Excellent results have been obtained by departing wholly from the traditional method of preparing greenhouse soils, making up a purely artificial soil from such materials as sifted coal-ashes, sand, commercial peat, etc., which contain little plant-food, and treating the mixture with commercial plant-food materials.

In the absence of farm manure or other desirable organic matter, any rich garden loam can be successfully used for ordinary greenhouse work by mixing it thoroughly with commercial plant-food materials at the time of filling the benches or pots and then keeping it well supplied with additional plant-food materials after the plants are well started.

However the greenhouse soil is prepared, it is essential to keep plenty of organic matter in the soil. When the soil is made up, the use of one or two pounds of fine-ground limestone or marl for 100 square feet may be found advantageous.

Fertilizer features to consider are: (1) Farm manure, (2) wood-ashes and bone-meal, (3) commercial plant-food materials, (4) when to begin feeding, (5) overfeeding, (6) relations to light, temperature and moisture, (7) liming.

(1) *Farm manure* is used in three ways: (*a*) As part of the greenhouse soil; (*b*) as a mulch after plants are well rooted; (*c*) as water-extract. The first method of use has been dwelt upon. When used as a mulch, well-decomposed horse, cow, sheep, or mixed manure is applied to the surface in a layer not over half an inch deep; it is generally best to sterilize manure by steam before using it as a mulch. Commercial dried animal manures are frequently used. When the manure mulch has largely disappeared, a second application may be made, and with some crops, further additions. When used as a mulch, manure rich in urinary nitrogen (p. 32) may give off ammonium carbonate to such an extent as to injure

the mulched plants, especially when used in considerable amounts and in a very warm place.

Water-extract of animal manures, commonly known as *liquid manure,* has been the chief source of plant-food in traditional forcing-house methods. Several ways of preparing it are used, but the product is, of necessity, variable in composition, owing to variation in the composition of the material extracted. The following method is in common use: For 50 gallons of water use 20 pounds of comparatively dry, well-decomposed (not fresh) horse, or cow, or mixed manure, or 10 to 12 pounds of dried pulverized sheep manure, or 5 to 8 pounds of air-dried hen manure. For convenience, the manure, placed in a coarse gunny sack, is hung in the water near the surface for two or three days. This largely avoids sediment in the solution. Larger amounts of manure may be used for the same amount of water and the solution diluted before application, enough to make it the proper strength. This water extract is used on soils once in a week or ten days, the treatment depending upon the crop.

Equally effective results are obtained more conveniently and uniformly by dilute solutions of nitrates, ammonium compounds and urea.

(2) *Wood-ashes and bone-meal.*—In addition to farm manure, wood-ashes and bone-meal have been used more or less extensively for greenhouse crops. These are applied to the surface and worked in lightly. Wood-ashes furnish potassium, a small amount of phosphorus and considerable calcium carbonate. The good effect of wood-ashes, usually ascribed to potassium, is probably more often due to the results of neutralizing soil acidity by both the potassium and calcium carbonate and making conditions more favorable for nitrification (p. 139). Their uncertain composition and high price in relation to plant-food value, often make wood-ashes unsatisfactory to use.

Bone-meal has been largely used in greenhouse soils. Raw bone-meal is less quickly available than the steamed (p. 183), though it contains more nitrogen. Probably, steamed bone-meal or bone-flour gives the best results, all things considered. Bone products should be free from salt for greenhouse use. They have the following advantages: (1) They are lasting on account of their moderate availability; (2) they furnish phosphorus, and some nitrogen; (3) they can be used in fairly large amounts without burning or otherwise injuring crops; (4) they are in a convenient form to apply. Bone products have the disadvantage of high cost relative to the nitrogen and phosphoric acid they contain.

(3) *Commercial plant-food discussion includes:* (*a*) various materials, (*b*) plant-food mixtures, (*c*) application.

(*a*) Materials of various kinds are already described (pp.162-

191), and tabulated (p. 158). From these, several compounds can be selected which, from a chemical point of view, have ideal properties for plant feeding: Ammonium nitrate, concentrated ammonium phosphate, potassium nitrate and potassium ammonium nitrate. In addition to the reasonable cost of the plant-food constituents, these compounds have two marked advantages: (1) Easy solubility, and (2) freedom from constituents not required by plants. In most of the plant-food materials used we are compelled for the sake of economy to employ compounds that contain constituents not needed—chlorine in potassium chloride, sulfuric acid in potassium sulfate, sodium in sodium nitrate, gypsum in superphosphates, etc. These are largely left in the soil and may or may not be desirable, according to special conditions. In using compounds like those mentioned above, the constituents are all plant-food so no residues of useless or objectionable materials are left in the soil. They can all be applied in solution and their amounts closely regulated.

Because so-called chemicals were formerly little understood and were often applied too strongly, prejudice against them was general. These materials are highly concentrated so must be applied only in dilute form.

(*b*) Plant-food mixtures for greenhouse crops.—Several fertilizer mixtures are in use for growing flowers and vegetables under glass. These vary widely even for the same plant.

Two grades or analysis are suggested for general use, 4—8—8 and 8—6—6. The former is used on soils that contain enough animal manure to furnish considerable nitrogen; the latter on soils that contain materials that furnish less in readily available form; for example, a soil in which the organic matter is supplied by commercial peat. These mixtures can be made either of materials applied dry or of others entirely soluble in water and applied in solution. Using materials adapted to greenhouse work, we suggest the mixtures or formulas given below. It should be stated that we use steamed bone-meal in larger amounts than called for by the analysis, on the ground that only one-half of the 22 per cent. of phosphorus and 2.5 per cent. of nitrogen become available to the growing crop. The amounts of materials are given on the basis of one ton, but can easily be reduced to a basis of any amount as one may desire to make.

Plant-food Mixture No. 7.—*For use with forcing-house crops: Analysis:* 4—8—8; *formula;* potassium nitrate, 365 pounds (47 pounds N and 160 pounds K_2O); ammonium sulfate, 100 pounds (20 pounds N); bone-meal, steamed, 600 pounds (15 pounds N and 80 pounds available P_2O_5); superphosphate (16%), 500 pounds (80 pounds available P_2O_5); drier, 435 pounds.

This mixture is applied dry on the surface and worked lightly into the soil at the rate of 16 to 24 ounces for 100 square feet of soil. It is used when only moderate rapidity of action is desired.

For more rapid action, a readily soluble mixture can be conveniently prepared from equal weights of concentrated ammonium phosphate (about 11 per cent. of nitrogen and 48 to 52 per cent. of phosphoric acid) and potassium nitrate (about 13 per cent. of nitrogen and 44 per cent. of potash), as follows: **Plant-food Mixture No. 8.**—*Applied in solution for use when quick action is desired: Analysis:* 12—24—24; *formula;* ammonium phosphate, 10 pounds (1.1 pounds N and 4.8 pounds P_2O_5); potassium nitrate, 10 pounds (1.3 pounds N and 4.4 pounds K_2O).

This mixture contains nitrogen, phosphoric acid and potash in approximately the 1—2—2 or 4—8—8 ratio but in greater concentration, having an analysis of about 12—24—24. For use one ounce of this mixture is dissolved in 4 gallons of water and 2 quarts of this solution are applied to one square yard of soil surface at intervals indicated further on.

For greenhouse soils that contain only small amounts of readily available nitrogen, the following mixture can be used at the rate of 16 to 24 ounces for 100 square feet, applied dry on the surface and worked lightly into the soil: **Plant-food Mixture No. 9.**—*For use with forcing-house crops: Analysis:* 8—6—6; *formula;* sodium nitrate, 550 pounds, (110 pounds N); potassium nitrate, 275 pounds, (36 pounds N and 120 pounds K_2O); bone-meal, 550 pounds, (14 pounds N and 60 pounds P_2O_5, available); superphosphate, 375 pounds, (60 pounds P_2O_5, available); drier, 250 pounds.

A mixture corresponding to the preceding for use in liquid form is prepared as follows: **Plant-food Mixture No. 10**; *Applied in solution for quick action: Analysis:* 20—15—15; *formula:* ammonium phosphate, 10 pounds (1.1 pounds N and 4.8 pounds P_2O_5); potassium nitrate, 11 pounds (1.4 pounds N and 4.8 pounds K_2O); ammonium nitrate, 11 pounds (4 pounds N).

In place of ammonium nitrate 9 pounds of urea could be used. This mixture has an analysis of 20—15—15, having the same ratio of constituents as the 8—6—6 analysis but in greater concentration. The analysis would be approximately the same if equal weights of the three materials were mixed. When used, one ounce of this mixture is dissolved in 4 gallons of water and 2 quarts of the solution applied to one yard of soil at intervals of one to three weeks.

(*c*) Application of plant-food mixtures.—In making applications of soluble compounds to greenhouse soils, one is partly guided in respect to the amount of application by the fact that the feeding roots of plants are liable to injury when the solution is more concentrated than 1 pound of soluble matter in 500 pounds of water (approximately equal to 1 ounce in 4 gallons of water). There is likelihood of harm in case of soils, especially light soils, deficient in organic matter or humus. This relation has been taken into consideration in giving the amounts stated above. In the use of any kind of plant-food in solution, care must be taken to begin with a weak solution. In forcing-house work, it is easy to make application of plant-food, especially in solution, so the principle is carried out of making several applications of a dilute solution instead of stronger ones at longer intervals.

It is best not to make applications of plant-foods on hot or on cloudy days; in one case, the plant-food constituents are absorbed too rapidly; in the other there is danger of suffocation of roots.

Frequency of application is governed largely by the kind of crop one is growing. Ordinarily, when vigorous feeding is desired, applications are made at 7, 10 and 14 day intervals as long as necessary to secure the desired results.

Feeding of forcing-house crops is usually begun only after growth is well started, and the soil well filled with roots. At any time, if plants show signs of overfeeding with nitrogen (p. 278), the quickest way to remedy this condition is to apply carefully a solution containing 16 ounces of superphosphate (or 6 to 8 ounces of concentrated superphosphate) and 4 ounces of potassium sulfate, dissolved in 50 gallons of water, applying 2 quarts per yard of surface at intervals of a week to three weeks.

Roses.—(1) *Soil.*—For forcing-house purposes, roses may be grown on a great variety of soils. Some varieties do best on light loams, others on heavier soils. The soil is prepared as described (p. 428). Incorporation of added plant-food may be effected in various ways. By one method, one pound of finely-powdered potassium sulfate and 10 to 15 pounds of steamed bone-meal (free from common salt) or about 5 pounds of 16 per cent. superphosphate, are added to each cubic yard of compost at the time of the last forking over, one or more weeks before the compost is to be used. Another method is to apply to 100 square feet of the compost after it is in the benches 16 to 24 ounces of the 4—8—8 **Mixture No. 7**, working it well into the soil, or, in case of soils prepared with commercial peat instead of manure, the 8—6—6 **Mixture No. 9** (pp. 432, 433).

(2) *Fertilizers.*—When plants are well started, filling the soil with roots, feeding may begin, using a manure mulch, if obtainable, a couple of times as already described (p. 430). Later, usually in December or January, mulching is discontinued and feeding begun with one of **Mixtures No. 7-10.** Applications are made at intervals of

10 days to 2 weeks or less often, according to the behavior of the crop. When the plants have made a heavy growth, frequency of application may be lessened. At any time when growth is slow, due to cold or cloudy weather, the applications should be withheld until active growth calls for more food. The application of a solution of a nitrate or of urea is often used with good results, but a complete plant-food mixture should enable one better to control plant growth than when one depends on the exclusive use of nitrogen compounds. If at any time too much concentrated fertilizer has been used, the effects can be counteracted by sowing some quick-growing grass or other plant; or, when conditions allow, the soil can be thoroughly soaked with water, thus reducing the concentration of the soil solution.

Violets.—(1) *Soil.*—Violets grow well either in a sandy, gravelly or in a good clay loam like that preferred by most roses. Care must be taken in sandy soils to have plenty of partly decomposed organic matter, furnished by 1 part of manure to 4 parts of soil. In general, a compost may be made as for roses, except that other fertilizing materials are to be omitted. Violets must not have soil too rich in nitrogen. It is important that the compost be well treated with slaked lime as it is made up. Sterilization of soil is useful in killing insects and disease germs.

When the soil is properly made, violets are not given additional feeding; but, when necessary, liquid manure prepared as described or a solution of urea (1 ounce in 10 gallons of water) is applied not more frequently than once in three weeks, and then only during active growth or when vigorous development of leaf is desired. Lack of phosphorus is indicated by reddening of leaves along the veins and by a delayed development of flowers. When these occur, phosphorus is supplied as superphosphate (16 to 20%) at the rate of 8 ounces for 100 square feet, sprinkled on the soil between the rows and lightly worked in. Generally, two or three applications of superphosphate at intervals of two or three weeks are enough. It is suggested that experiments be made with **Mixture No. 7 or 8,** using about one-half or one-third of the amount used for roses at 15 or 20-day intervals (pp. 432, 433).

Carnations.—(1) *Soil* adapted to carnations is preferably a good clay or a sandy loam, one that drains rapidly and has good ventilation. In preparing soil in large amounts for carnations, the field method already described (p. 429) is used. It is usually well to apply 2,000 pounds an acre of pulverized limestone (p. 195). When the soil is ready for the benches, it is well to add 600 to 800 pounds an acre of a fertilizer like **Mixture No. 7** (p. 432) and, in case of light soils, a liberal coating of well-decomposed manure in addition to the fertilizer mixture. This soil is removed to the depth of 6 inches and placed in the greenhouse benches. When a sod soil of good character is not obtainable, the next best thing is to sow in spring clover or some other suitable legume to be turned under before going to seed. The organic matter thus furnished is thoroughly worked into the soil by plowing and harrowing. It will decompose more rapidly if treated with 3 or 4 tons of manure immediately before plowing. In case of light soils, 600 to 800 pounds of **Mixture No. 7** can be used an acre. If a compost is prepared, the same method is used as with roses (p. 434). If, however, a soil mixture is made in the fall, one uses 3 parts of loam to 1 part of manure and 20 to 25 pounds of bone-meal or 10 pounds of superphosphate to each cubic yard. When the compost is turned for the first time, one can add 15 to 20 pounds of pulverized limestone.

(2) *Fertilizers.*—When the plants, started out of doors, have been removed to the greenhouse, they are not treated with any fertilizer until the roots fill the soil. Then they are given a manure mulch, as in case of roses. When the days become shorter and colder, one can begin applying fertilizer as with roses. It should be stated that the carnation is even more sensitive than the rose to overfeeding, especially during cloudy weather or semi-dormant periods, or when roots are not well developed. Accumulation of plant-food in the soil is removed, as in case of roses (p. 435). The common practise has been to rely largely on liquid manure and animal manure mulches but excellent results have followed complete plant-food mixtures, such as 4—8—8 (**No. 7 or 8**) or 4—12—4. Beginning in October, the first application can be made for two months and then again once a month from February to May, using 16 to 24 ounces of **No. 7** for 100 square feet of soil, or one ounce of the concentrated **No. 8** dissolved in 8 gallons of water and applied at the rate of two quarts of solution to one square yard of soil twice a month.

Chrysanthemums.—(1) *Soil.*—The best soils for chrysanthemums are such as drain readily and are well ventilated. To insure this, charcoal and sharp sand are used when needed. A suitable compost can be made as already described. The field method used for preparing soil for carnations can be followed, except that the ground limestone should be omitted. **Mixture No. 7** (p. 432) (600-800 pounds per acre) is incorporated in the soil thoroughly before placing in benches. An inch-thick mulch of decomposed manure or commercial peat is applied when the plants are well started.

(2) *Fertilizers.*—It is the rule not to give plants any special feeding until the buds begin to appear. Then **Mixture No. 8** (p. 433) can be applied in the way indicated every 10 or 15 days. It is believed that a complete fertilizer produces better results than

heavy feeding with nitrogen. Overfeeding with nitrogen produces in chrysanthemums too great succulence and softness of stems, leaves and petals. Red and other dark varieties are likely to show effects of burning as the result of overfeeding with nitrogen. With them solutions of nitrogen compounds should not be used after buds are well formed and of good size. In no case, should liquid manure be used after the flower is about three-quarters expanded. Precautions in feeding chrysanthemums are much the same as with other plants so far considered.

Tomatoes.—(1) *Soil.*—In the case of tomatoes grown in a forcing-house, best results are usually obtained with a light sandy loam that permits good drainage. A compost with these properties, may be prepared as already described.

(2) *Fertilizers.*—When the plants are set, a 4—12—4 fertilizer (p. 451) is applied at the rate of 16 to 24 ounces for 100 square feet of soil and well worked into the surface. When the plants grow normally and fruit begins to mature, additional applications of this fertilizer in the same amount may be made at intervals of two or three weeks. Exclusive or excessive use of manure or of nitrate does not give as satisfactory results as a complete fertilizer.

Lettuce.—(1) *Soil.*—Lettuce is as sensitive to soil conditions as is the chrysanthemum. A mellow, light or medium clay loam with good aeration, drainage and water-holding power gives best results. Lettuce grows better on soils slightly acid than on neutral or alkaline soils. Compost may be used, prepared as previously described.

(2) *Fertilizers,* on fertile, compost soils, may be applied as in the case of tomatoes.

XXXIV

NURSERY CROPS AND ORNAMENTAL PLANTS

THESE include young fruit and ornamental trees, bushes, vines, flowering plants, etc., used largely for out-of-door decoration. The extensive nursery interests of the United States are located in various sections where soil and climate combine to furnish the most favorable conditions. The best nursery soils are of the more distinctly clay-loam type, commonly known as "strong" soils. These crops are grown almost, if not entirely, without the use of plant-food material except farm manure or of various composts. In many cases, nitrogenous organic matter is applied so liberally in the form of partly decomposed manure, that the available phosphorus and potassium furnished by the soil have not been sufficient to produce the kind of normal growth desired. Under these conditions, the growth of leaves and wood is rapid and extreme and is prolonged late in the season, so that the texture of the tree is soft and succulent instead of solid or firm.

Such trees easily succumb to injury by cold weather and are believed to be subject to attack by plant diseases.

It is highly desirable to keep a liberal amount of organic matter in the soil used for growing nursery stock, but it is also essential that the nitrogen do not exceed the proper balance between it and the phosphorus and potassium. Therefore, when large amounts (over 20 tons an acre) of farm manure are used, it is desirable to apply some form of phosphorus and potassium. Some growers, without knowing just how the results were brought about, have used wood-ashes to produce firmer stock, but this material is practically unobtainable now at reasonable prices.

Probably in most cases where manure must be purchased and drawn 3 or 4 miles it is not an economical source of nitrogen or of organic matter. Leguminous cover-crops can usually be made to take the place of manure drawn long distances. The land can be treated, if necessary, with phosphorus and potassium compounds to secure a good yield of the cover-crops, so comparatively little commercial nitrogen in addition will be called for. The use of cover-crops will also correct the unwise practice of leaving the soil practically bare during winter.

Nursery ground is occupied a few years and then put into some ordinary farm crop, sooner or later seeded to grass and used in this way 3 or 4 years before again planted with nursery stock. Some nurserymen make the mistake of depending wholly upon the period intermediate between nursery crops for restocking the soil with organic matter and do not use either farm manure or cover-crops to maintain the supply of organic matter during the growing period of the next nursery crop. The result is that the soil often fails to furnish enough potassium or phosphorus or, when these are supplied artificially, the lack of organic matter prevents the plants from utilizing them to advantage.

When a nursery crop grown under such conditions is removed, it is likely to leave the soil in poor mechanical condition. Clay soil suffers seriously in its ability to promote the vigorous growth of crops if its supply of organic matter is allowed to diminish too much. In addition, the method of removing a crop of trees is such as to add greatly to impairment of the mechanical condition of the soil. The trees are dug in fall or early spring when the soil is generally wet, so manipulation seriously injures its structure and causes puddling (p. 74), the effects of which require time and labor to overcome.

Fertilizers.—In considering the need of fertilizers for nursery crops, the question should be first raised as to the character of product desired. From the nurseryman's standpoint, as well as from that of the orchardist a young tree should be vigorous and hardy; that is, it should be able, when transplanted, to grow reasonably fast under favorable conditions and to withstand extreme climatic changes to which subjected. This result is reached when the physical, chemical and biological conditions of the soil all contribute and favorable climatic conditions furnish the trees abundant plant-food at the right time and in sufficient amounts. Chief among them are good drainage (p. 108), abundance of organic matter (pp. 87-100), a good supply of calcium carbonate (pp. 205-211), good soil structure (p. 76), moisture and abundance of available plant-food. Another feature of producing young trees, desirable from the nurseryman's point of view, is that the growth shall not only be hardy but that it shall be rapid and continuous, so the trees shall not occupy the soil longer than necessary. Hardiness and extreme rapidity of growth are not consistent. The aim is, therefore, to hasten growth as rapidly as consistent with hardiness and vigor. This is done by observing the conditions requisite for growing most crops. To meet these needs we make the following suggestions:

(*a*) Calcium carbonate.—If there are any indications that the soil is acid, apply once in 4 or 5 years half a ton of quicklime or a ton or more of ground limestone, preferably in connection with clover or other leguminous crop. Trees use large amounts of calcium in the growth of leaves and wood.

(*b*) Organic matter and nitrogen can best be furnished by sowing clover or some other legume as a cover crop between the rows of nursery stock in July or August. Clover can also be grown as a rotation crop during the interim between the removal of one nursery crop and the planting of another. In case too much nitrogen in the soil is shown by excessive growth prolonged late into the season, use some non-leguminous crop such as rye or buckwheat as a cover crop and apply 300 or 400 pounds of superphosphate or basic-slag phosphate an acre.

(*c*) Potassium.—On clay soils it will not usually be necessary to supply potassium, provided the soil is limed occasionally and kept supplied with organic matter.

(*d*) Phosphorus probably in most cases can be amply and most cheaply furnished for both nursery crops and those that follow by incorporating once in 5 years 500 to 1,000 pounds of ground phosphate rock (floats), along with manure or green manure.

(*e*) Amounts of plant-food used by nursery crops to grow a crop of trees are not large, compared with the requirements of many field crops. Ordinary fruit trees in the nursery use annually less than 20 pounds of nitrogen, 6 of phosphoric acid (2.6P) and 10 of poatsh (8.3K) an acre.

(*f*) Effects of nursery crops on soils not due to exhaustion of plant-food.—Nurserymen do not commonly plant nursery crops in quick succession because they think that the crop exhausts the soil of elements especially needed for that particular kind of crop. When we consider the small amounts of plant-food used it is evident that this explanation is inadequate. The real explanation lies largely in the change of mechanical structure of the soil due to frequent cultivation for a prolonged period without maintaining the supply of organic matter, together with the abuse to which the soil is often subjected in removing the trees when the soil is wet.

(*g*) Prolonged seasonal growth.—The maximum, but not the most hardy, growth is most easily obtained by furnishing nitrogen as manure in large amounts. It becomes gradually available during the season and prolongs the growth of foliage and wood longer than advisable, producing a soft growth which lacks hardiness. This is prevented preferably by using smaller amounts of manure, and corrected by applying phosphorus and potassium compounds.

(*h*) Light soils.—When nursery stock is grown on light soils, it may be well to apply 200 to 300 pounds of one of the mixtures suggested for fruit crops (p. 450).

(*i*) Drainage and root growth.—The absolute necessity of good drainage for nursery crops can hardly be emphasized too much. It is essential that a tree, in order to be vigorous, have a strong root system. No condition is so disastrous to root development as a soil loaded with stagnant water.

ORNAMENTAL PLANTS

Outdoor, hardy, ornamental plants, whether trees, bushes or vines, were formerly seldom given commercial plant-food, dependence for nutrients being placed upon the liberal use of manure when the plants were set, and afterwards an occasional mulch with manure. When obtainable, enough manure should be worked in to keep the soil well supplied with organic matter and supplemented with a complete fertilizer. In the absence of manure, commercial peat or such materials as leaves, straw, etc., can be used. While much benefit is shown by annual application of nitrates, ammonium compounds, urea or calcium cyanamid in the spring, better results can be expected from complete fertilizers, such as **Mixture No. 7** or **No. 9,** applied at the rate of 16 to 24 ounces for 100 square feet. In the case of small areas, **Mixture No. 8** or **No. 10** can be applied in solution, one ounce in four gallons of water, using two quarts of this solution on one square yard (pp. 432, 433).

Roses probably give more pleasure for the amount of care required than other plants. They well repay care, especially liberal feeding. For best growth they require deep, well-drained, mellow loams rich in organic matter, whatever the general character of the soil otherwise, clayey or sandy; soils reasonably retentive of moisture are desirable.

A soil in which rose-bushes are to be set should be made deep (15 to 20 inches), mellow and fine; a liberal application of manure, preferably well-decomposed, or commercial peat or similar material if the soil is light, is spread uniformly and worked thoroughly into the soil. When bushes are set in spring and a rapid early start is desired, one can work into the surface soil about one ounce of **Mixture No. 7** or **No. 9** (pp. 432, 433) around each plant.

When a rose-bush has been established the first season, it can be treated liberally the following spring with manure or peat, which should be well worked into the soil to furnish organic matter. Then applications of nitrate or of a complete fertilizer can be made at intervals during the season. A good plan will be to top-dress with about 1 ounce of nitrate to each bush early in spring, and then 1 to 2 ounces (according to size and age of bush) of **Mixture No. 8** or **9** when flower-buds begin to show; another similar application after the first flowering is well along will promote continuance of blossoms (on monthly blooming varieties) through the season. A final application early in August will also help and enable the plant to store food to give it a vigorous start the following season. The amount of such a plant-food mixture that can be applied without doing harm is surprising to one who has not tested it. It is possible, however, to use such large amounts of manure, especially when fresh, as to promote growth of leaves and branches at the expense of flowers. The same is true of nitrate, if used alone and in large amounts. But if the complete mixture is used after the first spring application of manure and of nitrate, there need be no fear of overgrowth of foliage and branches at the expense of flowers. Roses require abundance of water, so a mulch of well-decomposed manure or peat in summer promotes water-holding power.

A liberal dressing of pulverized limestone may be applied once in four years.

House plants, whether flowering or non-flowering, are rarely given plant-food, and yet no form of vegetation responds more favorably to feeding. Plants grown in boxes for window or veranda decoration can be made to flourish much more vigorously than ordinarily by judicious feeding. To feed house plants, one can use either of the **Mixtures No. 8** or **10** (p. 433). These may be prepared in small amounts, taking one ounce of each material, and the mixture dissolved 1 ounce in 4 gallons of water. About one-fourth of a pint of this solution is used once a week or once in two weeks on a plant in a 6-inch flower-pot and in corresponding proportions for smaller and larger areas. The solution is poured onto the soil not on the foliage. If the mixture of liquid has stood long it should be stirred from the bottom before dipping out to put on the soil. This treatment is discontinued when the plant is in flourishing condition and renewed when the plant appears to need feeding. The surface of the soil is kept loosened by occasional stirring. A kitchen fork makes a good cultivator. This treatment will not insure good flowers or foliage unless the plants are properly watered and have good air, abundant sunlight, not too high temperature, nor too dry air.

Cut flowers will retain their freshness longer than usual when given proper care. The writer has found no single substance more effective for this purpose than ammonium nitrate (pp. 31, 166). About one level teaspoonful of the powder in four quarts of water. The stems of the flowers are placed in this dilute solution. Cutting off the ends of the stems and renewing the water daily will prolong the keeping-power of flowers for a surprising length of time. If one wishes to make up the solution in large amounts, dissolve one level tablespoonful of the powder in 3 gallons of water.

The data available in regard to the amounts of plant-food constituents used by greenhouse and nursery crops are meager, as indicated in the following table:

TABLE 55—APPROXIMATE AMOUNTS OF PLANT-FOOD CONSTITUENTS USED ANNUALLY BY NURSERY TREES PER ACRE

Crop	*Number per acre*	*Portion of plant*	*Nitrogen*	*Phosphoric acid (P_2O_5)*	*Potash (K_2O)*
			Lbs.		
Apple trees	8000	Wood	10.0	3.5 (1.5P)	7.0 (5.8K)
Peach "	5000	"	7.5	2.0 (0.9P)	4.0 (3.3K)
Pear "	5000	"	8.0	2.6 (1.1P)	4.5 (3.7K)
Plum "	5000	"	6.5	1.5 (0.7P)	4.0 (3.3K)
Rose bushes	15000	"	13.0	7.0 (3.1P)	12.0 (10. K)
		Leaves	17.0	4.0 (1.8P)	19.0 (16.3K)
		Flowers	3.0	1.0 (0.4P)	4.0 (3.3K)

XXXV

FRUIT CROPS

For our purpose fruits can be divided into three general classes: (1) Orchard, (2) bush and vine, and (3) strawberries; the two latter classes are commonly called *small fruits.*

GENERAL CHARACTERISTICS OF ORCHARD CROPS

Facts of a general character that bear on the subject include: (1) Differences between orchard and field crops, (2) relations of plant-food constituents to orchard crops, (3) application of fertilizers, (4) relations of soils to orchard crops, (5) tillage of orchards, (6) use of cover-crops.

Differences between orchard and field crops.—Study of the plant-food supply of orchards suggests fundamental differences between trees and farm crops, the more prominent points being:

(1) *Preliminary growth.*—Most farm crops begin and end their life cycle in a few months. Trees go through a period of development that lasts for several years, during which they merely increase in size and vigor without coming to fruitage.

(2) *Seasonal growth* of most field crops is shorter than that of trees, whose activity begins early in spring, before most field crops have been seeded, and continues until late in fall after most field crops have been harvested. Another point of difference is the character of growth. In field crops, the growth of leaves and stems stops when seed or fruit formation begins, but, in fruit trees, the growth of leaves and wood continues at the same time as does the growth of fruit.

(3) *New wood and bud growth.*—The growth of one season in fruit trees stores up nutrition in new wood and buds, which form in part the source of supply for the following season in the development of fruit. The growth of new wood is influenced largely by the amount of nitrogen supply. An insufficient supply of nitrogen compounds at the time of blossoming may limit or prevent formation of blossom buds for the following year, or, if blossoms are formed they may fail to develop into fruit.

(4) *General character of plant-food.*—When we consider the shorter period in which field crops get their food, we can appreciate the fact that their supplies must be in somewhat more concentrated condition and must also be in forms either immediately available or capable of becoming so rapidly enough to meet crop demands during the relatively short period of growth. In trees, comparatively small amounts of plant-food are removed from the soil in any one season, either during preliminary growth or during seasons, when, owing to insects, frosts, drouth, etc., fruit crops fail. Even in the case of bearing fruit-trees, the plant-food requirements are, as a rule, not often more in amount than in case of many field crops that give yields comparable. Therefore, owing to the long-continued feeding season of fruit-trees, the plant-food compounds required to supply their needs can be present in the soil solution in smaller amounts and can be in forms that become available more gradually than in the case of field crops. Another point in this connection is that in the case of some important field crops, like wheat, barley and other quick-growing crops, the important period of early growth comes when the processes of change in the soil are least active and when quickly available forms of plant-food are likely to be less than at any other part of the growing season; but, in the case of fruit-trees, there is ample opportunity to benefit by the result of all biochemical and other changes that occur during summer and contribute to increase the supply of available plant-food.

(5) *Distribution of feeding roots.*—Tree roots have a more extended feeding ground than do most field crops, because they go deeper into the soil and extend laterally in all directions. In relation to the amount of new material actually produced (fruit, leaves and new wood), the root system is in general more extensive in well-grown trees than in most field plants.

(6) *Transpiration of water.*—When comparison is made on a uniform basis, it is believed that during the growing season more water passes into trees and out of the leaves than in the case of field crops. Hence the soil solution that carries plant-food into trees can be more dilute than in case of field crops and still carry sufficient amounts to furnish nutrition as required.

(7) *Plant-food constituents.*—Fruit crops in general contain a larger percentage of water and, therefore, a smaller percentage of plant-food constituents than most field crops at the same stage of maturity. Speaking of the fruit alone, and not including leaves or new wood, fruit crops use less nitrogen and phosphorus than most field crops, and less potassium than many, when we consider the amounts used to the acre by average crops.

(8) *Rotations.*—Field crops and soils get the benefit of properly

planned rotations. Orchard crops occupy the ground continuously and represent the single-crop system, except as some superficial variation (cover crops) can be introduced. Orchard crops call year after year for the same kinds of plant-food in about the same relative proportions.

Relations of plant-food constituents to orchard crops.—The inorganic constituents most largely used by orchard crops are nitrogen, phosphorus, potassium, calcium and magnesium. In this connection we will consider (1) forms of plant-food materials that can be used, (2) amounts and relative proportions, (3) plant-food suggestions, and (4) relation of plant-food to quality of fruit.

(1) *Forms of plant-food materials to use.*—Organic nitrogen can be depended on mainly to supply this element. The high cost of such materials as dried blood, cottonseed-meal, fish-scrap, ground tobacco stems, meat-tankage, etc., makes their use in large amounts impracticable under usual conditions. However, by the use of appropriate cover crops it will be easily possible to incorporate into the soil much of the nitrogen needed. Manure should be utilized when easily obtainable, especially if one does not make a practice of growing cover crops. Light dressings of manure are useful. The nitrogen of cover crops and manure can be supplemented, if needed, by application of nitrate, ammonium compounds or urea. Excessive amounts of nitrogen in proportion to phosphorus and potassium must be avoided, because it makes the fruit ripen more slowly, affects its color, making it greener, and produces new wood of soft texture, lacking in hardiness and subject to injury by freezing.

Phosphorus can be furnished most economically as superphosphate.

Potassium can be applied as nitrate, high-grade chloride or sulfate, or in the lower-grade form of kainit which has the advantage of containing magnesium compounds. When wood-ashes are obtainable at less than $10 a ton, they can be profitably used on soils lacking in calcium and magnesium carbonate and potassium.

(2) *Amounts and relative proportions.*—An investigation made by the writer with trees in full vigor of bearing shows that fruit trees use per acre in one crop of fruit, foliage and new wood the following (Bul. 265, N. Y. Agr. Expt. Sta., 1905) amounts of plant-food:

Nitrogen	30 to 75 lbs.
Phosphoric acid	7 to 18 "
Potash	33 to 72 "
Calcium oxide (lime)	38 to 114 "
Magnesium oxide	11 to 35 "

The amounts of nitrogen and of potash vary within about the same limits in any one kind of tree, while the amount of phosphoric acid is only about one-fourth to one-third that of nitrogen or potash. The amounts used by various kinds of trees vary greatly, as shown in Table 56.

TABLE 56—AMOUNTS OF PLANT-FOOD USED PER ACRE

Variety	*Number of trees per acre*	*Nitrogen*	*Phosphoric acid (P_2O_5)*	*Potash (K_2O)*	*Lime (CaO)*	*Magnesia (MgO)*
		Lbs.	*Lbs.*	*Lbs.*	*Lbs.*	*Lbs.*
Apple	35	51.5	14.0	55.0	57.0	23
Peach	120	74.5	18.0	72.0	114.0	35
Pear	120	29.5	7.0	33.0	38.0	11
Plum	120	29.5	8.5	38.0	41.0	13
Quince	240	45.5	15.5	57.0	65.5	19

It is evident that the amounts of plant-food constituents given in Table 56 were present in the soil in available form during the growing season and within reach of the root systems. If one knew that there were no available plant-food in the soil, the above quantities would be the smallest amounts that one should apply for a season's growth. On the other hand, if one knew how much available plant-food was present in the soil, then it would simply be necessary to supplement this supply, if it were less than the season's requirements. At present we can ascertain what we want to know about the amount of available plant-food the soil can furnish only by rather crude experimenting.

When we consider the relative proportions of plant-food constituents fruit trees use we find that various species do not vary widely. This statement is supported by the data embodied in Table 57, in which 1 pound of nitrogen is used as a basis of comparison:

TABLE 57—RELATIVE PROPORTIONS OF PLANT-FOOD CONSTITUENTS USED

Variety	*Nitrogen*	*Phosphoric acid (P_2O_5)*	*Potash (K_2O)*	*Lime (CaO)*	*Magnesia (MgO)*
	Lbs.	*Lbs.*	*Lbs.*	*Lbs.*	*Lbs.*
Apple	1	0.27	1.07	1.10	0.45
Peach	1	0.25	0.97	1.53	0.47
Pear	1	0.24	1.10	1.30	0.37
Plum	1	0.29	1.29	1.40	0.52
Quince	1	0.33	1.25	1.44	0.42
Average of all.	1	0.27	1.14	1.35	0.45

The results embodied in Table 57 mean that, under like conditions of soil fertility, a mixture of nitrogen, phosphorus and potassium compounds that would meet the requirements of one variety would approximately meet those of other varieties, so far as the supply of these plant-food constituents is concerned. What proportions are best adapted to supplement most economically the supply of any particular *soil* can be determined only by special experiment.

(3) *Plant-food suggestions.*—Nitrogen from leguminous cover crops should be depended upon as an important source of supply for this plant-food constituent, supplemented by nitrate or ammonium compounds. In many soils phosphorus, though used by fruit trees in relatively small amounts, is the constituent most largely needed to be supplemented. Potassium is required in relatively large amounts but does not need large applications on most clay soils kept well supplied with calcium carbonate and organic matter. On light, sandy soils, it can be supplied in larger amounts.

It may be suggested as a general rule that as long as trees continue to make satisfactory growth of wood and produce average crops of well-colored fruits, commercial fertilizers need be supplied only in moderation and according to the results of observation. When other conditions prevail, the fruit-grower should ascertain by specific fertilizer experiments what combination and amount of plant-foods give best results.

FERTILIZERS FOR FRUIT-TREES

In deciding upon the use of fertilizers in orchards, one is governed by such conditions as age of trees and method of treatment, whether cultivated or under sod-mulch.

(1) Young trees.—In the development of fruit-trees during the years previous to fruiting, the main purpose is to stimulate growth as much as possible, yet to avoid a growth of wood late in the season. For this purpose, one uses under careful regulation nitrogen compounds (nitrate, ammonium compounds, urea) or a complete fertilizer high in nitrogen, such as the 8—6—6 analysis (p. 351).

(2) Bearing trees.—In the case of bearing trees it is essential to insure each year proper growth of wood, a plentiful formation of fruit-buds, abundance of blossoms and complete development of fruit, without causing excessive growth of leaves and twigs. These conditions are largely controlled by regulation of the plant-food supply. An excessive amount of nitrogen compounds in relation to phosphorus compounds must be avoided in the use of fertilizers.

A mixture that contains 2 per cent. of nitrogen, 5 of phosphoric acid (P_2O_5) equal to 2.2 per cent. of P, and 10 of potash (K_2O),

How fertilizing affects the yield of apples. The eight unfertilized trees of Rome Beauty at the left yielded nine barrels; the eight at the right fertilized for the first time in April yielded twenty-one barrels.

equal to 8.3 per cent. of K, furnishes in 1 ton 40 pounds of nitrogen, 100 of phosphoric acid (or 44 of P), and 200 of potash (or 165 of K). These amounts are furnished approximately in each of the mixtures given below. They are suitable for fruit trees.

PLANT-FOOD MIXTURES FOR FRUITS

No. 12 A

400 lbs. cottonseed-meal
400 " bone-meal
100 " acid phosphate
150 " potassium chloride
950 " kainite

No. 12 B

600 lbs. cottonseed-meal
700 " basic-slag phosphate
300 " potassium chloride
400 " kainite

No. 12 C

1000 lbs. tankage (4% nitrogen and 17% phos. acid)
200 " potassium chloride
800 " kainite

No. 12 D

600 lbs. tankage (6½% nitrogen and 9% phos. acid)
300 " acid phosphate
150 " potassium chloride
950 " kainite

No. 12 E

700 lbs. ground tobacco stems
500 " bone-meal
175 " potassium chloride
625 " kainite

No. 12 F

200 lbs. ground tobacco stems
200 " bone-meal
400 " tankage (6½—9)
200 " acid phosphate
200 " potassium chloride
800 " kainite

No. 12 G—*Soluble mixture for use on orchards in sod.*

125 lbs. sodium nitrate
100 " ammonium sulfate
725 " acid phosphate
180 " potassium chloride
870 " kainite

Relation of plant-food to quality of fruit.—Under the expression, "quality of fruit," are included such properties as color, flavor (including sweetness and acidity), keeping power, etc. While plant-food constituents unquestionably have more or less influence in determining quality in fruit, we are far from knowing fully what the relations are and still less how to control them. We knew that, when properly fed, fruit-trees produce larger yields of fruit and of higher quality than when underfed. It has been satisfactorily demonstrated that excessive feeding with nitrogen produces fruit of greener color. We know the general relations of nitrogen, phosphorus, potassium, calcium, etc., to many of the physiological processes of plants, but much remains to be learned in detail about their relations to the quality of fruits. The view most generally held is that conditions, such as sunshine and moisture and temperature, have much more to do with determining the qualities of fruit than do regulation of plant-food supply.

Application of fertilizers for fruit-trees.—Both method and time of applying fertilizers are important, for they have an important bearing on the location of the feeding-roots. Roots tend to develop most where the most plentiful supply of available plant-food is. Top-dressing of orchard soils with fertilizers tends to attract the feeding-roots to the surface; in that location, trees cannot resist drouth as well as when their roots are well down in the soil. Therefore, in order to keep the feeding-roots of trees located in the lower layers, fertilizers should be worked into the soil as deeply as practicable, and especially when applied during the growth of young trees. Where cultivation is practiced, fertilizers can be distributed on the surface and then plowed in. For orchards in sod, it is better to apply only soluble forms of plant-food which are carried down into the soil promptly by rain.

It is desirable to apply all fertilizer by the time the leaf buds begin to open or earlier. In some cases it may be well to apply part of the fertilizer early and the rest two or three weeks later. Use of fertilizer, especially of nitrogen compounds, is to be avoided late in the season, because there may be delay in the proper ripening of wood and consequent danger of injury by freezing during winter.

Relations of soils to orchard crops.—An ideal soil for fruit-trees possesses qualities common to all soils adapted to raising good crops; namely, abundance of plant-food, good physical condition, abundance of organic matter, calcium carbonate and good drainage. Medium sand, gravel and clay loams are preferable to soils either very heavy or light. On soils well suited for orchard crops and properly managed, the application of fertilizers may not be profitable until years have passed. On soils more or less deficient in qualities characteristic of fertile soils, fertilizers will generally be necessary from the start to insure good yield and quality.

Tillage of orchards.—Two systems of soil management have been commonly practiced with fruit orchards, tillage and continuous sod (sod-mulch). On hilly and rough lands, where tillage is not practicable, orchards are kept in grass, which is mowed and left as a mulch to decompose on the ground. The sod method formerly prevailed, but cultivation has become general with all fruits excepting apples. Cultivation tends to hold soil moisture by keeping down weeds and, to some extent, by lessening evaporation. Cultivation decreases the amount of organic matter in the soil, and should, therefore, be accompanied by green manures. Cultivation should be thorough, beginning as early as practicable in spring and continuing at intervals until the seeding of the cover crop.

Use of cover crops.—An essential part of the system of tillage is the cover crop, seeded in June or July and plowed under early the following spring. Cover crops perform several useful functions (pp. 256-265) chief of which are to supply organic matter and also, in case of legumes, organic nitrogen as a source of plant-food supply. Cover crops are especially needed where farm manure is not obtainable in sufficient amounts.

Cover crops which can be used to best advantage are red and sweet clover in the North and cowpea or velvet bean in the South. When a cover crop is started late in the fall, oats or rye may be used, being turned under in spring.

Growing some farm crop between the rows of fruit-trees during the first years of an orchard is called *inter-cropping*. It permits some crop rotation and usually calls for commercial fertilizers the same as when grown by themselves. The following are illustrations of rotations practicable for inter-crops:

(*a*) Fall potatoes, roots or market-garden crops, 1 year; corn, 1 year; clover, or vetch in fall or spring.

(*b*) Corn, 1 year; cotton, 1 year; cowpea or velvet bean, 1 year.

(*c*) Strawberries, plowed after once fruiting and cabbage plants to follow for the balance of the season; (2 years) clover, 1 year, plowed early for strawberries again.

When inter-crops are used, they should be kept far enough from the trees to avoid interference, one with the other. Cultivation of such crops keeps down weeds and makes the condition of the soil favorable for the growth of the trees.

SOD CULTURE

Sod culture has several advantages, being less expensive than cultivation and, in the case of apples, permitting use of the drop fruit. If an efficient grass cover is secured and well fertilized, the organic matter of the soil will be well maintained. For permanent sod, blue grass serves well. Alfalfa and clover are used as cover crops more or less permanent and are efficient in furnishing nitrogen. In case a grass or a leguminous sod runs down, it is plowed up and, after the orchard is cultivated for a season, it is again seeded. This should be done also when sod becomes so heavy that the usual application of nitrogen compounds fails to have effect.

Apples can be grown on a great variety of soils but give best results on porous, limestone soils and clay loams.

Fertilizers.—Recent experiments have shown that on fertile soils, especially of heavier types, fertilizers, continued for years even in large amounts, do not show appreciable effects, whether in case of old orchards or young ones, provided cover crops, including legumes, are used and up-to-date methods of tillage practiced. Even applications of nitrogen large enough to be regarded as excessive have not appeared to have any marked influence, favorable or otherwise. However, when the so-called sod-mulch method is used, the application of fertilizers to orchards grown on similar fertile soils gives increased yield of fruit. Under such conditions, it is better to use a fertilizer that is easily soluble, such as that already suggested (No. 12 G, p. 450), applying to bearing trees not less than 300 pounds an acre yearly and using with it a mixture of 50 pounds each of sodium nitrate and ammonium sulfate, or top-dressing with a light application of manure. Such an application will furnish less than one-half the amount of nitrogen used, and about one-half the potassium, while adding about the needed phosphorus.

On light soils known to be deficient in available plant-food, fertilizers should accompany the starting of the orchard and continue regularly. An application of not less than 300 to 600 pounds a year of any of the orchard mixtures (p. 450) is suggested, together with the regular utilization of leguminous cover crops. On such soils, applications of 1,000 pounds or more of a fertilizer have given profitable returns in case of mature trees in full bearing.

On what might be called good average soils, fertilizing may begin when the orchard comes into bearing, with applications of not less than 300 pounds an acre.

In orchards in which foliage shows poor color in spring, deficiency of available nitrogen is suggested and calls for a top-dressing of 100 pounds of sodium nitrate.

The question of amounts of plant-food materials to apply must be, as we have repeatedly stated, a matter of individual experimenting carried on for a series of years. As trees increase in bearing, the amounts of plant-food supplied should be greater.

Pears respond to fertile soils somewhat more noticeably than apples. They are adapted to heavier soils than are apples and peaches, doing best on stiff clay loams or even hard well-drained clays. They are, however, successfully grown on a great variety of soils, including lighter loams or sandy soils, but the tendency on such soils is to make too rapid growth and to produce wood that winter-kills.

Fertilizers.—Statements that apply to apples apply also to pears in regard to the use of fertilizers, except that pears do not use such large amounts of plant-food per acre (Table 56), so somewhat smaller applications should be sufficient.

Peaches are grown on a great variety of soils from heavy to light, but are best adapted to well-drained, sandy loam, even on fairly poor soils. Heavy, compact clay soils are least desirable. On soils rich in nitrogen, there is danger that growth may be prolonged too far into fall, and produce buds and wood that easily winter-kill.

Fertilizers.—Peaches differ from apples and pears in respect to several features which bear upon plant-food supply: (1) Compared on the acre basis, peach crops are larger consumers of plant-food. They use about one-third more nitrogen, phosphorus and potassium than do apples and twice as much calcium and considerably more magnesium, while they use two to three times as much of each of these constituents as pears do (Table 56). (2) The growing-stage, preliminary to fruit bearing, as well as the life cycle, is shorter with peach-trees. (3) Peach-trees are more sensitive to large nitrogen supply in respect to prolonging growth of foliage and branches late in the season. If nitrate is applied late, or if too large amounts of organic nitrogen are used early, conditions are favorable to prolong the normal growing season of leaves and new wood so late that the buds and wood are easily winter-killed.

For the foregoing reasons, the application of fertilizers to peach orchards should observe the following general rules: (1) Plant-food constituents should be applied in larger amounts than in case of apple or pear-trees under the same soil conditions. (2) The use of nitrogen should be so regulated as to furnish a moderate amount of quickly available nitrogen in spring and enough organic nitrogen to carry the trees through the season without unduly prolonging their growth. The use of leguminous cover crops must be carefully controlled so as not to overstock the soil with nitrogen. This can be regulated in accordance with the behavior of the trees.

On soils of good average fertility, use 300 to 600 pounds an acre of any one of the orchard mixtures (p. 450) after trees begin fruiting. On poor soils, these amounts may be used when the trees are planted and continued as an annual application. When trees begin bearing, one may use, in addition, 200 to 300 pounds an acre of **Mixture No. 12 G.**

Because peaches are so often grown on poor sandy soils and, therefore, require generous feeding, the impression has become common that the crop must always be heavily fertilized. On good apple soils, peaches require only light feeding in addition to the use of cover crops and judicious use of farm manure.

Apricots call for the same treatment, in general, as peaches.

Quinces closely resemble apples in their plant-food requirements.

Plums resemble pears more nearly than the other fruits in respect to the amounts of plant-food used.

Cherries do not need heavy applications of fertilizers. An annual application of 300 pounds or more for bearing trees on average soils will meet usual requirements. The crop matures earlier than the other orchard fruits, so to supply quickly available plant-food one may use equal parts of **Mixture No. 12 G** along with any of the other orchard mixtures. Excessive supply of nitrogen should be avoided.

Citrus fruits, orange, lemon, grapefruit or pomelo and lime require essentially the same general methods of management (cultivation, cover crops, etc.) and the general principles developed in connection with other orchard fruits, though special adaptations occur to meet specific local conditions. Details of methods practiced in California differ from those of Florida.

Fertilizers.—The quality of citrus fruits is more profoundly affected by the character of the plant-food furnished the trees than in case of northern fruits. Excess of nitrogen, especially of organic nitrogen, tends to coarse texture, thick rough skin, and inferior flavor. Potassium results in the production of thinner skin, larger proportion of pulp and sweeter juice.

Inorganic materials are largely used, especially for bearing trees; sodium nitrate, ammonium sulfate for nitrogen; superphosphate for phosphorus; potassium sulfate or the double manure salt (potassium and magnesium sulfate) in preference to potassium chloride or to kainite. Organic matter is supplied by cover crops instead of farm manure and care must be taken to avoid an excess of nitrogen accumulation as the result of using leguminous cover crops too extensively.

During the stage of tree growth preliminary to fruit bearing, such a mixture as the following can be used on light soils: 500 pounds of sodium nitrate, 200 of cottonseed-meal, 500 of superphosphate and 125 of potassium sulfate. On heavy soils cottonseed-meal is omitted. The amounts usually applied per tree are as follows: 1st year, 3 pounds; 2d year, 4½ pounds; 3d year, 6 pounds; 4th year, 9 pounds; 5th year, 12 pounds; 6th year, 15 pounds. It is well to divide the applications and put on one half in winter the other in summer.

When trees come into bearing, a fertilizer of the following composition, for illustration, can be used: 400 pounds of sodium nitrate; 1,150 pounds of superphosphate; 450 pounds of potassium sulfate.

The amounts applied are on the basis of the yield of fruit. Thus, for a tree that bears 3 boxes of fruit, 10 pounds of the mixture may be applied; for one that bears 6 boxes, 20 pounds, etc., up to a yield of 10 boxes per tree.

The applications are preferably made in three portions at different times of year as, for example, February, June and September.

The method of application is to broadcast beneath each tree, putting the bulk nearer the center. After distribution, the fertilizer is thoroughly worked into the soil.

SMALL FRUITS

Under this head come grapes, blackberries, raspberries, currants, gooseberries and strawberries as the most important representatives; as a matter of convenience a brief discussion of pineapples is added. These crops possess certain characteristics which resemble those of orchard fruits, but in their general management they are closer to garden crops. Like orchard trees, the plants of small fruits pass through a stage of preliminary growth before they reach the bearing stage, but the length of this period is much shorter than in case of fruit-trees. Unlike fruit-trees, the duration of their bearing life is generally shorter. Grapevines are comparatively long-lived, and gooseberries and currants with proper treatment can be kept in bearing many years, while blackberries and raspberries may continue 8 to 12 years but a strawberry plant does not bear well for more than 2 years. They resemble garden crops in that they are usually subjected to intensive culture and generous feeding. The period of development from the beginning of the season to the fruiting is short in most cases and therefore calls for treatment that takes advantage of this short growing season to get the largest fruitage.

The statements made in connection with orchard fruits with reference to the relations of soils to crops, tillage, use of cover crops, and inter-crops, apply in modified form to most of the small fruits.

Grapes are widely distributed, and the crop is one of the best-bearing of our fruits. The period of preparatory growth before the vines begin to bear is 1 to 3 years from the time of planting.

(1) *Soils.*—Grapes grow well on most well drained soils when given good cultivation. Soils of varied character appear equally good; gravelly loams containing considerable clay, open clays, some sandy loams, limestone soils and some others produce excellent crops.

(2) *Fertilizers.*—Grapes use somewhat larger amounts of plant-food than any of the other small fruits, but not as large as orchard crops. The constituent used in relatively large amounts is potassium. Nitrogen can be largely supplied in the form of leguminous cover crops. On light soils, one can use 800 to 1,200

pounds per acre of one of the fruit-tree mixtures (p. 450) along with an additional 100 pounds of potassium chloride or sulfate when the vines are in bearing. On fertile soils well managed, one-half the above amounts can be tried. Excessive use of farm manure should be avoided, and generous applications of lime should be made once in four or five years. In Germany and France it is generally believed that the wine-making qualities of grapes are much modified by the methods of feeding.

Blackberries thrive on deep clay loams well supplied with organic matter, but grow well on a great variety of other soils; however, soils that lose moisture too easily near fruiting time are undesirable.

Fertilizers.—It must be kept in mind that blackberries bear fruit on canes which grew from the roots during the season preceding. The main object of directing the growth is to secure a vigorous development of canes. For this purpose, the less active forms of nitrogen can be used, but should not be applied in such large amounts, relative to phosphorous and potassium, as to prolong the growth of the canes too late to ripen properly, for this makes them liable to injury by cold weather. An annual application in spring of 300 to 600 pounds of one of the fruit-tree mixtures (p. 450) will furnish abundant plant-food in case of good soils. On light soils, additional nitrogen should be provided by leguminous cover crops. Good results have been obtained especially on light soils by applying 75 to 100 pounds an acre of sodium nitrate after the plants blossom. While this treatment promotes the development of the fruit, the applications should not be such as to produce weak overgrowth of new canes as above explained.

Raspberries.—Deep moist soils well underdrained and supplied with humus furnish best conditions of growth. Red raspberries are better adapted to light loams and black-caps to heavier loams. Fertilizers recommended for blackberries apply equally well to raspberries.

Currants grow well on any soil that produces good general farm crops and best in a cool climate like that of the Northern States. They prefer a cool location and do well in partial shade. When properly cultivated, fed and pruned, the bushes continue in bearing for many years.

Fertilizers.—Farm manure was formerly used exclusively with the frequent result of an accumulation of nitrogen in the soil, a condition which promotes mildew. When farm manure is used, there should be an application of phosphorus and potassium compounds, especially on light soils. The best plan is to provide most of the nitrogen through leguminous cover crops and to use a fertilizer that supplies relatively large amounts of phosphorus and potassium. Any of the fruit-crop mixtures (p. 450) will meet these conditions, using 300 to 600 pounds on heavier soils and larger amounts on light, sandy ones.

Gooseberries are treated like currants.

Strawberries are the most important of the small-fruit crops, largely because they come as the first fresh fruit of spring. They grow well on any soil that raises good crops of corn. While clay loams usually produce larger crops, sandy loams ripen their crops earlier. The soil should contain an abundance of organic matter.

Fertilizers.—Strawberry plants have one year of growth preparatory to the fruiting stage. During this preliminary period the chief object is to develop the plants to the highest state of vitality possible to enable them to store feeding material sufficient to produce vigorous growth in early spring. Upon the vigor of the plant secured during the preparatory year of growth, especially during the autumn, when buds form, depends mainly the crop of the following year. It is obvious, therefore, that the soil in which strawberries are set should contain liberal supplies of readily available plant-food. On soils in good condition, an application of 300 to 500 pounds of **Mixture No. 4** (p. 406) can be used previous to setting out the plants; on poor, light soil more can be used. In the spring following the setting of the plants, an early application of the same fertilizer should be used in the same or larger amounts, put as near the row as practicable and worked into the soil lightly. The practice among some growers is to apply a top-dressing of 100 to 200 pounds of sodium nitrate an acre in spring after the foliage is well started and before blossoms appear, but in such cases there should be plenty of phosphorus and potassium within reach of the plants. The practice of applying the complete fertilizer as given above, and then a top-dressing of 100 pounds or more of sodium nitrate about the time of blossoming or immediately after is on the increase. Large applications of nitrate increase the vigor of the plants and the yield of fruit, provided phosphorus and potassium are also supplied in liberal amounts. Applications of too much nitrogen in relation to phosphorus and potassium usually result in softer berries that contain a larger percentage of water, impairing shipping and keeping quality. In applying care must be taken to keep nitrate from burning the plants.

Pineapples.—The growth of pineapples in the United States is largely confined to a portion of Florida, and to our island possessions, notably Hawaii.

The soil used in Florida is a fine sand containing small amounts of plant-food. Practically all of the plant-food used by the crop must, therefore, be applied as fertilizer. The annual application of 2,000 to 4,000 pounds of fertilizer to the acre containing 5 per cent. of nitrogen, 4 per cent. of available phosphoric acid and 10 per cent. of potash is practiced. The fertilizer is divided into two to four applications a year; 5 or 6 applications are made during the first 18 months and then 2 a year. It has been found that pineapples are sensitive to certain forms of fertilizing materials which produce injury to growth or to quality of fruit or to both. Thus, acid phosphate is not used unless the soil is kept suplied with calcium carbonate, additions of this being made to the soil every year or two at the rate of 1,000 pounds an acre. Sodium nitrate may be used for the first six months, or, with caution, for the first year, but after

Fertilizer experiment with pineapple. *Left,* cottonseed-meal, potassium, magnesium carbonate and bone-meal; *right,* cottonseed-meal, kainite and superphosphate.

that only organic nitrogen in the forms of dried blood, cottonseed-meal, castor-pomace, tobacco stems or dust, etc. Potassium sulfate, preferably the double manure salt, potassium and magnesium sulfate, are used, not the chloride or kainite. As soon as possible after the plants are set out, about a tablespoonful of cottonseed-meal, or of a mixture of 3 parts of cottonseed-meal and 1 part of tobacco dust, is dropped into the bud. About 6 weeks later the first regular application of fertilizer mixture is put on broadcast and well worked in. Details of application vary with different growers, but the following treatment serves as a illustration: In October or early November a mixture of about 700 pounds of blood and bone and 500 pounds of potassium-magnesium sulfate may be applied, and again the same amount in February. Before the beginning of the rainy season, a third amount is used containing 1,000 to 2,000 pounds of blood and bone and 750 to 1,500 pounds of potash salt.

TABLE 58—APPROXIMATE AMOUNTS OF PLANT-FOOD CONSTITUENTS USED IN ONE CROP OF FRUIT

Kind of crop	*Yield per acre*	*Trees per acre*	*Nitrogen* Lbs.	*Phosphoric acid* (P_2O_5) Lbs.	*Potash* (K_2O)	*Lime* (CaO)
APPLE						
Fruit	300 bu.	30	7.5	3.0 (1.3P)	15.0 (12.5K)	1.5 (1.1Ca)
Leaves	1000 lbs.	30	10.0	1.5 (0.7P)	3.5 (2.9K)	17.0 (12.1Ca)
New wood ...	100 "	30	0.5	0.2 (0.1P)	0.3 (0.2K)	1.5 (1.1Ca)
Total			18.0	4.7 (2.1P)	18.8 (15.6K)	20.0 (14.3Ca)
PEACH						
Fruit	400 bu.	100	22.2	11.0 (4.8P)	45.5 (37.8K)	2.8 (2. Ca)
Leaves	5300 lbs.	100	47.7	8.0 (3.5P)	42.0 (35. K)	120.8 (86.2Ca)
New wood ...	1500 "	100	8.6	2.0 (0.9P)	2.5 (2. K)	6.4 (4.6Ca)
Total			78.5	21.0 (9.2P)	90.0 (74.8K)	130.0 (92.8Ca)
PEARS						
Fruit	300 bu.	100	7.5	3.0 (1.3P)	15.0 (12.4K)	2.0 (1.4Ca)
Leaves	2400 lbs.	100	16.8	2.9 (1.3P)	9.6 (8. K)	25.0 (18. Ca)
New wood ...	600 "	100	1.8	0.6 (0.2P)	1.5 (1.2K)	4.0 (2.8Ca)
Total			26.1	6.5 (2.9P)	26.1 (21.6K)	31.0 (22.2Ca)
PLUMS						
Fruit	200 bu.	120	15.3	5.5 (2.4P)	21.3 (17.6K)	5.0 (3.6Ca)
Leaves	2000 lbs.	120	15.2	3.5 (1.5P)	16.0 (13.3K)	31.0 (22. Ca)
New wood ...	700 "	120	3.5	1.5 (0.7P)	2.0 (1.6K)	11.5 (8.2Ca)
Total			34.0	10.5 (4.6P)	39.3 (32.6K)	47.5 (33.8Ca)
QUINCES						
Fruit	200 bu.	160	12.0	5.5 (2.4P)	25.0 (20.8K)	2.0 (1.4Ca)
Leaves	1500 lbs.	160	13.5	2.7 (1.3P)	6.4 (5.3K)	30.0 (21.4Ca)
New wood ...	400 "	160	2.0	0.8 (0.3P)	1.6 (1.3K)	10.0 (7.2Ca)
Total			27.5	9.0 (4.0P)	33.0 (27.4K)	42.0 (30. Ca)
Cherries	8000 lbs.	120	16.0	4.0 (1.8P)	20.0 (16.6K)	......
Blackberries	4000 qts.	...	11.0	3.0 (1.3P)	12.0 (10. K)	......
Red raspberries .	4000 "	...	10.5	4.5 (2.0P)	12.0 (10. K)	......
Black raspberries	3000 "	...	10.0	4.0 (1.8P)	13.0 (10.8K)	......
Strawberries	5000 "	...	7.5	3.0 (1.3P)	12.5 (10.4K)	......
Currants	3200 "	...	12.0	5.0 (2.2P)	12.0 (10. K)	......
Gooseberries	3200 "	...	6.5	3.0 (1.3P)	12.0 (10. K)	......
Grapes	6000 lbs.	...	9.0	6.0 (2.6P)	18.0 (15. K)	......

P, phosphorus. K, potassium.

In the Hawaiian Islands, the conditions of growth are wholly different from those in Florida, only fertile soils being used and the amounts of applied fertilizers being less. An interesting experience in connection with the growing of pineapples in Hawaii shows that, in some of the soils that contain manganese oxide in amounts as high as 5 per cent., the young plants start well, but in a few months the leaves turn yellow and many of the plants never bear fruit, or, if any is produced, it is inferior in both size and quality. The trouble has been found to be due to action of manganese dioxide in preventing assimilation of iron by the plants. Apparently the iron is prevented from being converted into a form sufficiently soluble for absorption. Direct application of small amounts of ferrous sulfate to the plants results in normal growth and bearing.

XXXVI

SPECIAL CROPS: COTTON, TOBACCO, SUGAR-CANE, HOPS, FLAX, PEANUTS

SOME important crops which do not come under any of the previous groups, and are more or less closely localized, have been reserved for this closing chapter.

Cotton has long been the chief dependence of agriculture in the South, because it is the crop from which the farmer obtains most of his income. The cotton-belt of America includes about one-fourth of the area of the United States, extending from Florida and Texas to southern Virginia and northern Oklahoma, though small amounts are grown in scattered spots outside of this area.

All parts of the cotton plant serve some useful purpose. Next to the lint, the seed possesses high value; the oil is widely used in cooking and in manufactures, while the portion remaining after the oil is removed has high value as a fertilizer and as an animal food. The hulls serve as fuel. The roots of the plant furnish a useful drug, while the stems and leaves may be used as fodder or fuel.

Profitable cotton-growing depends on climate, quality of soil, proper soil management and good seed. The cotton crop has been extensively studied by various experiment stations in the cotton-belt and a large amount of literature has been published on the methods of growing it. The results of this scientific study have raised cotton-growing from a haphazard, soil-exhausting, unprofitable industry to one that can be made to yield satisfactory returns and, at the same time, maintain the soil in good productiveness.

(1) *Crop-rotation for cotton.*—Crop-rotation is, if possible, even more essential to profitable growing of cotton than of cereals. Numerous rotations are possible, but one of two or three years appears to give best results. The use of a leguminous crop, usually cowpeas, is essential. The following serve as illustration:

(*a*) Two years: Oats or corn with cowpeas, 1 year; cotton, 1 year.
Corn and cowpeas, 1 year; cotton and crimson clover, 1 year.

(*b*) Three years: Corn, 1 year; cotton, 1 year; oats with cowpeas or velvet beans, 1 yr.
Corn and cowpeas, 1 year; oats and cowpeas, 1 year; cotton, 1 year.
Cotton and crimson clover, 1 year; corn, 1 year; wheat and cowpeas, 1 year.

(2) *Soil and preparation.*—While good crops of cotton can be raised on a great variety of soils, clay loams and silty loams are superior to sandy loams. The best type of land is a well-drained, light clay loam or medium heavy sandy loam, well provided with organic matter, and a subsoil not too heavy and compact.

In preparing the soil, the object should be a well-pulverized, loose seed-bed. The cotton plant has a long tap-root which will go well down into the soil if it has a chance. On soils not subject to severe washing, plowing in early winter is practiced, while on lighter soils and those liable to washing, the plowing is done just

This outfit is putting in fertilizer, bedding and planting two rows of cotton at one operation.

before planting. Depth of plowing varies from 4 to 8 inches according to the character of the soil. If necessary, the soil is harrowed to make it fine. Two or three weeks before planting, it is laid out in rows varying from 3 to 4 feet wide, leaving a furrow in which fertilizer and seed are placed.

(3) *Fertilizers* for cotton-growing have been extensively studied, and some fairly definite facts have been developed. Conditions are favorable for obtaining plant-food materials to advantage, since leguminous crops supplemented by cottonseed-meal can be used as the chief source of nitrogen, and acid phosphate, produced in the South, is the source of phosphorus. The only materials that require to be brought from long distances are sodium nitrate, which is used to some extent, and potassium compounds, which are commonly applied either as kainite or chloride. It is stated that kainite is a specific against cotton-rust and blight. On heavy clay soils, little potassium appears to be needed, provided green-crop manures

and calcium carbonate are in good supply. Well-decomposed farm manure applied early in the season gives excellent results, especially when supplemented with 100 to 200 pounds of acid phosphate for each ton of manure. Ground rock-phosphate or "floats" applied with farm manure or with a green-crop manure at the rate of 1,000 pounds an acre once in several years appears to offer promise as a cheap source of gradually available phosphorus, but the effectiveness of this practice has not been sufficiently demonstrated yet.

Cotton is far from making heavy drafts on plant-food in soils, and does not require heavy fertilization when only the lint is taken from the farm. The various recommendations indicate an application of 200 to 700 pounds of mixtures containing cottonseed-meal, acid phosphate and potassium chloride. The following formulas are typical of those published for use in fertilizing cotton:

TABLE 59—PLANT-FOOD MIXTURES FOR COTTON CROPS

Materials	*For clay loam soil*	*For sandy loam soil with clay subsoil*	*For sandy soil with clay subsoil*	*For coarse, sandy soils*	*For deep, sandy soils*
Cottonseed-meal .	750	850	925	900	1000
Acid phosphate ..	1200	1100	1000	1000	875
Potassium chloride	50	50	75	100	125

In connection with these mixtures sodium nitrate is likely to prove useful when 50 to 75 pounds an acre are applied as a top-dressing about the middle of May. In some states, an application of 100 to 150 pounds of cottonseed-meal and 100 to 200 pounds of superphosphate, without any potash, is advised; in others, the application of 120 to 160 pounds of cottonseed-meal, 80 to 120 pounds of superphosphate and 80 to 120 pounds of kainit per acre is used.

The method of application commonly followed is distribution in the furrow at seeding time, while some planters prefer two weeks ahead of planting. Application in the row is more generally advised than broadcasting. One application is preferably made, in general, rather than part at the time of planting and the rest later.

On lands that lack calcium carbonate, an application of 1,000 to 2,000 pounds of slaked lime an acre once in 4 or 5 years has proved beneficial.

(4) *Planting and Cultivation.*—The heaviest seed is found to give best results. Planting varies from the middle of March to the middle of May, according to locality and season. The rows are 3 to 4 feet apart and the distance between the plants 12 to 24 inches, being closer on poor soils.

Cultivation commences early and is as frequent as practicable, depending on weather conditions. Cultivation after rains is important.

Tobacco growing is a specialty. Not only is the industry localized according to soil or climate or a combination of both, but special varieties are further localized. Moreover, the methods of culture differ widely according to the variety and the purpose for which the crop is grown.

Tobacco growing in sandy loam and peat mixtures. *Left to right,* Loam, no treatment; loam, plus fertilizer; peat, no treatment; one part peat, three parts loam, plus fertilizer; one part peat, four parts loam, plus fertilizer.

(1) *Soil.*—Probably no plant is so easily affected by character of soil as tobacco. The variety of tobacco in a particular locality is grown there largely because of the special type of soil. Physical properties of the soil appear to be more influential than the chemical. Soil types vary from coarse, sandy pine barrens to heavy limestone clays.

(2) *Crop-rotations for tobacco.*—The following illustrate rotations practiced in various localities for tobacco growing:

Two years: Tobacco, 1; cowpeas, 1.

Three years: Tobacco, 1; rye or wheat, 1; clover, 1.

Three years: Tobacco, 1; wheat and clover, 1; clover grazed, 1.

Three years: Corn, rye cover crop, 1; rye plowed under, and tobacco, 1; grass, 1 or 2.

Four years: Clover, 1; corn and cowpeas, 1; tobacco, 1; wheat and clover, 1.

(3) *Fertilizers.*—Tobacco in its relations to plant-food is characterized in two ways: (1st) It uses nitrogen, potassium and cal-

cium in amounts larger than most other crops; (2d) its quality is influenced as in the case of few other crops by the kind of plant-foods applied. In addition, the finest qualities are grown only on light, sandy soils, which naturally contain only small amounts of plant-food, and which, for this reason, must be heavily fed.

The quality of the crop as well as the yield must be considered. Certain plant-food compounds injure quality. It appears to be generally agreed that chlorine in a tobacco fertilizer impairs the quality for smoking. When burned a tobacco that contains a chloride produces an ash that fuses, on which account it is unsatisfactory

The preceding crop may check the benefit of fertilizer. These plots received the same fertilizer treatment but the effect of a timothy sod plowed under two years previous was still checking the growth of the tobacco on the left.

for smoking, especially in the form of cigars. For this reason, potassium should not be applied wholly in the form of chloride, but mainly as sulfate or carbonate. Large applications of nitrogen tend to produce thick, heavy leaves, loaded with gummy material, a condition desirable in manufacturing tobaccos, but unsuitable for those uses which call for thin-textured leaves or fine, bright tobaccos. It is generally believed by investigators that nitrogen in any form is suitable, whether as nitrate, ammonia or in organic form. Plant-food mixtures for tobacco can be made from the usual fertilizing materials, except that compounds or materials that contain chloride must be excluded.

The following points should be observed in growing tobacco (1) The soil should be kept well supplied with calcium carbonate. (2) Organic matter should be in abundance, preferably by means of leguminous crops, which will also furnish much of the nitrogen needed. (3) In addition to nitrogen furnished by such crops, nitrogen in any commercial form most easily obtainable will answer to furnish the balance needed. In the South, cottonseed-meal is extensively used to advantage. (4) Phosphoric acid is most conveniently furnished as superphosphate. (5) Potassium should be used only in the form of sulfate or carbonate, never as chloride. (6) When farm manure is used, it should be well decomposed. (7) Nitrogen, phosphorus and potassium should all be applied, but excessive nitrogenous fertilization avoided.

The following are representative tobacco fertilizers:

TABLE 60—PLANT-FOOD MIXTURES FOR TOBACCO CROP

Materials	*1*	*2*	*3*	*4*
	Lbs.	*Lbs.*	*Lbs.*	*Lbs.*
Sodium nitrate	...	200	200	150
Dried blood	800	600	...	...
Cottonseed-meal	...	...	...	900
Fish-scrap	...	...	800	...
Acid phosphate	800	800	600	600
Potassium sulfate	400	400	400	350

The amount of fertilizers of the foregoing composition to apply varies greatly according to conditions of soil and general previous soil management; but the use of 1,000 pounds an acre is commonly regarded as a comparatively light application, while the application of 2,000 pounds is not uncommon.

Sugar-cane.—Sugar-cane in the United States is mostly grown in Louisiana; it is a prominent crop in the island dependencies, where it forms by far the largest industry in Hawaii, and where for many years it has been grown with much intelligent skill.

Sugar-cane requires a warm climate and a fertile soil, well-drained, cultivated and watered. The soil must also be well supplied with calcium carbonate. Occasional liming is usually found beneficial. The plowing is generally 18 inches or deeper. Good surface cultivation is practiced while the crop is growing. As sugar cane usually occupies the ground three years or more for each planting, frequent crop-rotations are impracticable. It is highly important that between plantings the soil be occupied one year with other crops, one of which should be leguminous. One method is to follow corn with cowpeas, plowing the leguminous crop under as green-manure. Any of the well-known plant-food materials can be used in fertilizing sugar-cane.

While the crop removes large amounts of plant-food constituents and would therefore impress one as an exhausting crop, it must be remembered that the one substance sold is sugar, a carbohydrate that contains only traces of nitrogen, phosphorus, potassium or calcium as impurities. Therefore, practically all of the plant-food material removed can be restored to the soil. This is true of phosphorus and potassium, but less true of nitrogen, since much of the extracted cane is used as fuel and the nitrogen thus lost. Experiments indicate that in case of good sugar-cane land in Louisiana, there should be an annual application of 25 to 50 pounds of nitrogen, about 35 pounds of phosphoric acid (15 P), and potash in accordance with the character of the soil, good clay soils rich in organic matter and calcium carbonate often not requiring any.

Commercial fertilizers have been extensively employed in the growing of sugar-cane and much study has been given to their use. The fertilizer may be applied broadcast and worked into the soil before planting, or it may be placed under the plant-cane and covered with soil at the time of planting. Some growers apply part of the fertilizer in the row and part after the plants are well started, distributing it on each side of the row. After a crop has been cut, the stubble cane, when allowed to grow another crop, is not fertilized until about the time each sprout has sent out its own rootlets.

In Hawaii so great variation of soils and rainfall is found that each plantation is likely to have fertilizing problems peculiar to itself. In the districts where excessive rains fall, nitrates are not used much on account of leaching, but organic nitrogen is preferred in forms such as tankage and fish-scrap. Ammonium sulfate can be safely used. Tankage, containing about 8 per cent. of nitrogen and 10 of phosphoric acid (4.4 P), is applied in the furrow at planting time. After the cane is growing well, a top-dressing of a high-grade, easily available fertilizer, especially high in potash is given, the sulfate being preferred where the rainfall is heavy, because less leaching of calcium follows its use than that of chloride. In districts where irrigation is practiced, nitrate nitrogen is extensively used. Fertilizers are applied where the irrigation water dissolves and carries the plant-food down to the plant roots.

Hops.—Formerly hop-raising was confined to New York, it now forms a prominent agricultural industry in the Pacific coast states where it can be carried on under more favorable climatic conditions and cheaper methods.

Rich, sandy loams, moist but not wet, are best suited to hops. The crop does well on a great variety of well drained soils.

No extensive systematic studies have been made of the feeding of hops, largely because in the Pacific states fertilizers are not yet used much, while in the East the interest in hop culture has been decreasing. In using plant-food the custom has been to use 12 to 20 tons of farm manure per acre in starting a crop. After the plants are established the crown of the plant each fall is given a generous forkful of manure which in the spring is worked into the soil about the plant. Manuring between rows is also practiced. The use of fresh manure in too large amounts is likely to cause excessive growth of leaf and green hops of undesirable quality, as the result of oversupply of available nitrogen.

Commercial fertilizers have been successfully used, sometimes in alternation with farm manure. Wood-ashes when more plentiful were extensively used and gave excellent results, especially in combination with ground bone. The most successful practice appears to be to apply manure on the crown in the fall and some commercial fertilizers in the spring previous to the first cultivation. Fertilizers with a composition about like those applied to potatoes are commonly used for hops. A fertilizer like **Mixture No. 9** (p. 433) is well adapted to supply the needs of the hop crop, at the rate of 400 to 800 pounds an acre. In case of soils rich in potash, the amount given in this mixture may be reduced one-half.

Flax, while valuable as a fiber crop, is grown in the United States more extensively for its seed, the source of linseed-oil and linseed-meal (p. 168). The principal flax-growing states are Minnesota and the two Dakotas. Flax, when grown for seed, is adapted to about the same climate as wheat, while a cool, moist climate is best suited to the production of the best fiber. Flax has had a bad reputation agriculturally, having long been regarded as a robber and poisoner of soils with which it comes in contact; but the investigations of some of our experiment stations have proved the falsity of this old belief.

(1) *Crop-rotation for flax.*—Traditional rotations based on the belief that flax should be grown on the same soil only once in several years, have been rejected, and also the simple rotation of small grains and flax has been shown to be wholly undesirable. The prevalence of the flax-wilt disease, the germs of which persist in an infected soil for years, introduces a complication which greatly modifies any rotation based on plant-food requirements. In any rotation system, the elimination of disease germs must be kept in mind. So far as the crop itself is concerned otherwise, the main things to keep in mind are (*a*) the use of a leguminous crop, usually clover, one or two seasons preceding flax, and (*b*) soil management that will insure freedom from weeds. Experimental work on rotations indicates the marked value of one or more crops of cultivated corn in the rotation with the flax crop immediately preceded by hay and pasture for several years.

(2) *Soil and preparation.*—Generally speaking, good corn or wheat soils will grow flax well for seed; moist, deep, clay loams, well drained, are generally better than sandy loams for fiber pro-

duction. More, however, appears to be dependent on soil preparation than upon the particular type. For years it was supposed that successful flax growing requires land rich in plant-food constituents and that the crop is a particularly exhausting one. But it has been demonstrated that an average crop of flax uses less total plant-food per acre than a crop of potatoes, corn, wheat, or oats.

Another tradition is that soil is badly injured by flax, since it is a common experience that flax can not be successfully grown on the same soil continuously for many years without trouble from "flax-wilt," a condition ascribed to the soil being "flax-sick." It has been recently shown that this is due to a fungus disease, usually introduced into the soil with the seed, and that, therefore, so-called "flax-sickness" is in no sense due to any peculiar soil character or property. It is, therefore, now generally accepted that the character of the soil is of secondary importance, while the strains of seed used and climatic and seasonal conditions are of more importance.

The aim of soil preparation for the crop should be to furnish a well-worked lower layer to give a close structure for carrying water up to the surface and at the same time so work the surface layer as to provide a fine, mellow but shallow seed-bed about one inch deep. Deep plowing and working should precede seeding time as long as possible in order to make plant-food available for the coming crop.

(3) *Fertilizers.*—Flax uses twice as much nitrogen as potash and three times as much as phosphoric acid. In feeding flax, the habit of growth should be considered, especially that of the roots, and also the brief period required to complete crop growth. Flax has a delicate, threadlike taproot with a few slender branches; its tendency is to go down rather than horizontally; the roots have relatively few root-hairs. Its feeding-range, is therefore, much more limited than that of most crops, and it is commonly described as a "dainty feeder." The growing season is short, most of the plant-food being taken from the soil in the first 45 or 50 days. This somewhat unusual combination of limited feeding-range and short period of growth requires that the plant-food shall be in immediately or quickly available forms and well distributed through the soil within reach of the roots. Slow-acting fertilizers are not, therefore, suitable.

The application of fresh farm manure containing diseased flax straw has been shown to work great injury to the succeeding crop; when, however, such manure is sufficiently decomposed, the disease germs are killed. Large applications of available nitrogen are thought to produce too much wood in straw and to thicken the fiber. On light soils, a light top-dressing with nitrogen and potassium at about blossoming time prolongs the period of growth and is said to result in more pliable, longer straw.

Since it has been the custom in the Northwest to grow flax on virgin soils, the question of plant-food application has received slight attention. On used soils nitrogen can be provided mostly by means of farm manure and leguminous crops used one or two seasons before flax is grown. Additional nitrogen, if needed, can be provided by a top-dressing of 50 to 75 pounds of sodium nitrate to the growing crop. On soils that require addition of fertilizer, an application of 200 or 300 pounds of **Mixture No. 3** (p. 375) is suggested. The soil must be kept well supplied with organic matter, especially to hold moisture and keep the soil mellow, in addition to furnishing nitrogen.

(4) *Seeding.*—Strains of seed resistant to flax-wilt are used as far as possible, or else seeds that are carefully treated with some germicide such as formaldehyde. The seed is sown in spring as early as the work can be done without danger of injury to the young plants by frost. Growth during the rather cool weeks of spring and early summer tends to produce long and fine fiber. Prolonged dry weather hardens the fiber. Uniformity of depth in planting is essential and drilling appears to give best results.

Peanuts are now grown over an extended area from the South Atlantic states to California. The uses of the crop are varied and important, including use as human food, as stock-food and plant-food. The plant belongs to the leguminous family and is a vigorous gatherer of atmospheric nitrogen. The climatic conditions required are a long season without frost, a comparatively light rainfall during the growing period, plenty of sunshine and warmth.

(1) *Crop-rotation for peanuts.*—Among rotations used in peanut-growing, the following will serve as illustrations:

(*a*) Two years: Cotton, 1; corn with peanuts, 1.

(*b*) Two years: Corn with cowpeas or crimson clover, 1; peanuts, 1.

(*c*) Three years: Cotton, 1; corn, 1; peanuts, 1.

(*d*) Four years: Corn and cowpeas, 1; peanuts, 1; cotton, 2.

(2) *Soil and preparation.*—The best soil for the peanut crop is a well-drained, sandy loam, fairly well supplied with organic matter and an abundance of calcium carbonate. Like most leguminous plants, the peanut is sensitive to acidity. When it is grown for market, it is desirable to use light or grayish soils rather than dark soils which may stain the shells. When grown for forage or green-manure, larger crops are usually obtained on soils that contain considerable clay and retain water well.

The land is prepared as in case of any other leguminous crop

(p. 360); it should be moderately compact, mellow, well pulverized and free from weeds.

(3) *Fertilizers.*—Commercial fertilizers are extensively used in growing peanuts. With a good rotation, containing some other leguminous crops, and with the application of farm manure to some other crop in the rotation, little or no nitrogen need be used. **Mixture No. 2** (p. 363), modified in respect to its constituents, as shown in Table 61, can be used at the rate of 200 to 400 pounds an acre on soils in good condition:

Sodium nitrate..	100 pounds	(15 pounds	nitrogen)
Cottonseed-meal	350 "	(25 "	")
Acid Phosphate .	1150 "	(160 "	available phosphoric acid, or 70 lbs. P)
Potassium chloride	400 "	(200 "	potash, or 165 lbs. K)

On soils well provided with nitrogen by farm manure or green-crop manure, a mixture of 250 to 500 pounds an acre may be used, consisting of 1,200 pounds of kainite and 800 pounds of acid phosphate, or 500 pounds of potassium chloride and 1,500 pounds of superphosphate.

As in the case of other leguminous crops, abundance of calcium carbonate must be kept in the soil. Applications equal to 2,000 to 4,000 pounds of calcium carbonate are usually made once in four or five years.

TABLE 61—APPROXIMATE AMOUNTS OF PLANT-FOOD CONSTITUENTS IN ONE CROP

Crop	*Portion of Crop*	*Yield per acre Pounds*	*Nitrogen Pounds*	*Phosphoric Acid (P_2O_5) Pounds*	*Potash (K_2O) Pounds*	*Lime (CaO) Pounds*
Cotton .	Seed	550	19.3	7.7 (3.4P)	6.3 (5.2K)	1.7 (1.2Ca)
	Leaves ...	570	12.8	3.3 (1.5P)	6.3 (5.2K)	30.3 (21.6Ca)
	Stems	630	4.4	1.3 (0.6P)	6.3 (5.2K)	5.0 (3.6Ca)
	Burrs	350	3.5	1.7 (0.7P)	11.4 (9.5K)	4.0 (2.9Ca)
	Lint	300	0.6	0.3 (0.1P)	1.9 (1.5K)	0.3 (0.2Ca)
Total		....	40.6	14.3 (6.3P)	32.2 (26.6K)	41.3 (29.5Ca)
Tobacco	Leaves ...	1,000	44.0	5.0 (2.2P)	57.5 (47.7K)	55.0 (39.3Ca)
	Stalks	....	15.0	2.7 (1.2)	20.5 (17.0K)	9.0 (6.4Ca)
Total		....	59.0	7.7 (3.4P)	78.0 (64.7K)	64.0 (45.7Ca)
Sugar-cane..	Stalks, stripped	30,000	16.2	15.6 (6.9P)	18.3 (15.2K)	7.8 (5.6CaO)
	Leaves and tops ...	20,000	25.2	7.0 (3.1P)	18.0 (14.9K)	27.4 (19.5CaO)
Total		50,000	41.4	22.6 (10. P)	36.3 (30.1K)	35.2 (25.1CaO)
Hops ..	Cones ...	1,600	27.2	17.6 (7.7P)	24. (20. K)	....
	Vines and leaves ..	4,000	26.0	12.0 (5.3P)	54. (44.8K)	....
Total		....	53.2	29.6 (13. P)	78. (64.8K)	....
Flax ...	Seed	900	39.	15. (6.6P)	8. (6.6K)	3. (2.1Ca)
	Straw	1,800	15.	3. (1.3P)	19 (15.8K)	13 (9.3Ca)
Total		....	54.	18. (7.9P)	27. (22.4K)	16. (11.4Ca)
Peanuts	Peanuts ..	1,500	42.9	8.1 (3.6P)	6.8 (5.6K)	1.5 (1.0Ca)
	Vines and leaves ..	6,000	25.2	4.2 (1.8P)	40.0 (33.2K)	41.4 (29.3Ca)
Total		....	68.1	12.3 (5.4P)	46.8 (38.8K)	42.9 (30.3Ca)

APPENDIX

APPENDIX

In the following pages, we give, for convenience of reference, the approximate amounts of nitrogen, phosphoric acid (P_2O_5), and potash (K_2O), with equivalent amounts of phosphorus (P) and potassium (K) commonly present in materials used as sources of plant-food in fertilizers and also in farm products and other materials that may be of interest to crop growers:

TABLE 62.—COMPOSITION OF FERTILIZING MATERIALS USED AS SOURCES OF NITROGEN, PHOSPHORUS AND POTASSIUM

Material	*Pounds in 100*			
	Nitrogen	*Phosphoric Acid (P_2O_5)*		*Potash (K_2O)*
		Total	*Available*	
Ammonium sulfate (p. 166)	20			
Apatite (p. 175)		**33 to 44** (15 to 20P)		
Ashes				
anthracite coal		**0.1 to 0.15** (0.04 to 0.07P)		**0.1 to 0.15** (0.08 to 0.12K)
bituminous coal		**0.4 to 0.5** (0.15 to 0.20P)		**0.4 to 0.5** (0.3 to 0.4K)
lime-kiln (p. 190)		**0.5 to 1** (0.2 to 0.4P)		**0.1 to 1.5** (0.08 to 1.2K)
spent tan-bark		**1.5 to 2** (0.7 to 0.9P)		**1.5 to 2.5** (1.2 to 2K)
wood, leached (p. 190)		**1 to 1.5** (0.4 to 0.7P)		**1 to 3** (0.8 to 2.5K)
wood, unleached (p. 190)		**1 to 2** (0.4 to 0.9P)		**4 to 6** (3.3 to 6.6K)
Azotin. See meat-meal (p. 169)				
Basic-slag phosphate (p. 290)		**10 to 18** (4.4 to 8P)		
Blood, dried red (p. 169)	**13 to 15**			
" " **black**	**6 to 12**			
Bone (p. 183)				
" ash		**30 to 35** (13 to 15.5P)		
" black (p. 185)		**25 to 35** (11 to 15.5P)		
" black, dissolved		**13 to 16** (5.7 to 7P)	**12 to 14** (5.3 to 6P)	
" dissolved	**1 to 2**	**13 to 16** (5.7 to 7P)	**12 to 14** (5.3 to 6P)	
" **meal, steamed** (p. 183)	**2 to 3**	**23 to 25** (10 to 11P)		
" " **from glue-making**	**1 to 1.5**	**25 to 30** (11 to 13P)		
" " **raw**	**3 to 4**	**21 to 25** (9 to 11P)		
Carnallit				**13 to 14** (10.8 to 11.6K)
Castor-bean pomace (p. 168)	**5 to 6**	**2 to 2.5** (0.9 to 1.1P)		**1 to 1.25** (0.8 to 1K)
Cottonseed-hull ashes (p. 190)		**7 to 12** (3 to 5.3P)		**15 to 25** (12.5 to 20K)
Cottonseed-meal (p. 168)	**7 to 8**	**2 to 3** (0.9 to 1.3P)		**1.5 to 2** (1.25 to 1.65K)

Fish, dried and ground. (p. 170)				
" menhaden	**8 to 10**	**5.5 to 7** (2.5 to 3P)		
" acidulated	**5**	3 (1.3P)		
" dog-fish scrap	**7.5 to 9**	**3 to 6** (1.3 to 2.5P)		
" whale scrap	**11.5**	**0.8** (0.35P)		
Garbage rubbish (N. Y. City)	**3.4 to 3.7**	**0.10 to 1.47** (0.04 to 0.65)		**2.25 to 4.25** (1.8 to 3.5K)
Grape refuse from wine factory	**0.75**	**0.20** (0.09P)		**0.4** (0.3K)
Hair (p. 172)	**14 to 16**			
Hoof-meal and horn-dust (p. 171)	**10 to 15**	**1.5 to 2** (0.7 to 0.9P)		
Kainite (p. 189)				**12 to 13** (10 to 11K)
Leather, ground (p. 172)	**10 to 12**			
" acidulated	**7 to 8**			
Meat scraps (p. 169)	**13 to 14**			
Muriate of potash (see potassium chloride).				
Nitrate of soda (see sodium nitrate)				
" " potash (see potassium nitrate)				
Phosphates (p. 173)				
Florida hard rock (p. 174)		**36** (16P)		
Florida land pebble		**31** (14P)		
Peace River pebble		**28** (12P)		
South Carolina land (p. 174)		**27-28** (11 to 12P)		
Tennessee blue (p. 175)		**30** (13P)		
Tennessee domestic		**32 to 34.5** (14 to 15P)		
Potassium chloride, 80 to 85% (p. 188)				**50 to 53** (41.5 to 44K)
" carbonate, 90 to 95% (p. 189)				**61 to 64.5** (50 to 53.5K)
" " 96 to 98%				**65 to 66.5** (54 to 55K)
" nitrate (p. 165)	**13**			**44** (36.5K)
" sulfate, 90 to 95% (p. 188)				**48 to 51** (40 to 42K)
" double manure salt 48 to 53% (p. 189)				**26 to 29** (22 to 24K)
" manure salt, 20% (p. 189)				**11** (9K)
Powder-works waste	**2 to 3**			**16 to 18** (13 to 15K)
Sodium nitrate (p. 165)	**15 to 16**			
Superphosphate (p. 176)		**13 to 18** (5.7 to 8P)	**12 to 16** (5.3 to 7P)	
Tankage, general range (p. 169)	**4 to 10**	**2 to 18** (1 to 8P)		
" concentrated	**10 to 12**	**2 to 4** (0.9 to 1.8P)		
" crushed	**4 to 9**	**3 to 18** (1.3 to 8P)		
" garbage (p. 171)	**1 to 2**	**0.5 to 1** (0.2 to 0.4P)		**0.5 to 1** (0.4 to 0.8K)
Thomas slag (see basic slag).				
Tobacco waste (p. 190)	**2 to 4**	**0.5 to 1** (0.2 to 0.4P)		**5 to 10** (4 to 8K)
Wool waste (p. 172)	**5 to 6**	**2 to 4** (0.9 to 0.18P)		**1 to 3** (0.8 to 2.5K)

COMPOSITION OF FARM MANURES. See Tables 25 to 32, pp. 216 to 225.

TABLE 63—AMOUNTS OF NITROGEN, PHOSPHORUS AND POTASSIUM IN FARM PRODUCTS, FEEDING MATERIALS, BY-PRODUCTS, ETC.

		POUNDS IN 100	
Material	*Nitrogen*	*Phosphoric acid (P_2O_5)*	*Potash (K_2O)*
Ajax flakes	5.20	0.70 (0.31P)	0.20 (0.17K)
Alfalfa, green	0.60	0.15 (0.07P)	0.50 (0.42K)
" hay	2.45	0.50 (0.22P)	2.10 (1.74K)
Alsike clover (see clover)			
American poultry food	2.20	1.20 (0.53P)	0.90 (0.75K)
Animal meal	6.10	....	....
Apple:			
Fruit	0.05	0.02 (0.01P)	0.10 (0.08K)
Leaves	1.00	0.15 (0.07P)	0.35 (0.30K)
New wood	0.50	0.15 (0.07P)	0.25 (0.21K)
Pomace	0.20	0.02 (0.01P)	0.15 (0.12K)
Apricot	0.20	0.05 (0.02P)	0.30 (0.25K)
Artichoke	0.35	0.15 (0.07P)	0.45 (0.37K)
Asparagus, young shoots	0.35	0.10 (0.04P)	0.25 (0.21K)
" full-grown plant	0.40	0.12 (0.05P)	0.75 (0.62K)
" berries	1.10	0.35 (0.15P)	0.85 (0.70K)
Atlas meal, distillery feed	5.50	0.25 (0.11P)	0.15 (0.12K)
Balm, fresh stems and leaves	0.50	0.20 (0.09P)	0.70 (0.58K)
Barley, bran	1.75	0.90 (0.40P)	0.85 (0.70K)
" brewers' grains, dry	3.50	1.25 (0.55P)	0.20 (0.16K)
" " " wet	0.90	0.30 (0.13P)	0.05 (0.04K)
" chaff	1.00	0.25 (0.11P)	1.00 (0.83K)
" feed	2.30	1.30 (0.57P)	0.90 (0.75K)
" flour	1.60	0.85 (0.37P)	0.60 (0.50K)
" grain	1.75	0.75 (0.33P)	0.50 (0.42K)
" green forage	0.40	0.15 (0.07P)	0.50 (0.42K)
" ground	1.50	0.65 (0.30P)	0.35 (0.30K)
" malt	1.60	0.95 (0.42P)	0.45 (0.37K)
" malt-sprouts	3.75	1.65 (0.73P)	1.80 (1.50K)
" meal	1.55	0.65 (0.30P)	0.25 (0.21K)
" middlings	2.00	1.75 (0.77P)	0.70 (0.58K)
" straw	0.60	0.20 (0.09P)	1.10 (0.91K)
Bean, field, hay	3.00	0.65 (0.30P)	2.00 (1.65K)
" " meal	3.75	1.00 (0.44P)	1.50 (1.25K)
" " seed	4.00	1.20 (0.53P)	1.30 (1.10K)
" " shells	1.70	0.30 (0.13P)	0.35 (0.30K)
" " straw	1.30	0.25 (0.11P)	1.90 (1.60K)
Beans, garden, beans and pods	0.25	0.08 (0.03P)	0.30 (0.25K)
" " leaves	0.50	0.10 (0.04P)	0.50 (0.40K)
" " stalks	0.40	0.10 (0.04P)	0.55 (0.45K)
Beets, red, roots	0.25	0.10 (0.04P)	0.50 (0.42K)
" " tops	0.35	0.10 (0.04P)	0.55 (0.45K)
" young roots and tops	0.35	0.10 (0.04P)	0.65 (0.55K)
" mangolds or mangels	0.15	0.10 (0.04P)	0.35 (0.30K)
" sugar	0.20	0.10 (0.04P)	0.45 (0.37K)
" yellow fodder	0.20	0.10 (0.04P)	0.50 (0.42K)
Blackberries, fruit	0.22	0.06 (0.03P)	0.23 (0.20K)
" leaves	1.00	0.12 (0.05P)	0.77 (0.64K)

		POUNDS IN 100	
Material	*Nitrogen*	*Phosphoric acid* (P_2O_5)	*Potash* (K_2O)
Blackberries, new wood	0.55	0.15 (0.07P)	0.25 (0.21K)
" old wood	0.27	0.03 (0.02P)	0.15 (0.12K)
Brewers' grains (see barley).			
Buckwheat, bran	2.75	1.50 (0.66P)	1.10 (0.90K)
" feed	3.00	1.60 (0.70P)	1.00 (0.80K)
Buckwheat, flour	0.75	0.50 (0.22P)	0.15 (0.13K)
" grain	1.50	0.60 (0.26P)	0.30 (0.25K)
" green	0.40	0.08 (0.03P)	0.35 (0.30K)
" hay	0.80	0.60 (0.26P)	2.40 (2.00K)
" hulls	0.55	0.06 (0.02P)	0.75 (0.62K)
" in blossom	0.50	0.03 (0.01P)	0.40 (0.33K)
" middlings	4.75	3.00 (1.32P)	2.25 (1.85K)
" straw	1.25	0.15 (0.07P)	1.15 (0.95K)
" Japanese, green	0.25	0.15 (0.07P)	0.55 (0.45K)
" " hay	1.60	0.85 (0.37P)	3.30 (2.75K)
" silverhull	0.30	0.15 (0.07P)	0.40 (0.33K)
Butter	0.15	0.05 (0.02P)	0.03 (0.02K)
Buttermilk	0.50	0.15 (0.07P)	0.15 (0.12K)
Cabbage, head	0.30	0.10 (0.04P)	0.40 (0.33K)
" stump and outside leaves	0.37	0.10 (0.04P)	0.45 (0.37K)
Carrot, root	0.23	0.13 (0.08P)	0.53 (0.44K)
" tops	0.54	0.13 (0.08P)	0.11 (0.09K)
Cauliflower, head	0.28	0.10 (0.04P)	0.33 (0.27K)
" leaves	0.35	0.10 (0.04P)	0.47 (0.40K)
Celery	0.25	0.20 (0.09P)	0.75 (0.62K)
Cheese, whole-milk	3.75	0.75 (0.33P)	0.30 (0.25K)
" skim-milk	7.50	0.75 (0.33P)	0.30 (0.25K)
Cherry, fruit pulp	0.17	0.04 (0.02P)	0.20 (0.17K)
" stones	0.47	0.15 (0.07P)	0.25 (0.20K)
" stems	0.35	0.13 (0.06P)	0.60 (0.50K)
" leaves	0.60	0.11 (0.05P)	0.72 (0.60K)
" wood	0.45	0.15 (0.07P)	0.30 (0.25K)
Clover, alsike, green	0.50	0.12 (0.05P)	0.30 (0.25K)
" " hay	2.10	0.50 (0.22P)	1.25 (1.05K)
" crimson, green	0.45	0.12 (0.05P)	0.40 (0.33K)
" " hay	2.10	0.45 (0.20P)	1.50 (1.25K)
" mammoth, green	0.50	0.12 (0.05P)	0.40 (0.33K)
" " hay	2.20	0.55 (0.24P)	1.20 (1.00K)
" red, green	0.55	0.13 (0.06P)	0.50 (0.42K)
" " hay	2.10	0.50 (0.22P)	2.00 (1.65K)
" " seed	3.00	1.50 (0.66P)	1.40 (1.15K)
" sweet, green	0.55	0.15 (0.07P)	0.50 (0.42K)
" " hay	2.00	0.55 (0.25P)	1.85 (1.55K)
" white, green	0.50	0.20 (0.09P)	0.30 (0.25K)
" " hay	2.25	0.50 (0.22P)	1.25 (1.05K)
Cocoanut meal	3.90	1.60 (0.70P)	2.40 (2.00K)
Corn, bran	1.65	1.20 (0.53P)	0.70 (0.58K)
" broom, green fodder	0.30	0.15 (0.07P)	0.70 (0.58K)
" " seed	1.70	0.70 (0.30P)	0.50 (0.42K)
" " stalks	0.85	0.45 (0.20P)	1.85 (1.55K)
" cobs	0.40	0.06 (0.02P)	0.45 (0.37K)

Material	Nitrogen	POUNDS IN 100 Phosphoric acid (P_2O_5)	Potash (K_2O)
Corn products:			
" and cob meal	1.40	0.60 (0.26P)	0.50 (0.42K)
" fodder with ears	1.75	0.55 (0.24P)	0.90 (0.75K)
" " without ears	1.00	0.30 (0.13P)	1.40 (1.15K)
" grain	1.65	0.65 (0.30P)	0.40 (0.33K)
" husks	0.20	0.07 (0.03P)	0.25 (0.21K)
" green forage	0.30	0.13 (0.06P)	0.33 (0.27K)
" kafir, red, green forage	0.30	0.13 (0.06P)	0.45 (0.37K)
" " white " "	0.65	0.12 (0.05P)	0.50 (0.42K)
" meal	1.60	0.65 (0.30P)	0.40 (0.33K)
" middlings	1.40	0.35 (0.15P)	0.25 (0.21K)
" silage	0.30	0.12 (0.05P)	0.35 (0.30K)
" stalks	0.75	0.40 (0.18P)	0.90 (0.75K)
Cerealine feed	1.70	1.25 (0.55P)	0.65 (0.54K)
Germ meal	1.80	0.40 (0.18P)	0.20 (0.16K)
Germ oil meal	3.00	0.90 (0.40P)	0.10 (0.08K)
Gluten feed, grano	5.00	0.65 (0.30P)	0.20 (0.16K)
" " Queen	3.90	0.35 (0.15P)	trace
" " Star	3.70	0.35 (0.15P)	"
" " Warner's	2.80	0.30 (0.13P)	"
Gluten meal, Buffalo	4.20	1.15 (0.51P)	0.60 (0.50K)
" " Chicago	5.75	0.35 (0.15P)	trace
" " cream	6.70	0.30 (0.13P)	"
" " globe	4.15	0.60 (0.26P)	0.10 (0.08K)
" " Hammond	4.50	0.50 (0.22P)	0.10 (0.08K)
" " Iowa	4.70	0.45 (0.20P)	0.10 (0.08K)
" " King	6.00	0.65 (0.30P)	0.05 (0.04K)
" " Nebraska	3.15	0.45 (0.20P)	0.10 (0.08K)
Hominy feed	1.65	1.00 (0.44P)	0.50 (0.42K)
" meal	1.75	1.40 (0.62P)	0.75 (0.62K)
Maizeline	1.60	1.40 (0.62P)	0.80 (0.66K)
Sprouts	4.15	1.55 (0.70P)	1.85 (1.55K)
Starch feed, wet	0.80	trace	trace
Sugar feed	1.60	0.20 (0.09P)	0.10 (0.08K)
Corn, sweet, cobs	0.20	0.10 (0.04P)	0.26 (0.22K)
" " cobs and kernels	0.45	0.20 (0.09P)	0.30 (0.25K)
" " green fodder	0.30	0.12 (0.05P)	0.30 (0.25K)
" " husks	0.20	0.10 (0.04P)	0.35 (0.30K)
" " kernels	0.65	0.30 (0.13P)	0.35 (0.30K)
" " leaves	0.40	0.12 (0.05P)	0.55 (0.45K)
" " stalks	0.20	0.06 (0.03P)	0.55 (0.45K)
Cottonseed	3.15	1.25 (0.55P)	1.15 (0.95K)
" feed	1.50	0.45 (0.20P)	1.10 (0.90K)
" hulls	0.75	0.20 (0.09P)	1.10 (0.90K)
" meal	7.00	3.00 (1.30P)	1.75 (1.45K)
Cow-peas, green forage	0.45	0.12 (0.05P)	0.45 (0.37K)
" hay	2.50	0.55 (0.25P)	1.75 (1.45K)
" seed	3.10	1.00 (0.44P)	1.20 (1.00K)
Cream	0.50	0.15 (0.07P)	0.13 (0.11K)
Cucumbers, edible part	0.10	0.06 (0.03P)	0.20 (0.17K)
" leaves	0.40	0.08 (0.03P)	0.70 (0.60K)
" vines	0.20	0.08 (0.03P)	0.60 (0.50K)

Material	Nitrogen	POUNDS IN 100 Phosphoric acid (P_2O_5)	Potash (K_2O)
Currants, fruit	0.30	0.12 (0.05P)	0.30 (0.25K)
" leaves	0.70	0.20 (0.09P)	0.95 (0.82K)
" stems	1.00	0.35 (0.15P)	0.75 (0.62K)
" new wood	0.45	0.22 (0.10P)	0.35 (0.30K)
" old wood	0.57	0.23 (0.10P)	0.22 (0.18K)
Daisy, white, hay	0.25	0.40 (0.18P)	1.20 (1.00K)
" ox-eye, hay	0.30	0.45 (0.20P)	1.25 (1.05K)
Dandelion, green	0.35	0.10 (0.04P)	0.60 (0.50K)
Dewberries	0.22	0.06 (0.03P)	0.20 (0.17K)
Eggplant, edible part	0.20	0.05 (0.02P)	0.30 (0.25K)
" leaves	0.85	0.15 (0.07P)	1.00 (0.83K)
Eggs	2.25	0.40 (0.18P)	0.15 (0.12K)
Endive	0.30	0.07 (0.03P)	0.75 (0.62K)
Flax seed	4.35	1.60 (0.66P)	0.95 (0.80K)
" straw	0.85	0.20 (0.09P)	1.00 (0.83K)
Fox-grass hay	1.20	0.20 (0.09P)	0.95 (0.80K)
Germ-meal (see corn products).			
Glutten-meal (see corn products).			
Gooseberries, fruit	0.16	0.08 (0.03P)	0.03 (0.02K)
" leaves	0.70	0.23 (0.10P)	1.13 (0.94K)
" new wocd	0.40	0.21 (0.09P)	0.40 (0.33K)
" old wocd	0.50	0.25 (0.11P)	0.40 (0.33K)
Grape-pomace, fresh	0.95	0.14 (0.06P)	0.63 (0.52K)
Grapes, fruit	0.15	0.07 (0.03P)	0.30 (0.13K)
" leaves	0.45	0.10 (0.04P)	0.35 (0.30K)
" new wood	0.30	0.10 (0.04P)	0.30 (0.25K)
Hempseed	2.50	1.75 (0.77P)	1.00 (0.83K)
Herd grass	1.00	0.35 (0.15P)	1.50 (1.25K)
Hominy feed (see corn products).			
Hops, entire plant, air-dry	1.00	0.60 (0.26P)	1.40 (1.15K)
" cones or fruit, dry	1.70	1.10 (0.48P)	1.50 (1.25K)
" refuse	1.00	0.20 (0.09P)	0.10 (0.04K)
" vines, air dry	0.65	0.30 (0.13P)	1.35 (1.12K)
Horehound	0.70	0.15 (0.07P)	1.00 (0.83K)
Horse-radish, root	0.40	0.10 (0.04P)	1.00 (0.83K)
Hungarian grass, green	0.35	0.15 (0.07P)	0.50 (0.42K)
" " hay	1.20	0.40 (0.18P)	1.40 (1.15K)
" " seed	1.60	0.45 (0.20P)	0.40 (0.33K)
Hyssop	0.50	0.15 (0.07P)	0.55 (0.46K)
June grass, hay	1.05	0.35 (0.15P)	1.45 (1.20K)
Kentucky blue-grass, hay	1.20	0.40 (0.18P)	1.55 (1.30K)
Kohl-rabi	0.50	0.25 (0.11P)	0.45 (0.37K)
Lemon culls, Calif.	0.15	0.06 (0.02P)	0.26 (0.22K)
Lettuce	0.25	0.08 (0.03P)	0.45 (0.37K)
Linseed:			
" meal, old process	5.30	1.75 (0.77P)	1.30 (1.10K)
" " new process	5.70	1.85 (0.80P)	1.40 (1.16K)
Maize (see corn).			
Malt (see barley).			
Mangolds (see beets).			
Marsh rosemary, hay	0.85	0.05 (0.03P)	0.25 (0.21K)
Meadow, fescue, hay	0.95	0.40 (0.18P)	2.00 (1.65K)

Material	Nitrogen	POUNDS IN 100 Phosphoric acid (P_2O_5)	Potash (K_2O)
Meadow, foxtail, hay	1.50	0.45 (0.20P)	0.20 (0.17K)
" hay, mixed grasses	1.50	0.40 (0.18P)	1.35 (1.12K)
Milk	0.50	0.30 (0.13P)	0.18 (0.15K)
Millet, common, green	0.30	0.10 (0.04P)	0.50 (0.42K)
" " hay	1.25	0.45 (0.20P)	1.50 (1.25K)
" " straw	0.70	0.20 (0.09P)	1.70 (1.40K)
" " seed	2.00	0.95 (0.42P)	0.40 (0.33K)
" barnyard, green	0.25	0.15 (0.07P)	0.75 (0.63K)
" " hay	1.30	0.45 (0.20P)	2.75 (2.30K)
" Japanese, green	0.30	0.20 (0.09P)	0.70 (0.58K)
" " hay	1.10	0.40 (0.18P)	1.20 (1.00K)
" " seed	1.60	0.65 (0.54P)	0.40 (0.33K)
" pearl, green	0.20	0.15 (0.07P)	0.45 (0.37K)
" chaff	0.75	0.20 (0.09P)	0.45 (0.37K)
Molasses feed	3.00	0.85 (0.37P)	2.00 (1.65K)
Mushrooms	1.10	0.40 (0.18P)	0.70 (0.58K)
Muskmelons, fruit	0.22	0.08 (0.03P)	0.40 (0.33K)
" vines and leaves	0.35	0.08 (0.03P)	0.60 (0.50K)
Oak leaves	0.80	0.35 (0.15P)	0.15 (0.13K)
Oat products:			
Canada oat feed	0.70	0.30 (0.13P)	0.60 (0.50K)
Cream oat feed	1.15	0.55 (0.24P)	0.65 (0.54K)
Friend's oat feed	1.40	0.60 (0.26P)	0.65 (0.54K)
Iowa oat feed	1.70	0.50 (0.22P)	0.55 (0.45K)
Monarch oat chop	1.40	0.65 (0.30P)	0.45 (0.37K)
Royal oat feed	1.40	0.50 (0.22P)	0.70 (0.58K)
Vim oat feed	1.10	0.55 (0.24P)	0.70 (0.58K)
"X" oat feed	1.20	0.60 (0.26P)	0.75 (0.62K)
Oats and vetch, green fodder	0.30	0.15 (0.07P)	0.30 (0.25K)
" " " hay	2.00	0.60 (0.26P)	1.25 (1.05K)
Oats, bran	1.35	0.20 (0.09P)	0.70 (0.58K)
" chaff	0.75	0.15 (0.07P)	0.45 (0.37K)
" chop	1.35	0.65 (0.30P)	0.70 (0.58K)
" grain	2.00	0.80 (0.35P)	0.60 (0.50K)
" green fodder	0.60	0.15 (0.07P)	0.45 (0.37K)
" ground	1.85	0.75 (0.33P)	0.60 (0.50K)
" hay	1.50	0.65 (0.30P)	2.20 (1.80K)
" hulls	0.55	0.25 (0.11P)	0.50 (0.42K)
" middlings	2.40	1.35 (0.60P)	1.10 (0.90K)
" straw	0.60	0.20 (0.09P)	1.25 (1.05K)
" and peas, green fodder	0.35	0.15 (0.07P)	0.50 (0.42K)
" " " hay	1.65	0.60 (0.26P)	1.80 (1.50K)
" " " seeds	2.70	1.00 (0.44P)	0.90 (0.75K)
" " " straw	0.75	0.40 (0.18P)	3.20 (2.65K)
Olives, pomace	1.00	0.14 (0.06P)	0.26 (0.22K)
" refuse	1.22	0.18 (0.08P)	0.32 (0.27K)
Onions, bulb	0.23	0.09 (0.04P)	0.22 (0.18K)
" top	0.30	0.06 (0.03P)	0.55 (0.45K)
Oranges, California	0.18	0.05 (0.02P)	0.21 (0.17K)
" culls	0.20	0.13 (0.06P)	0.21 (0.17K)
" Florida	0.12	0.07 (0.03P)	0.48 (0.40K)

Material	Nitrogen	POUNDS IN 100 Phosphoric acid (P_2O_5)	Potash (K_2O)
Orchard grass, green fodder	0.45	0.15 (0.07P)	0.55 (0.45K)
" " hay	1.20	0.35 (0.15P)	1.75 (1.45K)
Parsnips, root	0.22	0.20 (0.09P)	0.65 (0.54K)
" top	0.42	0.15 (0.07P)	0.10 (0.08K)
Pasture grass	0.90	0.25 (0.11P)	0.75 (0.62K)
Peaches, fruit pulp	0.08	0.04 (0.02P)	0.20 (0.17K)
" stones	0.27	0.07 (0.03P)	0.08 (0.07K)
" leaves	0.90	0.15 (0.07P)	0.60 (0.50K)
" new wood	0.43	0.10 (0.04P)	0.22 (0.18K)
Peanuts, meal cake	1.40	1.30 (0.55P)	1.50 (1.25K)
" feed	7.50	0.25 (0.11P)	0.80 (0.66K)
" shells	0.80	0.15 (0.07P)	0.50 (0.42K)
" meal and shells	1.15	0.15 (0.07P)	0.60 (0.50K)
" seeds or kernels	3.60	0.70 (0.31P)	0.45 (0.37K)
Pears, fruit	0.05	0.02 (0.01P)	0.10 (0.08K)
" leaves	0.70	0.12 (0.05P)	0.40 (0.33K)
" wood	0.30	0.10 (0.04P)	0.25 (0.20K)
Peas, field, green forage	0.50	0.15 (0.07P)	0.50 (0.42K)
" " hay	2.00	0.40 (0.18P)	1.00 (0.83K)
" " meal	3.00	0.80 (0.35P)	1.00 (0.83K)
" " middlings	3.75	0.65 (0.30P)	0.75 (0.62K)
" " seeds	3.75	0.80 (0.35P)	1.00 (0.83K)
" " straw	1.40	0.35 (0.15P)	1.00 (0.83K)
" garden, green	1.15	0.30 (0.13P)	0.45 (0.37K)
" " leaves	0.55	0.07 (0.03P)	0.50 (0.42K)
" " pods	0.25	0.05 (0.02P)	0.20 (0.17K)
" " vines	0.25	0.05 (0.02P)	0.70 (0.58K)
Peppermint, garden	0.35	0.20 (0.09P)	0.65 (0.54K)
" wild	0.45	0.17 (0.07P)	0.55 (0.45K)
Plums, fruit pulp	0.12	0.05 (0.02P)	0.20 (0.17K)
Plums, fruit stones	0.50	0.15 (0.07P)	0.10 (0.08K)
" leaves	0.75	0.17 (0.08P)	0.80 (0.66K)
" new wood	0.50	0.18 (0.08P)	0.28 (0.23K)
" stems	0.35	0.12 (0.05P)	0.30 (0.25K)
Potatoes, leaves and stalks	0.60	0.15 (0.07P)	0.45 (0.37K)
" tubers	0.35	0.15 (0.07P)	0.50 (0.42K)
" sweet	0.25	0.10 (0.07P)	0.50 (0.42K)
Pot marjoram	0.50	0.16 (0.07P)	0.60 (0.50K)
Prickly comfrey	0.40	0.15 (0.07P)	0.70 (0.58K)
Prunes, refuse	0.18	0.07 (0.03P)	0.31 (0.26K)
Pumpkins, flesh	0.16	0.07 (0.03P)	0.26 (0.22K)
" vines and seeds	0.42	0.10 (0.04P)	0.57 (0.47K)
" seeds	0.87	0.50 (0.22P)	0.45 (0.37K)
Quinces, fruit	0.12	0.05 (0.02P)	0.25 (0.20K)
" leaves	0.90	0.17 (0.08P)	0.40 (0.33K)
" new wood	0.50	0.20 (0.09P)	0.40 (0.33K)
Radishes, root	0.15	0.06 (0.02P)	0.35 (0.30K)
" top	0.42	0.06 (0.02P)	0.45 (0.37K)
Rape, green	0.40	0.15 (0.07P)	0.45 (0.37K)
Raspberries, fruit	0.20	0.10 (0.04P)	0.25 (0.21K)
" leaves	1.35	0.27 (0.12P)	0.63 (0.52K)
" new wood	0.45	0.15 (0.07P)	0.25 (0.21K)

Material	Nitrogen	Phosphoric acid (P_2O_5)	Potash (K_2O)
	POUNDS IN 100		
Raspberries, old wood	0.30	0.06 (0.02P)	0.25 (0.21K)
Redtop hay	1.20	0.35 (0.30P)	1.00 (0.82K)
Rhubarb, leaves	0.30	0.08 (0.03P)	0.50 (0.42K)
" stems	0.10	0.04 (0.02P)	0.35 (0.30K)
Rice bran	0.70	0.30 (0.13P)	0.25 (0.21K)
" grain	1.10	0.20 (0.09P)	0.10 (0.08K)
" hulls	0.60	0.20 (0.09P)	0.15 (0.13K)
" polish	2.30	3.30 (1.45P)	1.20 (1.00K)
Roses, flowers	0.30	0.10 (0.04P)	0.40 (0.33K)
" leaves	0.40	0.10 (0.04P)	0.50 (0.42K)
" wood	0.15	0.07 (0.03P)	0.12 (0.10K)
Rue	0.55	0.15 (0.07P)	0.65 (0.54K)
Rutabagas	0.20	0.12 (0.05P)	0.50 (0.42K)
Rye, bran	2.25	2.15 (0.95P)	1.30 (1.08K)
" chaff	0.60	0.55 (0.24P)	0.50 (0.42K)
" feed	1.55	0.75 (0.33P)	0.45 (0.37K)
" flour	1.75	0.85 (0.37P)	0.65 (0.54K)
" grain	1.70	0.85 (0.37P)	0.60 (0.50K)
" green forage	0.45	0.20 (0.09P)	0.65 (0.54K)
" hay	1.05	0.40 (0.18P)	1.25 (1.05K)
" middlings	2.00	1.00 (0.44P)	0.75 (0.63K)
" straw	0.50	0.30 (0.13P)	0.85 (0.70K)
Rye grass, Italian, green	0.55	0.30 (0.13P)	1.15 (0.95K)
" " " hay	1.10	0.55 (0.24P)	1.20 (1.00K)
" " perennial, green	0.50	0.30 (0.13P)	1.10 (0.90K)
" " " hay	1.15	0.55 (0.24P)	1.50 (1.25K)
Sage	0.55	0.10 (0.04P)	0.60 (0.56K)
Sainfoin, green	0.70	0.20 (0.09P)	0.55 (0.46K)
" hay	2.20	0.50 (0.22P)	1.40 (1.16K)
Salsify, root	0.50	0.15 (0.07P)	0.55 (0.46K)
" top	0.50	0.10 (0.04P)	1.10 (0.90K)
Salt marsh hay	1.10	0.25 (0.11P)	0.75 (0.63K)
Sedge, creek, hay	1.30	0.05 (0.02P)	0.50 (0.42K)
" short, hay	0.15	0.15 (0.07P)	1.15 (0.95K)
Serradella, grass	0.40	0.15 (0.07P)	0.40 (0.33K)
" hay	2.40	0.75 (0.33P)	2.00 (1.65K)
Skim-milk	0.50	0.35 (0.15P)	0.20 (0.17K)
Sorghum, green fodder	0.30	0.12 (0.05P)	0.30 (0.25K)
" seed	1.50	0.80 (0.35P)	0.45 (0.37K)
Soy beans, green	0.50	0.15 (0.07P)	0.60 (0.50K)
" " hay	2.30	0.70 (0.31P)	1.10 (0.90K)
" " seed	5.30	1.80 (0.80P)	2.00 (1.65K)
" " straw	0.90	0.30 (0.13P)	0.75 (0.62K)
Spanish moss	0.60	0.10 (0.04P)	0.55 (0.46K)
Sphagnum moss	0.64	0.10 (0.04P)	0.26 (0.22K)
Spinach	0.50	0.15 (0.07P)	0.25 (0.21K)
Strawberries, fruit	0.15	0.06 (0.03P)	0.25 (0.21K)
" hulls	0.30	0.09 (0.04P)	0.40 (0.33K)
Street sweepings	0.38	0.40 (0.18P)	0.30 (0.25K)
Sucrene dairy feed	3.10	0.60 (0.26P)	0.25 (0.21K)
Sugar-beet feed, dry	1.30	0.25 (0.11P)	0.60 (0.50K)
" pulp, fresh	0.20	0.30 (0.13P)	0.10 (0.08K)

Material	*Nitrogen*	*Phosphoric acid* (P_2O_5)	*Potash* (K_2O)
		POUNDS IN 100	
Sugar beets (see beets).			
Sugar-cane	0.20	0.10 (0.04P)	0.45 (0.37K)
Sunflower seed	2.25	1.25 (0.55P)	0.75 (0.62K)
Sweet clover (see clover).			
Sweet corn (see corn).			
Tall meadow oat grass, hay	1.15	0.30 (0.13P)	1.70 (1.40K)
Tansy, garden	0.50	0.15 (0.07P)	0.75 (0.62K)
Tarragon	0.55	0.15 (0.07P)	0.75 (0.62K)
Teosinte grass	0.35	0.10 (0.04P)	0.25 (0.21K)
Thyme, French	0.40	0.15 (0.07P)	0.65 (0.55K)
Timothy, green	0.50	0.25 (0.11P)	0.75 (0.62K)
" hay	1.25	0.55 (0.24P)	1.00 (0.83K)
Tobacco, leaves	4.00	0.50 (0.22P)	6.00 (5.00K)
" stalks	3.70	0.65 (0.32P)	4.50 (3.70K)
" stems	2.50	0.90 (0.40P)	7.00 (5.80K)
Tomatoes, fruit	0.20	0.07 (0.03P)	0.35 (0.30K)
" leaves	0.35	0.10 (0.04P)	0 40 (0.33K)
" stalks	0.35	0.10 (0.04P)	0.50 (0.42K)
Turnips, root	0.25	0.10 (0.04P)	0.45 (0.37K)
" top	0.40	0.12 (0.05P)	0.55 (0.46K)
" rutabagas	0.20	0.12 (0.05P)	0.50 (0.42K)
Velvet bean, green	0.55	0.15 (0.07P)	0.55 (0.46K)
Vetch, hairy, green	0.50	0.10 (0.04P)	0.45 (0.37K)
" hay	2.80	0.75 (0.33P)	2.30 (2.00K)
" seed	3.70	1.40 (0.60P)	0.85 (0.71K)
" straw	1.00	0.25 (0.11P)	0.65 (0.54K)
" and oats (see oats).			
" kidney, green	0.45	0.10 (0.04P)	0.30 (0.25K)
" sand, green	0.55	0.15 (0.07P)	0.50 (0.42K)
" spring, green	0.35	0.10 (0.04P)	0.45 (0.37K)
Watermelons, fruit	0.17	0.06 (0.03P)	0.30 (0.25K)
" vines and leaves	0.50	0.07 (0.03P)	0.50 (0.42K)
Wheat, bran	2.65	2.90 (1.28P)	1.60 (1.33K)
" chaff	0.75	0.80 (0.35P)	0.45 (0.37K)
" feed	2.70	2.00 (0.88P)	0.55 (0.46K)
" flour	2.00	0.60 (0.26P)	0.50 (0.42K)
" grain	2.00	0.85 (0.37P)	0.50 (0.42K)
" middlings	2.50	1.35 (0.60P)	0.70 (0.58K)
" shorts	2.80	1.35 (0.60P)	0.65 (0.54K)
" straw	0.50	0.15 (0.07P)	0.60 (0.50K)
Whey	0.12	0.30 (0.13P)	0.15 (0.13K)
Winter savory	0.55	0.20 (0.09P)	0.70 (0.58K)
Wormwood	0.70	0.22 (0.10P)	0.85 (0.70K)

INDEX